WITHDRAWN
L. R. COLLEGE LIBRARY

Poisonous Plants and Venomous Animals of Alabama and Adjoining States

Whit Gibbons
Robert R. Haynes
Joab L. Thomas

With a Foreword by
Robert J. Geller, M.D.

Poisonous Plants and Venomous Animals of Alabama and Adjoining States

The University of Alabama Press

Tuscaloosa and London

Manufactured in the United States of America

∞ The paper on which this book is printed meets the minimum requirements of American National Standard for Information Science-Permanence of Paper Library Materials, ANSI A39.48-1984.

Library of Congress Cataloging-in-Publication Data

Gibbons, Whit, 1939–
Poisonous plants and venomous animals of Alabama and adjoining states / Whit Gibbons, Robert R. Haynes, and Joab L. Thomas.
p. cm.
Bibliography: p.
Includes index.
ISBN 0–8173–0442–8 (alk. paper)
1. Poisonous animals—Alabama. 2. Poisonous animals—Gulf States. 3. Poisonous animals—Southern States. 4. Poisonous plants—Alabama. 5. Poisonous plants—Gulf States. 6. Poisonous plants—Southern States. I. Haynes, Robert R., 1945– . II. Thomas, Joab L. III. Title.
QL100.G53 1990 88–34003
574.6′5′09761—dc19 CIP

British Library Cataloguing-in-Publication Data available

The drawing on the title page:
Poison Ivy (*Toxicodendron radicans*), by M. E. Jackson.

The color photographs are by the authors, Ted Borg, Trip Lamb, Lloyd Logan, Dave Scott, Rebecca Sharitz, and others.

To Our Wives

Carol,
Elizabeth,
and
Marly

Contents

Foreword

Every moment of our lives, we are interacting on a chemical level with the world around us. Most of these events are beneficial to us. However, discomfort or injury occasionally results from contact with plants or animals. These toxic effects are frequently preventable, however, with careful planning.

The authors of this book, all with field experience as professional biologists in the southeastern United States, have focused on potentially harmful plants and animals in Alabama. They have chosen Alabama as a paradigm for the southeastern United States, but their information is relevant to a much wider geographic area. Through discussion and the use of maps, diagrams, and tables, the reader is assisted in identifying plants and animals to be avoided or deserving of caution because of potential risk.

The Alabama Poison Center receives several hundred calls each year regarding human encounters with plants, insects, or snakes. Certainly some of these prove much more consequential than others. Each person's immune system is primed, from its genetic makeup and subsequent experiences, to respond differently. A bee sting to two brothers may cause one to have local pain and the other to develop life-threatening respiratory difficulty as a hypersensitivity reaction. This response pattern is rarely predictable, and medical management following exposure must therefore be cautious yet tailored to the individual.

Prevention of exposure is clearly applicable to all. This volume should prove useful not only to biologists and medical professionals, but also to the gardeners, outdoorsmen, hunters, and fishermen of our society seeking safe and relaxing pastimes.

Robert J. Geller, M.D.
Diplomate, American Board of Medical Toxicology
Fellow, American Academy of Pediatrics
Tuscaloosa, Alabama

Preface

Our goal in this book is to introduce the reader to the fascinating array of plants and animals having the common ability to harm human beings through some means of toxicity. We have chosen to focus on a geographic area—the state of Alabama, its adjoining states, and its nearshore coastal areas—although many of the species have widespread geographic ranges and the concepts themselves have no boundaries. The justification for such a book is that almost everyone has an inherent concern with the possibility of plant or animal poisoning. A poisoned person has at least a passing interest in how or why it might have happened, not to mention what might be done about it.

Our reason for choosing Alabama and adjoining areas as the region of interest is to limit consideration to a manageable group of species. With its coast, Coastal Plain, Piedmont, and mountainous regions, Alabama is an identifiable unit representative of southeastern flora and fauna. All toxic groups of plants and animals discussed occur in surrounding states; most are found throughout the Southeast, and many are transcontinental in their distribution. The selection of the particular geographic region of Alabama, then, has the advantage of generality while providing a boundary for a topic that could become open-ended. The geographic range maps used for the species accounts were based on a variety of sources including museum records and personal observations of the authors.

Any plant or animal species alive on the earth today can be characterized by the means with which it protects itself from the hazards of its environment, including the possibility of being eaten, trampled, or injured by other animals and plants. For many, this protection is in the form of poisonous or venomous substances.

The harmful properties of most venomous animals and poisonous plants are not directed toward humans, although we may indeed suffer the consequences. For example, the high concentrations of alkaloids in larkspur plants are probably an effective deterrent to certain insects or grazing animals that would otherwise make a meal of the plant. It so happens that humans are also affected by the chemical makeup and can thus be severely poisoned by ingestion. Even the waxy oils of poison ivy to which many humans are so susceptible are probably only byproducts of the plant's metabolic processes and did not develop as a protective mechanism. For example, deer will readily eat poison ivy, and we know of a wildlife manager who contracted a severe case of poison ivy during his examination, by hand, of the gut contents of a deer that had consumed large quantities of this plant.

Even venomous snakes, despite the inherent danger to us, use their venom primarily to obtain food, not to attack people. It is probably a safe estimate that, during their lifetime, the majority of venomous snakes use their venom hundreds of times to obtain prey for every one time they use it for protection.

We are fortunate in North America to live on a continent where there is far less danger from native species than is true of parts of Africa, Asia, Australia, and South America. No species of native wild animal in Alabama will normally attack a human unless provoked in some manner. Few plant species are harmful to the touch, and the greatest dangers from ingestion can easily be avoided by exercising reasonable caution about what we swallow or chew. Alabama has a wealth of wildlife, some outstandingly beautiful habitats, and much to offer those who enjoy the outdoors. Awareness of where the potential dangers lie, coupled with the knowledge that all plants and animals must protect themselves, should allow us maximum enjoyment of the natural wonders of this state.

Poison Paranoia

As seen in the hospital records below and in the species accounts themselves, most poisonous or venomous species in Alabama cause only minor problems. One way to gain perspective about the dangers of encounters with native plants or animals is to compare them with the number of deaths occurring from a variety of other causes. In the first half of the 1980s, fewer than a dozen people in Alabama died from consuming native poisonous plants, including mushrooms, and from the bites or stings of venomous animals, including poisonous snakes. During this same period in Alabama, a great number died from accidental causes: automobile accidents (5,225); gunshot wounds (3,829); drownings (605); household poisonings (7).

On a relative scale, our native flora and fauna give us little cause for worry. In addition, the plants and animals of the state have much to offer in the way of outdoor enjoyment and opportunities for learning. Our objective in emphasizing the harmful qualities of some species is to elevate the level of understanding of the protective mechanisms and to reduce fear of and intimidation by the plants and animals found in Alabama and adjoining states.

Whit Gibbons
Robert R. Haynes
Joab L. Thomas

Acknowledgments

A book such as this cannot be written without the assistance and cooperation of many people. We are particularly indebted to colleagues who have contributed knowledgeable advice about certain groups of organisms, in particular, Herbert Boschung, Rebecca R. Sharitz, and Trip Lamb. Portions of the manuscript were improved by the comments of Janie M. Gibbons, Michael W. Gibbons, Dr. Lamb, and Dr. Sharitz. Drs. Lamb and Sharitz and Dave Scott contributed their photographic skills and color slides. The photograph of the io moth caterpillar was graciously provided by Lloyd Logan. Miriam Stapleton, Susan Kinlaw, Karin Knight, Marianne Reneau, and Malinda Doherty were helpful in manuscript typing. We especially thank L. Marie Fulmer, Sarah Collie, Judith L. Greene, and Patricia J. West for their help with preparation of the manuscript. Mary E. Jackson made most of the drawings of the plants and animals. The stingray was drawn by Jean Coleman. Mary E. Jackson, Jean Coleman, and Linda Orebaugh drew the range maps. Research necessary to complete the book was supported by Contract No. DE-AC09-76SR00819 between the U.S. Department of Energy and the University of Georgia.

We owe a special thanks to several professional colleagues who are experts in their fields and were able to assist us concerning specialty groups of plants or animals. The following provided critical information: Charles Covell, University of Louisville (stinging caterpillars); David Jenkins, University of Alabama at Birmingham (mushrooms); Robert Matthews, University of Georgia (wasps); Douglas Rossman, Louisiana State University (scorpions); Robert Shipp, University of South Alabama (fishes); Rowland Shelley, North Carolina State Museum (centipedes); Fred Stehr, Michigan State University (stinging caterpillars); and David H. Nelson, University of South Alabama (marine invertebrates).

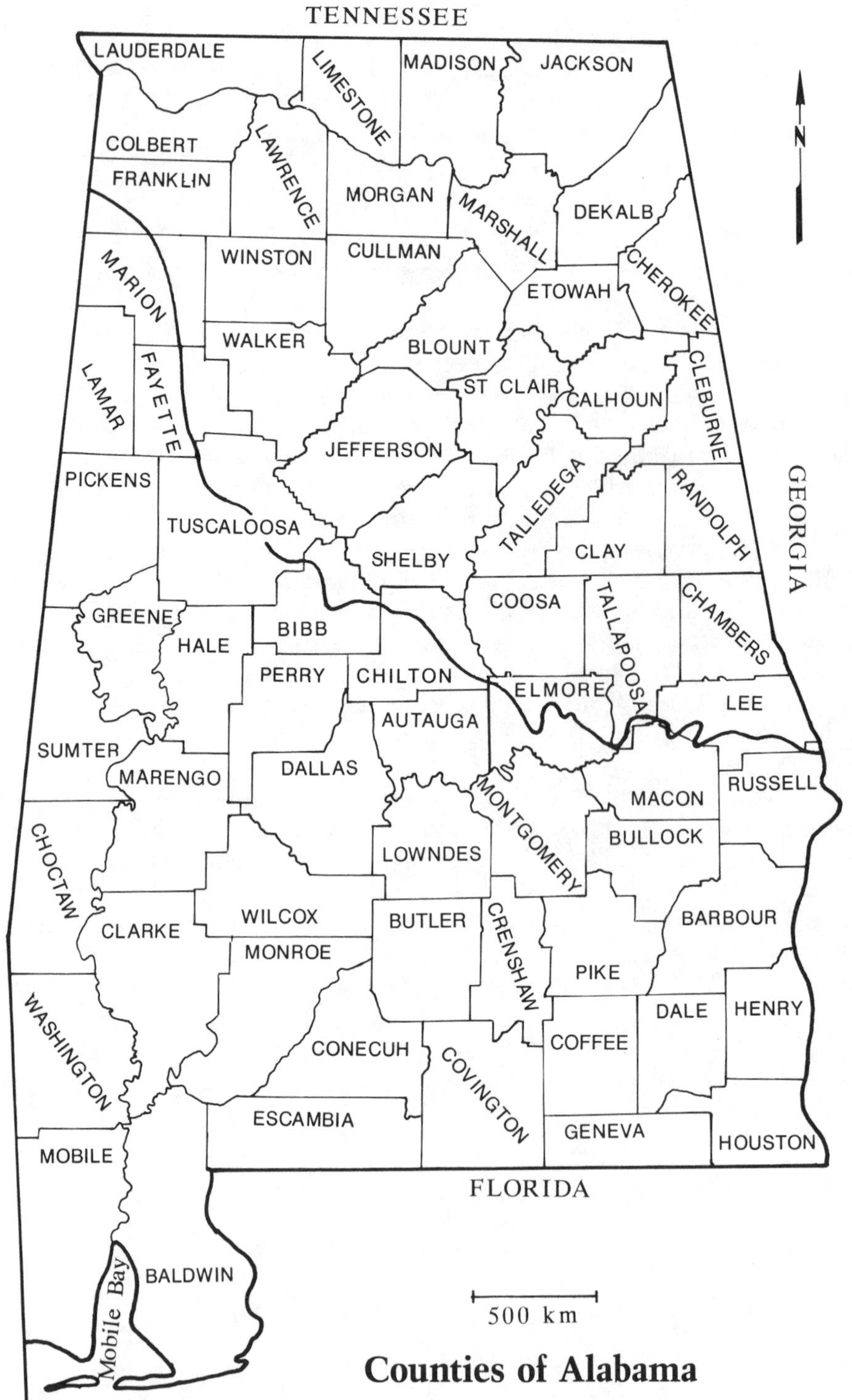

Counties of Alabama

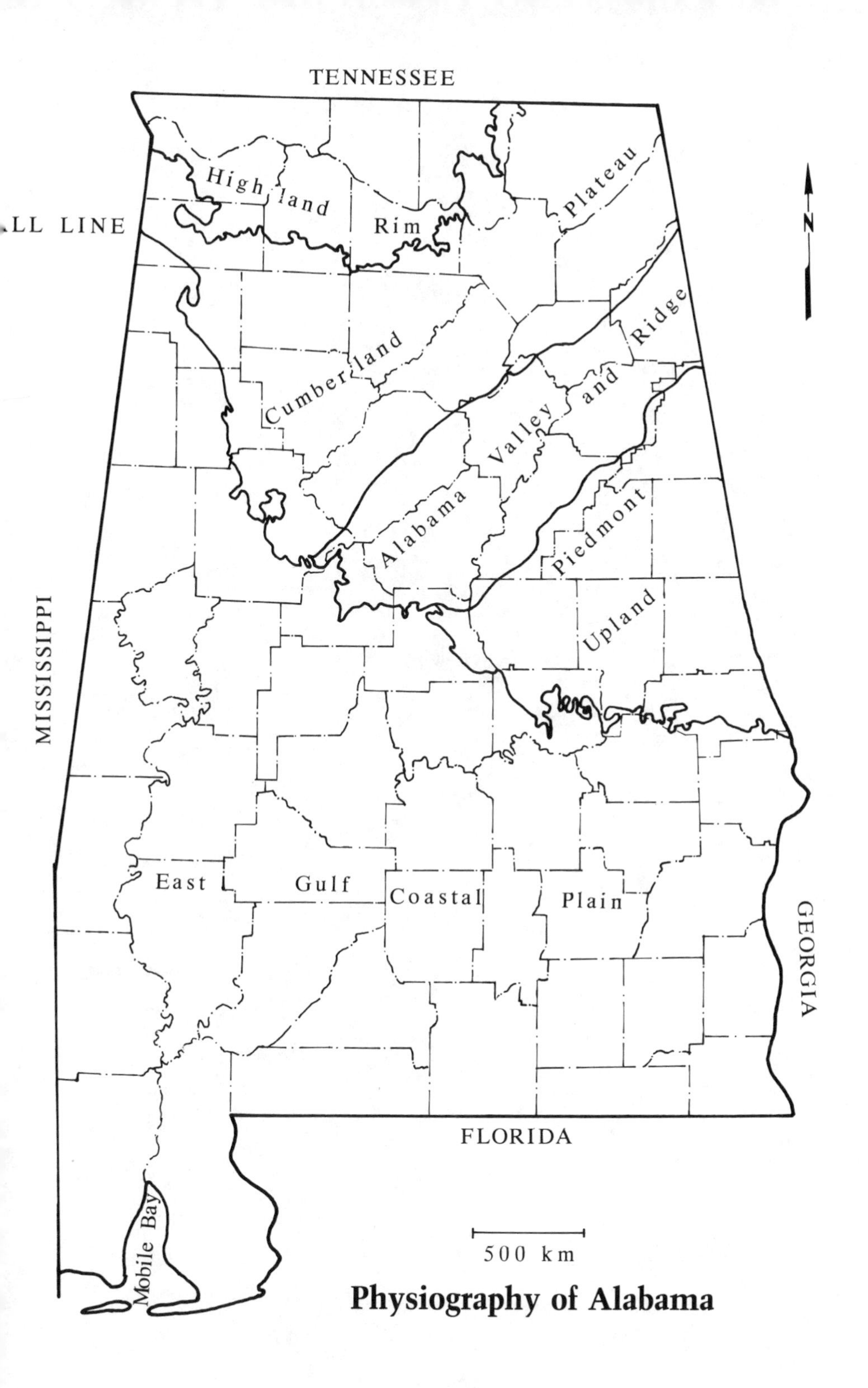

Physiography of Alabama

Chapter 1

Overview of the Poisonous Plants and Venomous Animals of Alabama

Alabama is similar to other southeastern states in the numbers and kinds of native plants and animals that can be poisonous in some form to humans. More than 400 species can be defined as technically poisonous and capable of having a harmful effect. However, the chance of serious harm from the majority of these species is minimal, and the probability of encountering a life-threatening species is very small for most people. Alabama does indeed have a few species, indicated below, that are truly dangerous under certain circumstances. But encounters with most poisonous flora and fauna of the Southeast usually cause no more than temporarily annoying, unpleasant, or slightly painful consequences. Alabama, like other eastern states, is a safe place to live with native creatures. A little knowledge of the biology of species that could potentially be harmful can make it even safer.

The toxicity of plants and animals has been known since the beginning of recorded history. The ancient Greeks obviously knew enough about some poisonous plants that Socrates ordered a glass of poison hemlock, and even before that Nicander actually wrote a book on poisonous plants of the world. As far as we know this was the first written word that recognized plants as being poisonous, but you can be sure that humans of any region have known that some plants and animals were dangerous. Cherokee Indians were either able to distinguish destroying angel mushrooms from the others or else they did not eat mushrooms.

More than 350 species of Alabama plants, including the mushrooms, have or are suspected of having poisonous properties at some stage in their life. Poisonous plants can be divided taxonomically into 3 primary groups: 1. Mushrooms that are known to harm people as a consequence of ingestion (Chapter 3). 2. Vascular plants that cause systemic effects; that is, the effect is internal, as a result of ingestion (Chapter 4). Records of harmful effects by many of these species are based solely on poisoning of livestock rather than on actual cases of human ingestion. Also, although the harmful effects from these plants are primarily from ingestion, the sap of some can irritate the eyes or nasal membranes. 3. Plants that cause dermatitis, an irritation or inflammation of the skin, as a consequence of contact with the plant or one of its volatile derivatives (Chapter 5).

Most of the venomous animals of the state are invertebrates (Chapter 6), which include all forms without backbones, such as the insects, spiders, and jellyfish. Although hundreds of species are technically capable of injecting poison in some manner, the number capable of causing harm to humans is relatively small. The remaining venomous animals in Alabama are vertebrates (Chapter 7), comprising primarily certain fishes and snakes.

What Are Poisonous Plants and Venomous Animals?

A poison is a substance that adversely affects and may even kill an organism through chemical action that can impair or destroy tissues, and that generally disrupts normal bodily functions. A toxin is a biologically produced substance; that is, one produced by a plant or animal and poisonous to another organism upon contact, either externally or internally. Venom is a toxin produced in certain glands of one animal and effectively transmitted to another by mechanical means, such as biting or stinging. All toxins and all venoms are poisons, but the converse is not true.

Innumerable plants and animals produce products that are poisonous if ingested by humans. Many of these are byproducts of the organism's natural metabolic processes and are not necessarily protective in nature. But some products are protective poisons that have been crafted by evolution to discourage other organisms, often specific ones, from adversely affecting the welfare of the individual or species. In seeking an appropriate limit for the subject matter of this book, we have chosen to consider any plants that are poisonous, by touch or through ingestion. Many of the plant species have only been known to poison livestock, not people, but have been included on the assumption that humans would be susceptible, too. As the title suggests, the coverage of animals is limited to those that poison primarily through injection of venom beneath the skin.

Why Do Plants and Animals Produce Toxins?

Most plants and animals have mechanisms to enhance their competitive position in the face of potential predators, browsers, or other encroachments. At the same time, individuals of each species must acquire and utilize energy in an effective manner to grow, to reproduce, and to maintain the metabolic processes necessary for day-to-day survival. To perform these functions, most organisms have developed into highly complex, living factories with intricate biochemical systems.

The chemical compounds produced by one organism are toxic when they are not compatible with the physiological makeup of another organism. This may be a part of the organism's defense mechanism or merely a byproduct of its metabolism.

In this book, the human is considered to be the receiving organism, and certain species of Alabama plants and animals that produce chemical compounds toxic to humans are considered to be the offending organisms. In nearly all instances, humans become victims of toxic chemical compounds that are produced by species whose evolutionary development in no way involved humans. It is probably coincidental that human systems react physiologically to certain chemical compounds as poisons. Surely, no one would be self-centered enough to suppose that the venom of the tiny madtom catfish or the poisonous oily sap of poison ivy was formerly or presently produced to cause grief to humans. Nonetheless, even though a compound was not intended for us, if we respond negatively to it, we should have an awareness of its capability and the foresight to avoid it.

Identification and Classification of Plants and Animals

A purpose of this book is to make it easy for anyone to determine whether a particular plant or animal might be a potential threat. Identification of species thus becomes important, and we have attempted to aid in the recognition process through descriptions, line illustrations, and photographs. A general understanding of the system by which related species are grouped is also helpful in learning to identify plants and animals.

Plants and animals are classified by biologists into natural groupings of species in a systematic manner from large groups to progressively smaller ones. The system was developed more than two centuries ago and is still the most effective approach for presenting and comparing the biological kinship and common ancestry of species. Each species in the plant or animal kingdom is a member of a **genus** that includes other closely related species, if there are any. Closely related **genera** (plural of genus) are grouped into a **family**. Families that have a common ancestry belong to an **order**, and closely related orders are placed in a particular **class**. Similar classes of animals are grouped into a **phylum** (plural is **phyla**). Similar classes of plants are grouped into a **division**. The classification scheme for grouping related organisms is indicated in Table 1.

This family tree arrangement serves as an indicator of the biological relationships of the countless species that inhabit the earth. Table 2 shows the families of poisonous plants and venomous animals of Alabama and the number of species found in each. The other taxonomic categories provide an indication of the taxonomic relationships of the families.

An attempt was made to use the classification system that is currently accepted for each major group of organisms. However, authorities sometimes disagree about taxonomic designations so that it was occasionally necessary to make our own judgment calls. Several different references were used in the determination of scientific and common names used in this book.

Format for Presentation of Species

The characteristics of the species of poisonous plants and venomous animals are presented in a similar format throughout most of the book, al-

Table 1. General Classification Scheme for Species of Plants and Animals

Poison ivy and the cottonmouth moccasin are used as examples of how biologists classify plants and animals.

SCIENTIFIC NAME (genus and species)	*Toxicodendron radicans*	*Agkistrodon piscivorus*
GENUS (may include several closely related species)	*Toxicodendron* (includes poison sumac)	*Agkistrodon* (includes copperhead)
FAMILY (may include several closely related genera)	Anacardiaceae (includes genus of cashews)	Crotalidae (includes two genera of rattlesnakes)
ORDER (may include several closely related families)	Sapindales (includes holly, maple, and horse-chestnut families)	Squamata (includes all families of snakes and lizards)
CLASS (may include several closely related orders)	Magnoliopsida (includes all orders of dicot flowering plants)	Reptilia (includes all orders of reptiles)
DIVISION (includes all closely related classes of plants)	Magnoliophyta (includes all flowering plants)	
PHYLUM (includes all closely related classes of animals)		Chordata (includes all vertebrates)

though some groups are handled differently because of special considerations. Rarely, a plant or animal specimen cannot be identified with certainty, even by an expert taxonomist using all of the scientific information and techniques available. Closely related species within a genus may appear almost identical, although the species are genetically different and normally do not interbreed. For practical purposes, only representative species of some such groups are discussed in any detail because of the similarity of appearance of several species within a genus (e.g., wild azaleas [genus *Rhododendron*] and bumblebees [genus *Bombina*]). For certain genera (e.g., pigweed [genus *Amaranthus*] and madtom catfish [genus *Noturus*]) only geographic range maps are provided for some species because of their similarity in appearance, ecology, and presumed toxicity. Some families (e.g., certain mushrooms [family *Boletaceae*] and ants [family *Formicidae*]) are discussed in an even more general manner, rather than by species accounts, because of taxonomic uncertainties, unresolved distribution patterns, or large numbers of species that look similar and about whose ecology little is known.

Measurements are given in metric units throughout the book. The follow-

Table 2. Classification and Higher Taxonomic Relationships of the Poisonous Plants and Venomous Animals of Alabama

The classification systems followed are indicated in the Selected References. An asterisk (*) indicates that the number of poisonous or venomous species is not designated because of uncertainty about the taxonomy, geographic distribution, or potential toxicity of species in the family. Parentheses enclosing species numbers indicate that the level of toxicity of some species is unknown, although closely related ones are toxic.

Plant Kingdom

Division	Class	Order	Family	Number of Poisonous Species in Alabama
BASIDIOMYCOTA	HOMOBASIDIOMYCETES	BASIDIOMYCETES (gilled mushrooms)	AMANITACEAE (Amanita family)	15+
			LEPIOTACEAE	2
			RUSSULACEAE	2
			TRICHOLOMATACEAE	5
			COPRINACEAE	2
			STROPHARIACEAE	2
			CORTINARIACEAE	2
			CLAVARIACEAE	1
			BOLETACEAE	2
ARTHROPHYTA	EQUISETOPSIDA	EQUISETALES (horsetails)	EQUISETACEAE (horsetail family)	2
PTERIDOPHYTA	PTERIDOPSIDA	PTERIDALES (ferns)	PTERIDACEAE (fern family)	1
			ASPIDIACEAE (sensitive fern family)	1
PINOPHYTA	PINOPSIDA	PINALES (conifers)	CUPRESSACEAE (cypress family)	1

Division	Class	Order	Family	Number of Poisonous Species in Alabama
MAGNOLIOPHYTA	MAGNOLIOPSIDA	MAGNOLIALES (magnolias)	CALYCANTHACEAE (spicebush family)	1
			ANNONACEAE (custard-apple family)	3
		ARISTOLOCHIALES (birthworts)	ARISTOLOCHIACEAE (birthwort family)	2
		RANUNCULALES (buttercups)	RANUNCULACEAE (buttercup family)	(30+)
			BERBERIDACEAE (mayapple family)	1
			MENISPERMACEAE (moonseed family)	1
		PAPAVERALES (poppies)	PAPAVERACEAE (poppy family)	3
			FUMARIACEAE (fumatory family)	3
		URTICALES (nettles)	MORACEAE (mulberry family)	1
			CANNABACEAE (hemp family)	1
			URTICACEAE (nettle family)	4
		JUGLANDALES (walnuts)	JUGLANDACEAE (walnut family)	2
		CARYOPHYLLALES	PHYTOLACCACEAE	1

	NYCTAGINACEAE (four-o'clock family)	2
	CHENOPODIACEAE (lamb's-quarter family)	6
	POLYGONACEAE (buckwheat family)	(20)
	AMARANTHACEAE (amaranth family)	(7)
BATALES (saltworts)	BATACEAE (saltwort family)	1
THEALES (teas)	HYPERICACEAE (St. John's wort family)	1
MALVALES (mallows)	MALVACEAE (mallow family)	1
ERICALES (heaths)	ERICACEAE (heath family)	(17)
ROSALES (roses)	FABACEAE (LEGUMINOSAE) (bean family)	32
	ROSACEAE (rose family)	4
	SAXIFRAGACEAE (saxifrage family)	2
MYRTALES (myrtles)	THYMELAEACEAE (mezereum family)	1
SANTALALES (sandalwood)	LORANTHACEAE (mistletoe family)	1
EUPHORBIALES (spurges)	EUPHORBIACEAE (spurge family)	(16)

Division	Class	Order	Family	Number of Poisonous Species in Alabama
MAGNOLIOPHYTA	MAGNOLIOPSIDA	RHAMNALES (grapes)	RHAMNACEAE (coffeeberry family)	2
			VITACEAE (grape family)	1
		SAPINDALES (soapberries)	SIMAROUBACEAE (quassia family)	1
			RUTACEAE (orange family)	1
			MELIACEAE (mahogany family)	1
			CELASTRACEAE (bittersweet family)	1
			ANACARDIACEAE (cashew family)	3
			ACERACEAE (maple family)	2
			HIPPOCASTANACEAE (hippocratea family)	5
		GERANIALES (geraniums)	OXALIDACEAE (oxalis family)	9
			BALSAMINACEAE (touch-me-not family)	2
		APIALES (parsley)	ARALIACEAE (ginseng family)	2

	APIACEAE (parsley family)	2
GENTIANALES (gentians)	LOGANIACEAE (logania family)	3
	APOCYNACEAE (dogbane family)	3
	ASCLEPIADACEAE (milkweed family)	(14)
POLEMONIALES (phlox)	SOLANACEAE (nightshade family)	7
	CUSCUTACEAE (dodder family)	6
LAMIALES (mints)	VERBENACEAE (verbena family)	2
	LAMIACEAE (mint family)	1
SCROPHULARIALES (snapdragons)	BIGNONIACEAE (trumpet-vine family)	1
	OLEACEAE (olive family)	2
CAMPANULALES (bellflowers)	CAMPANULACEAE (bellflower family)	11
RUBIALES (madders)	RUBIACEAE (madder family)	1
DIPSACALES (honeysuckles)	CAPRIFOLIACEAE (honeysuckle family)	5
ASTERALES (sunflowers)	ASTERACEAE (sunflower family)	10

Division	Class	Order	Family	Number of Poisonous Species in Alabama
MAGNOLIOPHYTA	LILIOPSIDA	NAJADALES (pondweeds)	JUNCAGINACEAE (arrow-grass family)	1
		CYPERALES (sedges)	POACEAE (GRAMINEAE) (grass family)	1
		ARALES (calla lilies)	ARACEAE (arum family)	3
		LILIALES (lilies)	LILIACEAE (lily family)	(28)
			HAEMODORACEAE (bloodwort family)	1
			IRIDACEAE (iris family)	6

Animal Kingdom

Phylum	Class	Order	Family	Number of Venomous Species in Alabama
COELENTERATA	HYDROZOA		PHYSALIIDAE	1
			PENNARIDAE	*
ECHINODERMATA				*
MOLLUSCA	GASTROPODA		CONIDAE	*
PLATYHELMINTHES				*
ANNELIDA				*

ARTHROPODA	CHILOPODA (centipedes)			*
	ARACHNIDA	SCORPIONIDA (scorpions)	VAEJOVIDAE	1
		ARANEAE (spiders)	LOXOSCELIDAE (brown recluse spider family)	1
			THERIDIIDAE (black widow family)	1
			LYCOSIDAE (wolf spider family)	*
	INSECTA	LEPIDOPTERA (moths and butterflies)	MEGALOPYGIDAE	*
			LIMACODIDAE	*
			ARCTIIDAE	*
			NOCTUIIDAE	*
			LYMANTRIIDAE	*
			SATURNIIDAE	2
			LASIOCAMPIDAE	*
		HYMENOPTERA (stinging insects)	VESPIDAE (wasp family)	7+
			SPHECIDAE (cicada killer family)	*
			MUTILLIDAE (velvet ant family)	*
			APIDAE (bee family)	*
			FORMICIDAE (ant family)	*

Phylum	Class	Order	Family	Number of Venomous Species in Alabama
CHORDATA	CHONDRICHTHYES	MYLIOBATIFORMES (rays and eagle rays)	DASYATIDAE (ray family)	2
			MYLIOBATIDAE (eagle ray family)	2
	OSTEICHTHYES	SILURIFORMES (catfishes)	ICTALURIDAE (catfish family)	18
			ARIIDAE (sea catfish family)	2
		PERCIFORMES (perches)	CARANGIDAE (jack family)	*
			SCORPAENIDAE (scorpionfish family)	3
			URANOSCOPIDAE (stargazer family)	*
	REPTILIA	SQUAMATA (snakes and lizards)	CROTALIDAE (pit viper family)	5
			ELAPIDAE (coral snake family)	1
	MAMMALIA	INSECTIVORA (insectivores)	SORICIDAE (shrew family)	1

ing table is provided to indicate the observations used and to simplify conversion between metric and English.

cm = centimeter	1 cm = 0.39 inches 1 inch = 2.54 cm	100 cm = 1 m
m = meter	1 m = 39.37 inches 10 m = 32.8 feet	1,000 m = 1 km
km = kilometer	1 km = 0.62 miles 1 mile = 1.61 km	

Toxic properties are sometimes considered at the species level in the accounts. However, because of the similarity of the toxins of all species within some genera or families or because of the lack of information about the distinction among the species within a group, it is sometimes more practical to consider the entire genus or family in this regard.

The geographic range is given for most of the species that are considered. The known ranges as used in this book do not reflect in every case actual museum records, but rather data from a variety of sources that provided evidence of the geographic range of the species. Future studies may serve to modify and refine the ranges presented in the maps. Sources used in the establishment of geographic ranges are listed in the Selected References.

A major attempt is made throughout the book to provide treatment recommendations for each species or toxic group when such information is known. However, the reader should be cautioned that plant and animal toxins are extremely complex, and effects are often unpredictable because of each individual's unique physiology, allergenic response, general health, and other circumstances. The treatments given are based on general success records, but none should be taken as medical advice. Victims are advised to consult a poison center, their private physicians, or an emergency treatment facility for medically based treatment recommendations.

Alabama's Potentially Lethal Species

Only a few species of native or introduced plants or animals are known to cause human death. Unfortunately, many of the statistics on deaths from plant and animal poisonings are suspect because hospital records of lethal poisonings are difficult to acquire and case histories are often complicated and confounded by extenuating circumstances. For example, there is a record of a forest ranger who was bitten by a copperhead and immediately had a heart attack, presumably from fright and excitement, before the snake's venom had any time to act. If he had died later from the effect of the venom, this could have resulted in a recorded death from the bite of a copperhead, one of the venomous snakes for which few if any human deaths have been documented.

There is another interesting example from the records of a local Alabama hospital. A boy had been bitten by a spider of some unknown identity, taken to the hospital for treatment, and released the same day with no serious ill effects. The hospital had a policy of making routine telephone calls to a released patient's home in such cases to inquire about the welfare of the pa-

Table 3. Alabama's Most Dangerous Plants and Animals

The species listed are those that can bring about serious medical consequences, even death in some instances, to humans. Although many other plant species in Alabama are poisonous to touch or eat, and numerous animals can cause noticeable irritation through envenomation, the species listed in this table are the potential offenders of which anyone in Alabama, its coastal waters and adjacent states, should be aware. All these species are believed to have caused human death, although for most there is no record of a death in the state of Alabama.

Scientific Name	Common Name
Delphinium sp.	larkspur
Aleurites fordii	tung oil tree
Cicuta maculata	water hemlock
Cocculus carolinus	coral beads
Conium maculatum	poison hemlock
Riccinus communis	castor bean
Solanum gracile	deadly nightshade
Phytolacca americana	common pokeweed
Kalmia latifolia	mountain laurel
Lathyrus hirsutus	everlasting pea
Phoradendron serotinum	American mistletoe
Melia azedarach	chinaberry
Nerium oleander	oleander
Datura stramonium	jimsonweed
Lobelia sp.	lobelia
Eupatorium rugosum	white snakeroot
Amianthium muscaetoxicum	fly poison
Zigadenus densus	black snakeroot
Amanita virosa	destroying angel mushroom
Physalia physalis	Portuguese man-o-war
Latrodectus mactans	black widow spider
Loxosceles reclusa	brown recluse spider
Order HYMENOPTERA	bees, wasps, hornets, ants
Dasyatis sabina	Atlantic stingray
Agkistrodon piscivorus	cottonmouth moccasin
Crotalus adamanteus	eastern diamondback rattlesnake
Crotalus horridus	timber (canebrake) rattlesnake
Sistrurus miliarius	pygmy rattlesnake
Micrurus fulvius	eastern coral snake

tient. The family apparently became annoyed after the second call, and on the third one, their answer to the question about the boy's condition was, "He died." Under the circumstances the hospital did not believe the answer, but some hospital records might attribute a relatively harmless spider with a human death.

Despite a mass of misinformation and unreliable data, certain species of plants and animals have been reliably reported in medical records to have been the definitive cause of death through the action of their poison. A list of the most dangerous species that occur in Alabama and are known to have caused death is shown in Table 3. In most instances the known deaths did not occur in Alabama.

Aside from its relatively few species known to cause death, Alabama has a large number of plant species that are technically poisonous and animal species that are technically venomous; the levels of effect may range from none to trivial to painful, but not be potentially lethal under normal circumstances. The majority can be placed in the first two categories—no effect or a trivial one; some of these will be discussed. However, a few species can be singled out because their effect is usually severe, even though the consequences are not considered life-threatening (Table 4).

Table 4. Nondeadly Poisonous Plants and Venomous Animals of Alabama That May Cause Noticeable Pain and Discomfort, but Would Not Be Lethal Under Normal Circumstances

Scientific Name	Common Name
Cnidoscolus stimulosus	stinging nettle
Toxicodendron radicans	poison ivy
Toxicodendron toxicarium	poison oak
Toxicodendron vernix	poison sumac
Arisaema triphyllum	jack-in-the-pulpit
Aralia spinosa	devil's-walking-stick
Urtica dioica	nettle
Amanita muscaria	fly amanita mushroom
Pennaria tiarella	stinging hydroid jellyfish
Vaejovis carolinianus	brown unstriped scorpion
Sibine stimulea	saddleback caterpillar
Family Mutillidae	velvet ants
Agkistrodon contortrix	copperhead
Noturus sp.	madtom catfishes
Bagre marinus	gaff-topsail catfish

Chapter 2

Categories of Biological Toxins and Suggested Treatments

One of the frustrations faced by the victim of any form of biological poisoning is the uncertainty of the level of danger. How many castor beans or how much destroying angel mushroom does one have to eat for it to be lethal? How big does a diamondback rattlesnake have to be to kill someone? The answers are not simple. They depend on numerous variables.

The complexity and variability in the types of toxins in plants and animals are overwhelming, even to the toxicologist. To most people a detailed biochemical accounting would be of little value. However, a general overview and categorization of some of the types of poisons that occur in organisms are provided. They will allow the reader to gain some familiarity with the general types of biological poisons that exist in nature.

Mushroom Toxins

The poisonous components of mushrooms have been classified into several groups, based on their chemical composition and their physiological effects. The highly complex biochemical nature of the toxins, however, has defied a precise identification of the primary toxic agent of particular species, in many instances, because of the difficulty in deciding whether one compound causes an effect or is simply associated with another compound that is the true culprit. Certain general toxin groups are identifiable in Alabama mushrooms.

Phallotoxins and Amatoxins

Several members of the genera *Amanita* and *Galerina* contain complex proteins known as phallotoxins and amatoxins that can have severe effects on the liver and gastrointestinal tract. Death or permanent organic damage can occur as a consequence of ingesting species containing these compounds.

Muscarine

Muscarine is an organic compound ($C_9H_{21}O_3N$) found in small amounts in the fly amanita and in greater concentrations in certain other mushrooms. Muscarine is believed to be responsible for the toxicity of some of the species. The effects include dilation of blood vessels and a reduction in rate of heartbeat.

Hallucinogenic Compounds

Varieties of organic compounds, including psilocybin, psilocin, muscimol, ibotenic acid, pantherin, and tricholomic acid, found in many mushroom species act on the central nervous system and can cause hallucinations. The effects of given concentrations and proportions are often dependent on the physiological and psychological condition of the victim so that it is difficult to predict the response one might have after ingestion. So, perhaps Lewis Carroll had as much experience as imagination when he had Alice's mind doing tricks as she ate the mushroom. However, hallucinogenic mushrooms should not be put in the category of recreational drugs since the consequences can be serious and potentially fatal.

Gastrointestinal Toxins

Because of the rarity of certain mushroom species and the infrequency of human poisoning, the toxic compounds of many species have not been thoroughly analyzed, although some component is known to cause poisoning. Thus, many gastrointestinal disorders, ranging from mild to serious, occur as a consequence of undescribed toxins in some mushroom species.

Vascular Plant Toxins

Numbers of compounds are produced by vascular plants that under certain conditions may cause toxic reactions to humans or domestic animals. The compounds mentioned most commonly can be arranged into 6 groups. These are alkaloids, glycosides, oxalates, phytotoxins, minerals, and those causing photosensitivity.

Alkaloids

Alkaloids in plants are byproducts of various metabolic pathways but have no apparent physiological role. These compounds are often toxic and almost always bitter. As a result, they possibly function for protection from grazing animals, including plant-eating insects.

Alkaloids are organic molecules containing nitrogen, which are usually present in a heterocyclic ring. The nitrogen will act as a base, since it has the ability to accept hydrogen ions. Most solutions containing alkaloids are basic. In general, alkaloids are white crystalline compounds and are only slightly soluble in water.

Over 3,000 alkaloids have been found in some 4,000 species of plants. The first alkaloid to be isolated and crystallized was the drug morphine, found over 150 years ago in the opium poppy. Other well-known alkaloids include nicotine, caffeine, quinine, strychnine, atropine, and colchicine.

Most alkaloids produce a strong physiological reaction when introduced into an animal, but a few produce no reaction at all. In most cases, activity is effected primarily via the nervous system by a mechanism that is at best poorly understood. Lesions are absent. Some types of alkaloids produce completely different syndromes; e.g., pyrrolizidine alkaloids cause severe liver damage.

GLYCOSIDES

Glycosides are compounds that yield one or more sugars and one or more other compounds (aglycones) when hydrolyzed. They are all phenolic compounds; that is, they contain one or more 6-carbon rings. Glycoside sugars are bound to one or more of the carbons. Although disaccharides do occur, the most common glycoside sugars are monosaccharides and include glucose, galactose, rhamnose, xylose, and arabinose. Hundreds of glycosides are known and new ones constantly are being discovered. The concentration of a particular glycoside in a plant depends on the genetics of the species, as well as ecological conditions in which the plant is growing. Therefore, a particular glycoside poison can occur in some specimens of a plant species but be totally absent in others, thus making it difficult to classify certain species as poisonous.

Toxic glycosides include cyanogenic (nitrile) glycosides, goitrogenic substances, irritant oils, coumarin glycosides, and steroid glycosides.

Cyanogenic Glycosides

Cyanogenic glycosides are composed mostly of one carbon ring and release hydrocyanic acid (HCN) upon hydrolysis. This class of compounds is found abundantly in the family Rosaceae. The intact glycosides are harmless, and the toxicity of these compounds is due entirely to the HCN released during hydrolysis.

Hydrocyanic acid acts by inhibiting cytochrome oxidase, an enzyme that is the terminal step in cellular respiration. This enzyme catalyzes the reaction linking atmospheric oxygen to the byproduct of respiration (hydrogen ions) in the formation of water. Therefore, HCN poisoning results in cellular asphyxiation. Death usually follows within 15 minutes to a few hours after a lethal dose is ingested.

Goitrogenic Substances

Goitrogenic substances, which include thiouracil, thiourea, cyanides, sulfonamides, thiocyanates, and L-5-vinyl-2-thiooxazolidone, prevent the thyroid from accumulating inorganic iodide. This inhibits the formation of the thyroid hormone.

Irritant Oils

These oils are present in the plants mostly as glycosides that upon hydrolysis yield a sugar and the irritant oil (an aglycone). Neither the glycoside nor the sugar is toxic, but the oil is, often in fairly low concentrations. Two such examples are mustard oils (of the Brassicaceae) and ranunculin (of the Ranunculaceae).

Coumarin Glycosides

The best-known coumarin glycoside is a coumarin derivative, dicoumarol, that is produced when *Melilotus* (sweet clover) spoils. Dicoumarol is a hemorrhagic agent that reduces blood prothrombin levels, and prevents blood from clotting.

Steroid Glycosides

Steroid glycosides have aglycones composed of cyclic chains of carbon atoms, usually numbering 20 or more. The large aglycone by itself is toxic, but is not, or is only slightly, soluble. The sugars increase the solubility, thus increasing the toxic effect.

This group can be divided by physiological means into two classes, those that stimulate the heart and those that do not. Cardiac glycosides, the first class, have been of medicinal use. They act directly on heart muscles to increase the force of contraction in systole, and on the vagus innervation to decrease the rate of heartbeat. Overdoses produce nausea, dizziness, blurred vision, and diarrhea.

Saponins, the second class, are large molecules that form colloidal solutions and produce a foam in stirred water. They are essentially nonsoluble in water. As a result, they are not readily absorbed into the bloodstream through healthy digestive tissue. However, when they possess, or are accompanied by, substances with irritant properties sufficient to injure the wall of the digestive tract, absorption occurs followed by severe gastroenteritis.

Oxalates

Oxalates occur in plants either in the soluble state, as oxalic acid, potassium acid oxalate, and sodium oxalate, or in the insoluble state, as calcium oxalate. Soluble oxalates that are absorbed into the bloodstream result in an immediate drop in ionic calcium in serum. A more serious effect, however, is the precipitation of oxalate crystals in the kidney tubules. Severe cases result in loss of kidney function.

In some plants, especially the Araceae, calcium oxalate occurs as crystals in all vegetative tissues. These crystals cause mechanical injury to tissues of the oral tract when plants containing the crystals are chewed and swallowed.

Phytotoxins

Phytotoxins are protein molecules that, like bacterial toxins, cause an antibody response in the individual. In addition, they function as enzymes and break down critical natural proteins, resulting in accumulation of ammonia. Unlike most bacterial and animal toxins, they can be absorbed through the digestive tract. Gastrointestinal irritation occurs, usually with inflammation and swelling of several organs.

Mineral Poisoning

Many species of plants concentrate minerals or elements in their tissues. These may be by-products from enzymatic activities or may be absorbed from the soil. Common minerals in plants include copper, lead, cadmium, fluorine, manganese, nitrogen, and selenium. Ingesting plants with high mineral concentrations may result in symptoms ranging from nausea to deformed offspring to death.

Photosensitivity

Chemicals causing photosensitivity are produced in only a few species. After vegetation containing such a compound is ingested, the compound is absorbed through the intestinal wall into the circulatory system, eventually reaching the peripheral circulation. Light is probably absorbed by this compound, resulting in an oxidation reaction. This oxidation causes the animal to become hypersensitive to light, resulting in the development of erythema and pruritus followed by edematous suffusions and often necrosis of the skin in affected areas.

Animal Venoms

Animal venoms are generally complex chemical compounds that may consist of a few or numerous proteins, including enzymes and polypeptides, as well as other components, such as steroids, amines, and lipids. By definition, a venom is produced in special glands or cells of one animal and injected into the body of another animal. The exact makeup of venom varies from one species to another. Venoms are so biochemically complex that only a tiny fraction of those that exist in the animal kingdom have been successfully surveyed and analyzed in their entirety. However, the painful or physiologically disruptive components have been identified for many venomous species. For example, peptide substances have been indicted as the chemical cause of some caterpillar stings. Histamine, serotonin, and acetylcholine are reported to be the primary pain producers from bee stings, whereas melittin is the most common substance in the venom and causes much of the physiological disorder. Most people are aware that formic acid is a major component and noxious agent in ant venoms.

Snake venoms are complex substances of which proteins, including many enzymes, are a significant part. Certain elements, such as zinc and calcium, are found in some snake venoms and have been suggested as possibly affecting certain physiological processes in victims. Peptides are considered to be the venom components responsible for the serious medical complications resulting from snakebite, but the complexity of the venom mixture (and the often unknown interaction of these constituents) makes it difficult to precisely identify the causal factor of a particular physiological response. Wide variability in physiological condition confounds predictions of how an individual will likely react to injection by a particular venom. Thus, not only are snake venoms mostly indecipherable chemical compounds, their effects are equally unpredictable.

Two traditional categorizations of snake venom that bear mentioning are the traditional *neurotoxic* and *haemotoxic* (or *hemotoxic*). The former is used to refer to venom that affects the nervous system and the latter to venom that affects the blood and body cells. The terms *myotoxic* (affecting the muscles) and *cardiotoxic* (affecting the heart) have also been used in connection with venomous snakebite. These terms are mentioned not to advocate their use but to cover terminology with which many readers may be at least vaguely acquainted. Many physicians and scientists experienced in venom biochemistry and snakebite treatment decry the use of any of these terms as being too simplistic and clinically misleading, but it would be an error to ignore the words and their standard definitions.

Treatment

One of the first things that a victim of poisoning by a plant or an animal wants to know is what to do about it immediately. People who even suspect the possibility of poisoning are likely to have a serious interest in the most appropriate treatment. The obvious expectation from a book on venomous animals and poisonous plants is general and specific advice.

The first appropriate step toward medical treatment following contact with a poisonous plant or venomous animal will vary with the situation. For someone who has ingested or come in physical contact with a suspected poisonous plant or who has been bitten or stung by an animal, but for whom no symptoms, or minor symptoms, have appeared, a telephone call to a poison center may be the most efficient approach. Advice can thus be obtained about what symptoms to look for and how to proceed if the victim's condition worsens. If the symptoms already seem serious, the victim should proceed or be taken to an emergency facility where stabilization treatment is likely to be immediately available. A trip to a poison center or emergency treatment facility may be more effective than one to the office of a physician because of the greater likelihood that the attending medical personnel will have had more experience with emergency treatment and that appropriate facilities and equipment will be present. No first-aid cure-all can be prescribed for biological poisoning. These recommendations are given as the least risky and most useful approaches to an unpredictable but potentially dangerous situation.

Treatment that works for some forms of toxicity may be totally inappropriate for others. For example, induced vomiting would be advisable for certain ingested plant materials to get them out of one's system; on the other hand, vomiting is not recommended for other plants, such as members of the aroid family (e.g., jack-in-the-pulpit), which produce calcium oxalate crystals. These crystals can severely cut the esophagus going down and would do the same thing coming up. So offering first-aid measures is an uncertain business because of the many variables.

A cautious effort has been made to recommend treatments under some species accounts in the book, but always with qualifications that should be carefully considered. This is not only because of lack of information but primarily because of the extreme complexity of the biochemical structure that constitutes the bewildering array of animal venoms and plant poisons. Most venoms are so complex biochemically that few have been analyzed with any satisfactory degree of certainty. It is not surprising, therefore, that the medical industry, which must also deal with the complexity of the human body and the variability among individuals, has yet to devise totally satisfactory means of treating venomous bites and stings. Few physicians are likely to give advice on the best treatment for animal bites and stings outside the medical arena of a doctor's office, clinic, or hospital. We will, of course, not try to second-guess the medical profession.

The consequences of poisoning from ingestion or envenomation are dependent on such a wide variety of factors that preparation can only be based on probability. The chances of assessing probable effects on a victim increase with the availability of information about certain variables. For example, if a physician knows that a patient under treatment for a diamondback rattlesnake bite has a history of heart ailments, then he knows that the chances of successful recovery are less than for a healthy person. Or, if the victim is allergic to horse serum-based antivenin, then this form of medical treatment could reduce his chances for speedy recovery. If the victim received venom from only one of the snake's two fangs, then his probability of ill effects from the venom is reduced by half.

Other variables that can influence the impact of biological poisoning include the size, age, and general health (both mental and physical) of the victim and the specific physiological state at the time of encounter. For example, someone with alcohol in the bloodstream or with a fast heart rate from jogging would respond differently to a snakebite than would an abstemious person calmly strolling through the woods.

Another personal factor that could influence response to a toxin is a previous encounter. For example, the sensitivity to some insect stings increases with each encounter with the toxin. Yet another factor is the place on the body where an animal stings or bites, or where a plant makes contact. Being brushed in the eye by poison ivy or a stinging caterpillar could result in far more extensive complications than contact on the forearm. Other factors are functions of the particular physiological conditions and past history of the offending plant or animal. As will be indicated in the species accounts, some portions of a plant can be extremely toxic whereas other parts are harmless. Toxicity can also vary with the age or developmental stage of the plant (e.g., old poke plants are more likely to be toxic than young ones), the season of the year, or even the soil type in which the plant is growing. The toxicity of some mushrooms is particularly difficult to evaluate because of their seasonal and regional variability.

Among venomous animals, the quantity of the poison injected can vary considerably among individuals of the same species. The amount of venom delivered depends on such factors as the size of the organism and the length of time since it last used its venom supply. In some species its sex makes a difference (for example, entomologists and physicians are unsure whether male black widow spiders can cause a serious bite or whether all male wasps are incapable of stinging). The condition of the structure used in envenomation is also important. A broken spine or fang will lessen the amount of venom injected.

A final little understood factor that is of utmost importance involves the level of control exercised by the offending animal. Some venomous snakebites have been reported that resulted in no injection of venom. Similar incidents have been observed with scorpions, centipedes, and spiders—a bite or sting that had no effect. Such "dry bites" or "dry stings" are thought by some zoologists to be controlled by the animal. Venom is a valuable commodity that is required for obtaining food by many species, and venom used for protection is no longer available for prey. So, some dry bites or stings are thought to be intentional attempts by a venomous organism to startle a large enemy (humans) without wasting the venom.

Besides the interaction between the condition of the victim and that of the toxic organism, another factor can have major influence on the level of severity. The treatment administered to the patient can sometimes make a difference in whether one leaves the encounter with practically no ill effects or whether one dies. Failure of treatment can not only result from improper or delayed first-aid measures but from improper or delayed treatment at a medical facility. Professional ignorance about the biochemical complexity of toxins is one reason for improper treatments. As an example, there is no logical and apparent reason why heat neutralizes stingray venom, whereas wasp stings are best assuaged by cold. Even a physician, if unfamiliar with a venomous species, could administer the wrong treatment because of the variability among types of venomous species.

Chapter 3
Poisonous Mushrooms

Mushrooms abound in Alabama, and the actual number of species found in the state is unknown. One reason the number remains in question is that many mushrooms are similar in appearance and difficult to identify. Thus, only a mushroom specialist can be assured of making proper identification of certain rare forms. This is precisely the problem that the average person faces in deciding whether a mushroom is safe to eat. And the problem can be more serious than whether one merely has a stomachache or sees visions. People have died from eating some of the mushroom species that are common in Alabama.

Harmful effects from mushrooms or other large fungi are caused by eating those that contain any of a variety of toxins that affect a human's physiology in different ways, including effects on the central nervous system. The safest way for the average citizen to eat mushrooms in Alabama or the United States in general is to buy them in the grocery store. Understand, however, that many edible species can be identified successfully by ardent mushroom collectors, and thousands are consumed annually by those who know precisely what they are doing, and what they are eating. For those who wish to enjoy the wonders of Alabama's out-of-doors and yet avoid plants and animals that could cause harm, the mushrooms pose no problems. None of them will attack, injure by contact, or harm a person in any way, as long as one keeps them off of the dinner table. It is not the intent of this book to train anyone to collect edible mushrooms, but a few references are listed in the Selected References that provide additional information on distinguishing some of the edible species from the poisonous ones.

Most books on edible mushrooms give the following advice in one way or another: There is no universal set of identifying characteristics that distinguishes edible from poisonous mushrooms. A species must be identified before one can be certain of its toxicity. So, even though most mushrooms are probably not harmful and many are excellent to eat, the deadly ones do not have any general characteristics that separate them from the others. A good rule is: Don't eat a wild mushroom because you know it is *not* a particular poisonous species; eat it because you know it *is* a particular nonpoisonous one.

The geographic distribution patterns of mushroom species have not been determined as precisely as they have for many species of animals and vascular plants. This may be due in part to the unpredictable time sequences in their life histories. Vascular plants often stay in the same spot for years, and animals have the courtesy to at least stay in the same general area. Many mushrooms appear, remain visible for a few days, seemingly disappear, and may not be seen again in the vicinity for years. Thus, clear definitions of exact geographic ranges have been difficult to establish for many of them. Therefore, this treatment will present in the narrative the general geographic distribution where particular species are known to occur, but it cannot provide detailed range maps.

The discussion will be confined to a few species that are known to occur or suspected of occurring in Alabama, that are documented as dangerous, and that should be kept out of one's digestive tract. But the reader must remember that even if certain species are not mentioned, one must not eat an Alabama mushroom unless one knows exactly what it is.

Family AMANITACEAE

AMANITA ***(Plate 1)*** AMANITA MUSHROOMS

Amanita is the premier genus of the poisonous mushrooms of North America. More people are poisoned by members of this genus than by any other because of its abundance and ubiquity and because of the many toxic species in the genus (certainly more than a dozen and possibly more than 3 dozen). Not a minor part of their reputation stems from the fact that the toxin of some is not only powerful enough to cause severe illness, but has been documented as lethal to humans.

As is true throughout nature, there are exceptions. Some *Amanita* are actually edible. However, as stated earlier, this book will have no recipes for mushrooms.

The genus *Amanita* includes more than 130, perhaps as many as 300, species of mushrooms that have a general similarity of certain characteristics. Mature *Amanita* are generally large mushrooms with flat caps and white gills and spores. Two other characteristics are a consequence of veils that enclose parts of the growing mushroom and result in 2 structures common to *Amanita*, the annulus and the volva. The annulus, present in the fully opened mushroom, is the lower remnant of a veil that covers the gills before the cap opens. Upon breaking from the pressure of an expanding cap, the only sign of the original veil is a ring of tissue around the stalk beneath the cap. The annulus is not always present, and certain other types of mushrooms can also have an annulus. The volva is a bulbous, often cuplike structure at the base of the stalk, a remnant of the "universal veil" that covers the entire mushroom as it pushes up through the soil. The membranous volva is characteristic of the genus *Amanita* but, as with the annulus, may sometimes be absent in some *Amanita* species or be present in other genera.

The habitat occurrence of all *Amanita* is basically the same. All forms grow in pine or hardwood forests, usually directly on the ground surface.

Some may be found in open pastures, along roadsides, or even in people's yards. Any notable habitat distinctions of particular species will be noted in the species accounts.

Toxic Properties: Among the active ingredients in *Amanita* that cause toxicity are the amatoxins, cyclic octapeptides that affect the liver and kidneys. One study reported that the amount of amatoxin present in 2 ounces of the most poisonous mushroom species can be fatal to an average adult. A variety of hallucinogenic compounds that are present in some species of *Amanita* can cause gastric disturbances. Ibotenic acid and muscimol (muscimol is produced upon drying of the mushroom) are found in some species.

The symptoms from *Amanita* poisoning are varied; they depend on the species involved, whether the specimen is cooked, raw, fresh, old, or dried, and whether the victim is a child or an adult. Unfortunately, the symptoms from the most dangerous *Amanita* may not be evident for several hours following ingestion, allowing the toxins to become incorporated into the bloodstream before the need for treatment is recognized. Death has been the final outcome of more than half of the victims who were poisoned by the most toxic species. It is worth repeating that no mushroom found in the wild in Alabama should be eaten unless the species is known for certain to be harmless. This is especially true for the *Amanita*.

Amanita brunnescens BROWN AMANITA

Species Recognition: The cap is up to 10 cm in diameter and is grayish brown with a number of white patches. The white stalk grows up to 12–15 cm and to almost 3 cm in diameter. This species is common during midsummer in some years.

Toxic Properties: Brown amanita is considered to be a poisonous species, although it apparently can be eaten safely by some individuals. The safest bet is to assume you are not one of them.

Geographic Distribution: Found throughout much of eastern United States except Florida and also the West Coast in Oregon and California. Found throughout Alabama.

Amanita citrina GREEN AMANITA

Species Recognition: This is a pale greenish yellow mushroom with a cap diameter of up to 12 cm. The stem may be more than 10 cm long and 4 cm thick.

Toxic Properties: One of the interesting aspects of this species is that some authors call it definitely toxic, others indicate that toxicity is suspected, and still others indicate that the species is edible. The simplest rule is, when in doubt, don't.

Geographic Distribution: Found throughout eastern North America into New England. Found throughout Alabama.

Amanita cokeri COKER'S AMANITA

Species Recognition: The 7–15-cm cap is white with white-to-brownish warts. The white stalk may be up to 18 cm long with a persistent annulus.

Toxic Properties: This species is of unknown edibility, and thus is not recommended for consumption.

Geographic Distribution: Mostly southeastern in occurrence and found throughout Alabama.

Amanita flavoconia ORANGE AMANITA

Species Recognition: *Amanita flavoconia* has been described as having a yellow-orange or chrome-yellow cap and stalk with an annulus of similar color. Yellow blotches may be present on the cap. The stalk may be almost white. The cap is up to 8 cm in width. The orange amanita is usually found singly, though in some species of *Amanita*, several are often found together.

Toxic Properties: This species has been described by most authors as probably poisonous, which means that no one has had the courage to eat it. We recommend that this cowardly attitude be maintained.

Geographic Distribution: Found throughout eastern United States and all of Alabama.

Amanita flavorubescens GOLDEN AMANITA

Species Recognition: The 5–10-cm cap is golden or yellowish with a yellow annulus and a yellowish volva. Yellow warts are present on the cap. The base of the stalk is often reddish. The species frequently occurs in hardwood areas and sometimes in pine forests.

Toxic Properties: The species is considered poisonous by most authorities or suspected of being poisonous by others. This would appear to be enough information to invite caution for anyone who would consider eating it.

Geographic Distribution: Eastern United States from Michigan to lower New England, south to South Carolina, and westward to Alabama.

Amanita gemmata GEMMED AMANITA

Species Recognition: The 3–11-cm cap is light yellow with several white warts. The whitish stalk can be 5–15 cm high and an annulus may be present.

Toxic Properties: This species should be considered poisonous because of possible hybridization with *Amanita pantherina*.

Geographic Distribution: Found throughout the United States and much of Europe. Found anywhere in Alabama.

Amanita muscaria FLY AMANITA

Species Recognition: *Amanita muscaria* growing in Alabama has a colorful cap that is orange or yellowish with lighter yellowish warts or scales. The inside of the mushroom cap is white. An annulus is normally present, and 2 or 3 rings may be seen above the volva. This is a large mushroom that may have a cap more than 20 cm in diameter and a stem that may be 15 cm long with a bulbous base.

Toxic Properties: This species is confirmed to be poisonous with hallucinogenic properties.

Geographic Distribution: Most of the United States from New England to Alaska and California and throughout the southeastern states except Florida and South Carolina. Found throughout Alabama.

Amanita pantherina PANTHER FUNGUS

Species Recognition: Panther fungus has a brown cap that varies from grayish to yellowish. The diameter is up to 11 cm. The volva is generally white with numerous warts. The stem is up to 4 cm in diameter,. and 13 cm high, and is generally white.

Toxic Properties: This species is considered poisonous and hallucinogenic based on accounts from Europe and should be presumed dangerous in Alabama.

Geographic Distribution: Varieties of this species are found throughout most of the eastern United States, on the West Coast, and in Europe. Found throughout Alabama.

Amanita rubescens BRONZE AMANITA

Species Recognition: The 5–20-cm cap is reddish brown with tan patches. The stalk is white or tan and may be up to 23 cm in height. The base of the stalk and the volva may be reddish, and the flesh of the mushroom slowly turns reddish where bruised. An annulus is generally present and may be white or tan.

Toxic Properties: This species is considered nonpoisonous by some authorities and by others is considered poisonous when it is not cooked. Hence, it qualifies as a poisonous variety.

Geographic Distribution: Found in most states in the eastern United States with the exception of Florida. Found throughout Alabama.

Amanita virosa DESTROYING ANGEL

Species Recognition: *Amanita virosa* is entirely white, with a cap up to 15 cm in diameter. The white stem can be almost 3 cm in diameter and 18 cm long. The annulus is large and hangs down below the cap.

Toxic Properties: *Amanita virosa* is considered to be deadly poisonous.

Geographic Distribution: Reported in most states from Maine to Florida and west to Minnesota and Iowa. Found throughout Alabama.

Other Species: Several other species of *Amanita* that occur in Alabama are suspected of being toxic or are definitely known to be so. These include *Amanita parcivolvata*, *A. polypyramis*, *A. bisporigera*, *A. verna*, and *A. aestivalis*. Some species of Alabama *Amanita* are nonpoisonous and edible, and more than 20 others are of unknown edibility. As indicated earlier, all mushrooms, whether cooked or uncooked, should be considered poisonous to humans unless one knows the exact species and is certain of its edibility. Mushrooms and snakes should be treated in a similar manner by the inexperienced. Look, but do nothing more, unless the identity is certain. A nice thing about *Amanita* and other mushrooms is that many are colorful, fascinating organisms, enjoyable to look at with no risk.

Family LEPIOTACEAE

The mushrooms in this family have gills and most have white spores. They lack the volva of *Amanita*.

CHLOROPHYLLUM

Chlorophyllum molybdites GREEN-SPORED MUSHROOM

Species Recognition: This can be a large mushroom (up to 30 cm in diameter) with a whitish cap that has brown scales. The gills are white when the mushroom is young, but turn green later. The spores are green. There is a large annulus beneath the cap.

Toxic Properties: This enigmatic mushroom is reported to be poisonous to some persons and nonpoisonous to others. It is suggested that Alabamians not become a part of the experiment to determine the percentage of people to whom this species is poisonous.

Geographic Distribution: Found in the South but occurring as far north as Michigan. Found throughout Alabama.

Habitat Occurrence: *Chlorophyllum molybdites* is found in open fields and lawns where it sometimes occurs in fairy rings. It is common in some years, from midsummer into late fall, and may be particularly abundant in years of heavy rains in late summer or fall.

LEPIOTA

Lepiota clypeolaria LEPIOTA

Species Recognition: The yellowish cap with brownish scales is 2–8 cm and is oval-shaped. The thin stem is up to 10 cm long and covered with white hairs.

Toxic Properties: Many of the species in this genus are poisonous or of unknown edibility.

Geographic Distribution: Eastern United States. Possibly throughout Alabama.

Habitat Occurrence: Mostly found in coniferous forests.

Family RUSSULACEAE

The members of this family have white spores, distinct gills, and grow in terrestrial situations. The species of Russulaceae do not have an annulus.

LACTARIUS

All members of this genus have a milky juice (latex).

Lactarius vellereus LACTARIUS

Species Recognition: This is a white mushroom that turns brown where it has been injured. It is not a particularly attractive mushroom, being short and coarse in appearance. Several other poisonous species of *Lactarius* may also be found in Alabama, including *L. piperatus*, *L. scrobiculatus*, and *L. chrysorheus*.

Toxic Properties: This is another puzzling species that apparently some people can eat with impunity whereas others cannot. Species of *Lactarius* with a milky latex that turns yellow or lavender are known to cause stomach upset. The latex of *L. piperatus* has a hot, pepperlike taste that should be a clue to anyone attempting to eat it. These are not deadly mushrooms but can cause gastrointestinal distress.

Geographic Distribution: Northern United States and the plains but considered common in many southeastern areas. Occurs in northern Alabama.

Habitat Occurrence: Found terrestrially along roadsides and in hardwood forests.

RUSSULA

Russula densifolia RUSSULA

Species Recognition: The 7–20-cm cap can range from white to gray. The white flesh turns red and then black when bruised. The thick stem may be up to 7 cm long. Other species of edible and poisonous russulas may be found in Alabama.

Toxic Properties: This poisonous species of *Russula* has a bitter taste and can cause gastrointestinal problems. The exact chemical makeup of the poisonous substance is unknown.

Geographic Distribution: Eastern United States and northern Alabama.

Habitat Occurrence: Most common in pine forests or mixed pine and hardwood. Found in summer and fall.

Family TRICHOLOMATACEAE

The species in this family of gilled mushrooms have white spores and are often found growing on decaying wood rather than on the ground. Most of the species are edible, but some cause gastric problems that can be unpleasant and sometimes are serious enough to require medical treatment. Members of this family, other than *Clitocybe*, that are known or suspected to be toxic in Alabama include *Omphalotus olearius*, *Omphalina chrysophylla*, *Collybia dryophila*, and *Tricholoma aurantium*.

CLITOCYBE

Clitocybe dealbata CLITOCYBE

Species Recognition: The cap, gills, and stem are a dull white. The cap may be flat and slightly more than 2 cm in diameter. The stem can be up to 5 cm long.

Toxic Properties: This particular species is known to be lethal.

Geographic Distribution: Eastern United States and possibly throughout Alabama.

Habitat Occurrence: Characteristically found growing in grassy areas, including lawns.

Family COPRINACEAE

The species in this family have dark purple or brown spores, and the cap is cone-shaped. The stem is thin, relative to the size of the mushroom. In Alabama the species in the genus *Panaeolus* are definitely poisonous. *Panaeolus separatus* is hallucinogenic and is generally found growing in horse dung. *Panaeolus foenisecii* usually grows in grassy areas.

Family STROPHARIACEAE

The 2 common poisonous species in this family have caps with distinctive colors and produce dark purple to brown spores.

PSILOCYBE

Psilocybe cubensis PSILOCYBE

Species Recognition: The cap is pale yellow, up to 8 cm in diameter. An annulus is present.

Toxic Properties: This species contains a hallucinogen that has been a popular drug in some subcultures.

Geographic Distribution: Florida and southern Alabama.

Habitat Occurrence: Most commonly associated with cow dung or other manure.

STROPHARIA

Stropharia aeruginosa STROPHARIA

Species Recognition: The 1–5-cm cap is bluish green, a trait that distinguishes it from most other mushrooms. An annulus is present.

Toxic Properties: The poisonous nature of this species is questionable, probably because few people have dared eat a bright green mushroom.

Geographic Distribution: Eastern United States, possibly throughout Alabama.

Habitat Occurrence: Characteristically in grassy areas, including lawns, or in wet forested habitats.

Family CORTINARIACEAE

The spores of species in this family are brown, and the gills are attached to the stem. The species in the genus *Cortinarius* are of unknown edibility, but representatives of at least 2 other genera are considered poisonous.

GALERINA

Galerina autumnalis AUTUMN GALERINA

Species Recognition: The 2–5-cm umbrella-shaped cap is brownish and sticky to the touch. This species often grows in clusters.

Toxic Properties: Amatoxins as well as phallotoxins have been reported from this and other species of *Galerina*. The species is potentially deadly.

Geographic Distribution: Found in many forested areas in North America. Can be expected throughout Alabama.

Habitat Occurrence: Characteristically grows on decayed wood, especially hardwood, but the wood may be beneath the surface. Most common in cool weather of early summer or fall.

INOCYBE

Inocybe fastigiata INOCYBE

Species Recognition: A small dark brown mushroom that smells like fresh green corn and has a pointed cap.

Toxic Properties: The poisonous substance in this mushroom is muscarine, which can cause severe poisoning if taken in large concentrations.

Geographic Distribution: Northern United States. Probably in northern Alabama.

Habitat Occurrence: Terrestrial in hardwood areas and may be numerous.

Family CLAVARIACEAE

The members of this family do not have the typical umbrella mushroom shape but instead look like ocean coral, growing upright with many fingerlike extensions or branches. Many are edible.

RAMARIA

Ramaria formosa CORAL RAMARIA

Species Recognition: Bright orange or red branches up to 25 cm high from a heavy whitish base. May be up to 15 cm thick.

Toxic Properties: The taste of this mushroom is bitter, and ingestion can cause digestive problems.

Geographic Distribution: Eastern United States and possibly throughout Alabama.

Habitat Occurrence: Grows most frequently on rich humic soil in hardwood areas.

Family BOLETACEAE

The members of this family look like typical mushrooms, but they do not have gills. Instead of the spores being released from gills, they are produced in tubes in the cap and are released from pores on the underside of the mushroom. Most species are edible, and some are highly preferred by eaters of wild mushrooms. At least 2 species in the genus *Boletus* are poisonous. Both are considered rare and unlikely to be found in most of Alabama. *Boletus satanus* occurs in the Smoky Mountains, and its range may extend into northern Alabama. It has a 10–20-cm cap, yellow or red spore tubes, and it turns blue when bruised.

Chapter 4

Vascular Plants Causing Systemic Poisoning

Family EQUISETACEAE

EQUISETUM HORSETAIL

Equisetum is a genus related to the ferns. It has green stems and whorls of scalelike brown leaves. The leaves are no more than 2 or 3 mm long, and there are upwards of 10 or 12 per whorl. The stems, which may or may not branch at the nodes, are hollow and vertically ridged. Two species occur in Alabama.

Toxic Properties: The toxic properties of *Equisetum* are not fully known, although silica and aconitic acid, as well as the alkaloids, nicotine, equisitine, and palustrine have been suggested at various times. Most recently, the enzyme thiaminase, a thiamine-destroying enzyme, has been suggested as the poisonous agent, but there is also evidence that this is not always the case. The toxic compound, whatever it is, is found in the green growing shoots of *Equisetum*. The compound normally is not destroyed when dried so that hay consumed by an animal up to several months later is still toxic. Symptoms in livestock include nervousness, unsuccessful attempts to stand up, muscular rigidity, inattentiveness, rapid and weak pulse, and cold extremities. These are followed by a period of quiescence, coma, and finally death.

Equisetum arvense FIELD HORSETAIL

Species Recognition: Field horsetail has many green branches at each node. The reproductive structures are produced on fertile branches that are nongreen stems. The green photosynthetic stems are sterile.

Geographic Distribution: Field horsetail occurs from Labrador to Alaska, south to Newfoundland, New England, New Jersey, Michigan, Wisconsin, North Carolina, and Alabama. The species is restricted to the northeastern quarter of Alabama.

Habitat Occurrence: Horsetails occur on woody slopes, usually calcareous substrate, roadsides, and edges of streams.

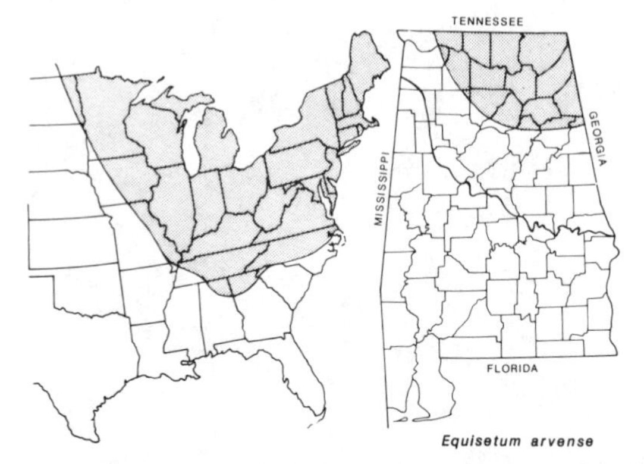

Range in eastern United States Known range in Alabama

Equisetum hyemale SCOURING RUSH; ROUGH HORSETAIL

Species Recognition: Scouring rush has green photosynthetic stems that are usually terminated by reproductive structures. There is no fertile nongreen stem. The stems arise from underground rhizomes, only rarely branch at the nodes, and are terminated by spore-producing strobili.

Geographic Distribution: Found from Newfoundland to Alaska, south to Florida, Texas, and New Mexico. Can be expected throughout Alabama.

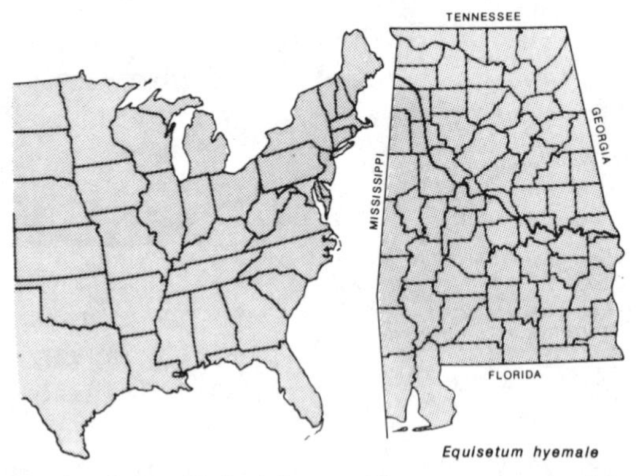

Range in eastern United States Known range in Alabama

Habitat Occurrence: Scouring rush occurs on dry or moist sandy shores, road embankments, roadsides, and in open woods.

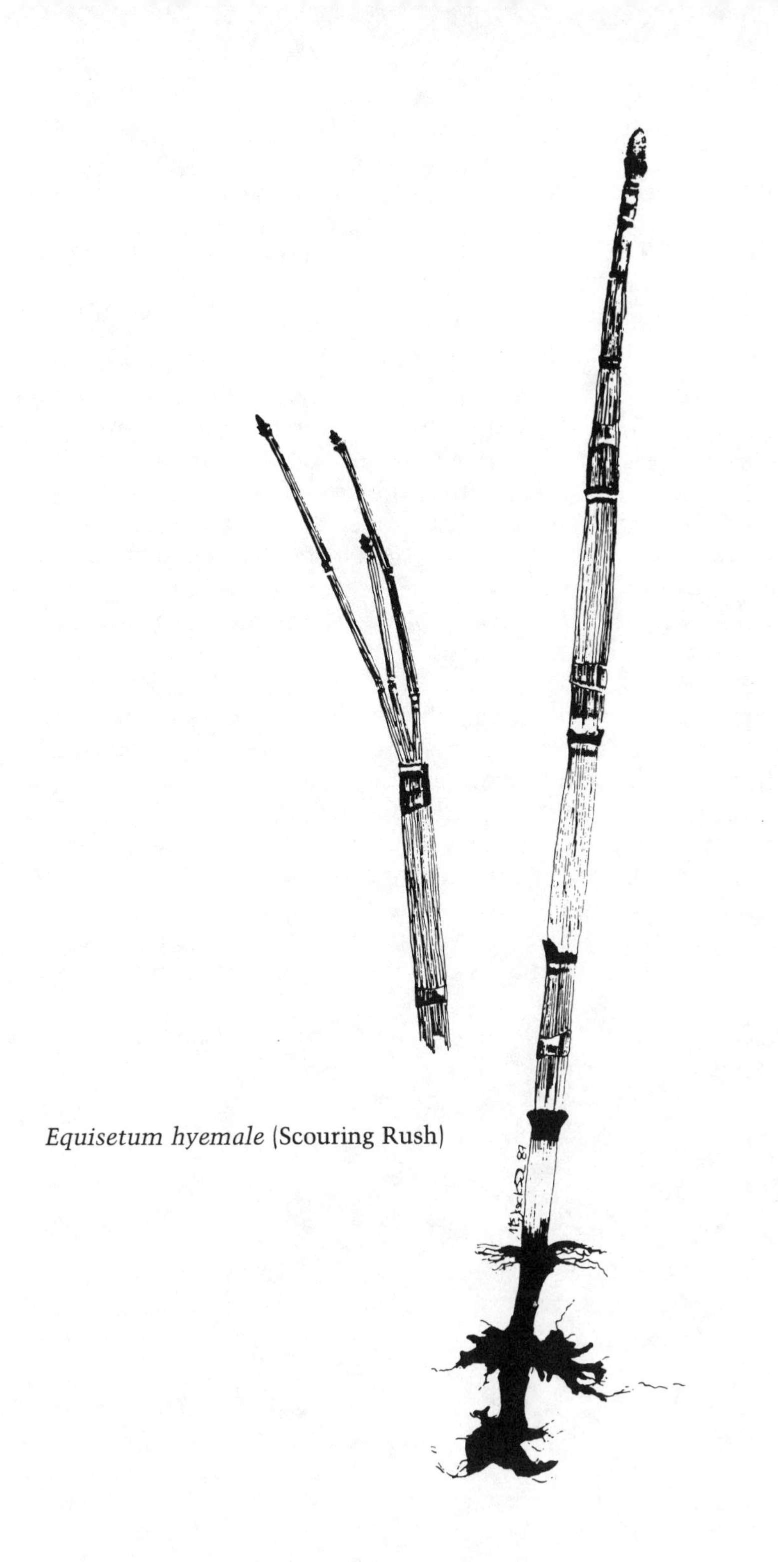

Equisetum hyemale (Scouring Rush)

Family PTERIDACEAE

PTERIDIUM BRACKEN FERN

Only one species occurs in Alabama.

***Pteridium aquilinum* (Plate 1)** BRACKEN FERN

Species Recognition: Bracken fern is an herbaceous perennial with stout blackish horizontal rhizomes often more than 1 meter in length. The leaves are erect from this underground rhizome. Narrowly or broadly triangular in shape, they are several centimeters to more than 1 meter in height. The leaves are divided into segments that are oblong in shape with entire margins. The sporangia are produced on the undersurface of the leaf segments and are covered by a very thin strip of tissue that looks as if the margin of the leaf is rolled back over on the underside.

Toxic Properties: All parts of the plants contain the enzyme thiaminase. When large amounts are consumed by livestock over a long period of time, toxic effects become evident—lack of coordination, lethargy, difficulty in standing, irregular heartbeat, and convulsions. Death will occur in a few days to a few weeks after the symptoms develop. The rhizome is considered to be at least 5 times as toxic as the leaves. Young leaves are sometimes cooked and used as a potherb by humans. However, the leaves contain a cancer-inducing agent that persists even after cooking, so they should not be eaten.

Geographic Distribution: Worldwide in distribution. Can be expected throughout Alabama.

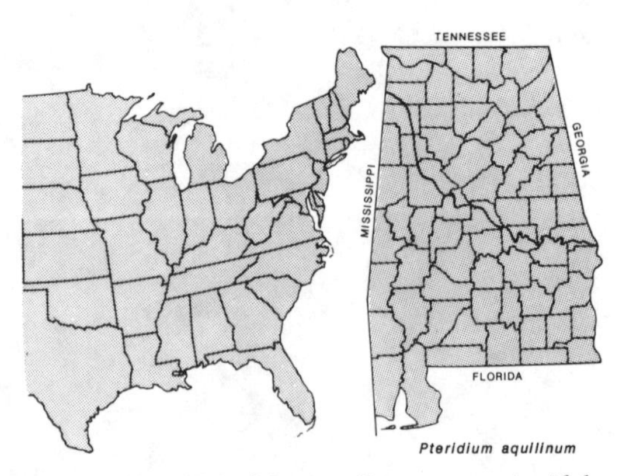

Range in eastern United States Known range in Alabama

Habitat Occurrence: Bracken fern occurs in old fields and secondary growth woodlands, especially in sandy soil and open areas.

Family ASPIDIACEAE

ONOCLEA SENSITIVE FERN

One species occurs in Alabama.

Onoclea sensibilis SENSITIVE FERN

Species Recognition: Sensitive fern is a perennial herb from a subterranean rhizome with pinnately lobed or compound leaves arising from the rhizome. Sori, the structures that produce the sporangia, are produced on a separate frond that has no chlorophyll and is completely dissimilar to the vegetative leaves. The margin of the vegetative leaves is crenate.

Toxic Properties: The principal poisonous compounds of the species are not known, but are seen primarily from livestock. Substantial quantities of sensitive fern must be ingested before symptoms appear. Animals fed hay composed of about 20 percent sensitive fern will develop the symptoms of incoordination in about 6 weeks. They eventually lose the ability to eat; they will fall down and remain unable to walk for several weeks.

Geographic Distribution: Throughout the eastern United States. Can be expected throughout Alabama.

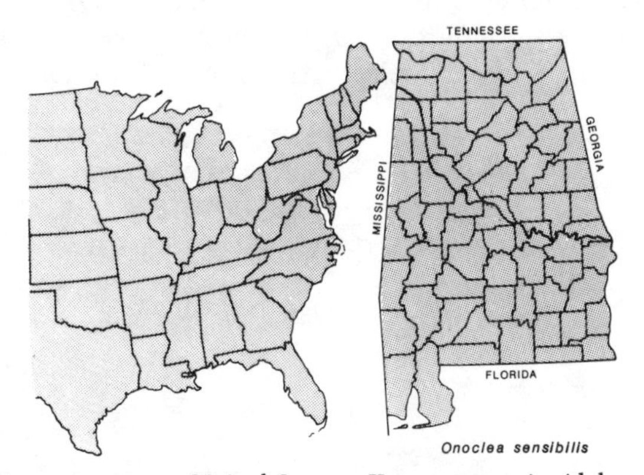

Range in eastern United States Known range in Alabama

Habitat Occurrence: Sensitive fern occurs in moist areas, especially in muddy ditches, marshes, swamps, and seepage areas.

Family CALYCANTHACEAE

CALYCANTHUS SWEETSHRUB

A single species occurs in Alabama.

Calycanthus floridus (Plate 2) EASTERN SWEETSHRUB

Species Recognition: Eastern sweetshrub is a shrubby plant that grows to a height of about 4 meters. The leaves are deciduous, opposite, entire, lanceolate to ovate-lanceolate, 5–18 cm long, 2–8 cm wide on a short petiole. The flowers have a sweet fragrance and are produced on short leafy branches when the leaves are partially expanded. The flowers are maroon and have sepals and petals that are similar in shape, texture, and color. The flowers have numerous stamens and carpels. The fruiting structure is about 8 cm long and 5 cm in diameter, fibrous, and formed from the enlarged receptacle. This structure contains numerous seedlike achenes (1-seeded fruits from the individual carpels), each about 10 mm long and 5 mm wide.

Toxic Properties: The active ingredients are the alkaloids calycanthidine and calycanthine, which are present in the seedlike achenes. No record of human poisoning has been found, but the species is suspected of poisoning livestock. Symptoms include strychninelike convulsions, myocardial depression, and hypertension. This plant is commonly cultivated, and the public should be made aware of its potential danger.

Geographic Distribution: Found from southern Pennsylvania to southern Ohio, south to northern Florida and Mississippi. Can be expected in all parts of Alabama except the northwest quarter.

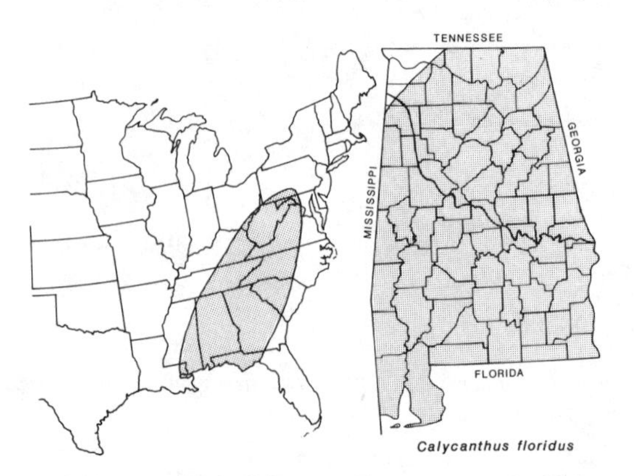

Range in eastern United States Known range in Alabama

Habitat Occurrence: Eastern sweetshrub occurs in deciduous forests, moist woodlands, clearings, and along stream banks. The species is most common in the Piedmont, but also occurs in the mountains and Coastal Plain.

Family ARISTOLOCHIACEAE

ARISTOLOCHIA BIRTHWORT

Aristolochia is a genus of erect herbs or sprawling vines, with leaves cordate at the base, palmate-veined, and alternate. The flowers have the sepals united into a tube. This tube is S-shaped or U-shaped, purple in color, and often densely pubescent. The fruit is a capsule that splits irregularly, either from the top or from the sides.

Toxic Properties: *Aristolochia* has been used medicinally as a reputed snakebite antidote. Large doses may cause gastroenteric irritation, vomiting, abdominal pain, dizziness, and diarrhea. The active ingredients are aristolochine, which is an alkaloid; aristolochin and serpentarine, both of which are resins; and aristinic acid. These compounds are found in the root.

***Aristolochia serpentaria* (Plate 2)** SNAKEROOT

Species Recognition: Snakeroot is an erect herb with flowers borne at the base of the stem. The capsule opens from the top with spreading valves.

Geographic Distribution: Connecticut to Illinois and Missouri, south to Florida and eastern Texas. Can be expected throughout Alabama.

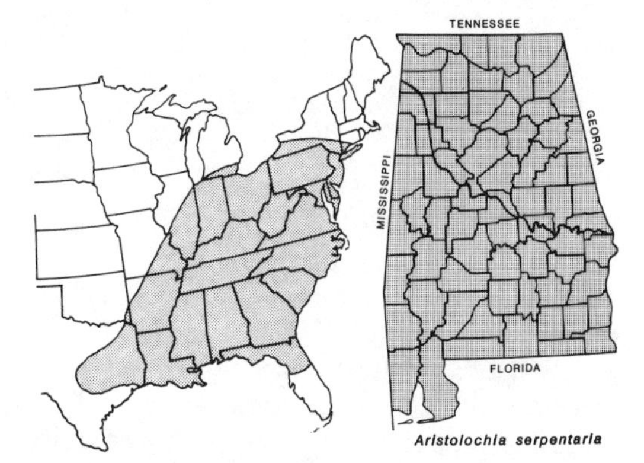

Range in eastern United States Known range in Alabama

Habitat Occurrence: Snakeroot occurs in mixed deciduous forests, woodland margins, and along stream banks. It occasionally occurs on ledges and wooded rocky slopes.

Aristolochia tomentosa PIPE VINE; WOOLLY DUTCHMAN'S PIPE

Species Recognition: Pipe vine is a high-climbing, woody, twining vine about 25 m long with dense white pubescence throughout. The leaves are cordate to rounded and often more than 10 cm wide. Flowers are borne in the leaf axis on the upper part of the stem, and the capsules open from the bottom or irregularly.

Geographic Distribution: North Carolina to Illinois and Missouri, south to Florida and eastern Texas. Can be expected throughout Alabama.

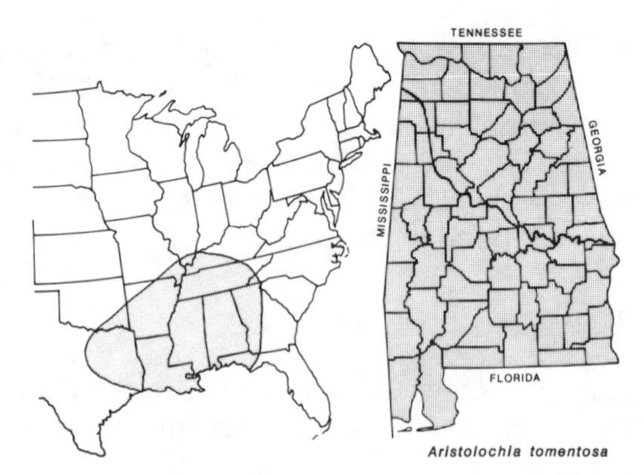

Range in eastern United States Known range in Alabama

Habitat Occurrence: Pipe vine occurs in shaded areas along streams and in bottomland forests, along rivers, and in coastal woodlands.

Family RANUNCULACEAE

ACONITUM MONKSHOOD

A single species occurs in Alabama.

Aconitum uncinatum SOUTHERN BLUE MONKSHOOD

Species Recognition: Southern blue monkshood is a weak-stemmed plant with a tuberous root. The stem may grow to 1.5 m and be either erect or sprawling over other vegetation. The leaves are divided into several linear segments. The flowers are blue and are produced in a loose raceme. One lobe of the petal-like structures has the shape of a hood, hence the common name. The fruits are produced mostly in clusters of 3, have several seeds, and are split along 1 margin.

Toxic Properties: This is an extremely toxic plant. Ingestion of a relatively small amount can cause death in humans within 1–6 hours. The toxic substances are various alkaloids, especially aconitine, which occur in all parts of the plant. Symptoms include nausea, vomiting, skin-tingling, respiratory difficulty, weakness, and irregular heartbeat.

Geographic Distribution: From Pennsylvania to Indiana, south to Georgia and Alabama. Extremely rare in Alabama and can be expected only in the mountains of the northeastern part of the state.

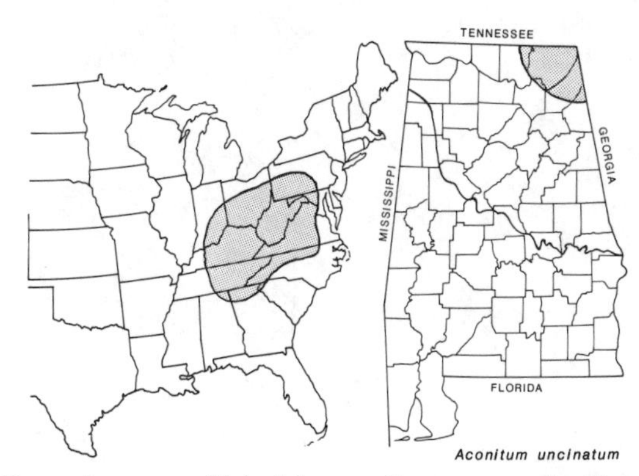

Range in eastern United States Known range in Alabama

Habitat Occurrence: Restricted to rich deciduous woods of the mountains and Piedmont.

ACTAEA BANEBERRY

A single species occurs in Alabama.

***Actaea pachypoda* (Plate 3)** WHITE BANEBERRY

Species Recognition: White baneberry is an herb that grows to about a meter in height. The leaves are mostly twice-pinnately compound with ovate leaflets that are coarsely serrate. The flowers are white and scattered along a raceme that grows to about 17 cm when in fruit. The fruit stalk becomes thickened and white and supports a red or white berrylike fruit.

Toxic Properties: In Europe, the ingestion of baneberry has been known to result in death of children. The species also can cause severe dermatitis after external exposure. The active ingredient is a glycoside that produces protoanemonin, an irritant oil. This glycoside is present in all parts of the plant, but is especially prevalent in the fruits and roots. Symptoms from ingestion include burning, vomiting, gastroenteritis, diarrhea, headache, dizziness, delirium, and, rarely, convulsions.

Geographic Distribution: From Prince Edward Island to Manitoba, south to northern Florida, Louisiana, and Oklahoma. Can be expected throughout Alabama, except possibly the extreme southern row of counties.

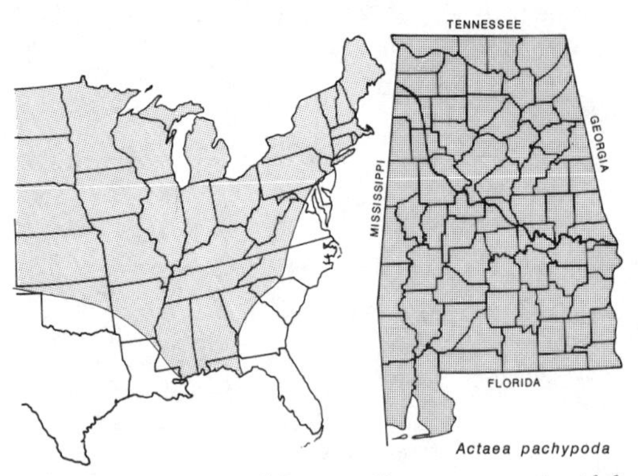

Range in eastern United States   Known range in Alabama

Habitat Occurrence: White baneberry occurs in rich deciduous woods of the Cumberland Plateau and, to a lesser extent, of the Piedmont and Coastal Plain.

ANEMONE WINDFLOWER

These are herbaceous perennials with tubers or rhizomes. The leaves are dissected and appear compound. The flowers are solitary on elongate stems without petals but with sepals that look like petals. The stamens are numerous, and the fruits are clustered into a cylindric or subglobose head. The individual fruits are hairy, often making the head look like a puff of cotton.

Toxic Properties: All parts of this plant contain a glycoside that forms protoanemonin, which is an irritant oil. When taken internally, this irritant oil causes burning and redness in the mouth and throat, blistered eruptions on the skin, gastroenteritis, dizziness, and, rarely, convulsions or death. Cooking or drying destroys the toxicity. Dermatitis can result externally from contact with the oil. Not all species have been tested for toxicity, but all should be assumed toxic until proven otherwise.

Anemone berlandieri SOUTHERN THIMBLEWEED

Species Recognition: The flowering stems are usually less than 40 cm tall with one flower. The plant grows from one tuber with 10–20 sepals that are often pink or blue. The achenes are densely pubescent, and the stem leaves are dissimilar from those at the ground. The stem leaves are divided into linear segments; those at the ground are divided into 3 segments.

Geographic Distribution: From Virginia to Oklahoma, and south to northern Florida and Texas. Most common in the Black Belt of Alabama but can be expected throughout the northern half of the state.

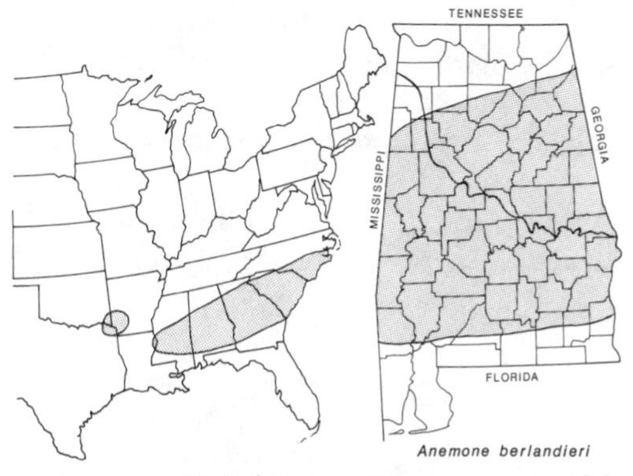

Range in eastern United States Known range in Alabama

Habitat Occurrence: Southern thimbleweed is found in open sun with calcareous or sandy clay soil.

Anemone caroliniana CAROLINA THIMBLEWEED

Species Recognition: The plant is less than 40 cm tall with one flower, with 10–20 pink or blue sepals and densely pubescent achenes. It is rhizomatous with a series of small tubers. Stem leaves are similar in shape to the basal leaves, the leaves divided into linear segments.

Geographic Distribution: From North Carolina to Indiana and South Dakota, south to Florida and Texas. Can be expected throughout the northern half of Alabama.

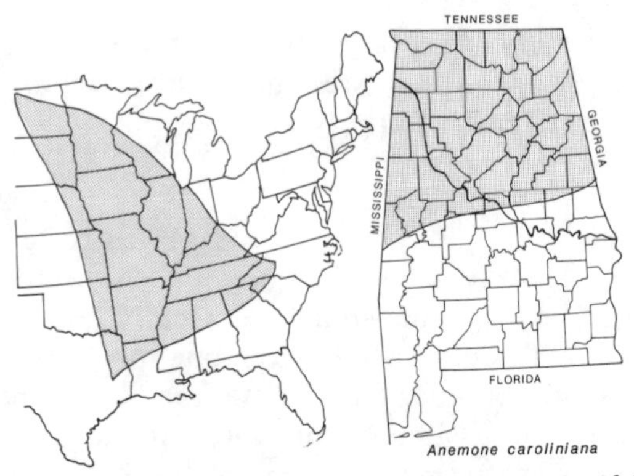

Range in eastern United States Known range in Alabama

Habitat Occurrence: Carolina thimbleweed occurs in sandy soils, in open fields, and along roadsides.

Anemone quinquefolia AMERICAN WOODLAND THIMBLEWEED

Species Recognition: The plant is less than 40 cm tall with 1 flower, usually white, with 5–8 sepals. Achenes are finely pubescent, the plant growing from long root stalks or rhizomes without tubers.

Geographic Distribution: From Quebec to Ohio, and south to North Carolina and Alabama. Can be expected in the northern half of Alabama.

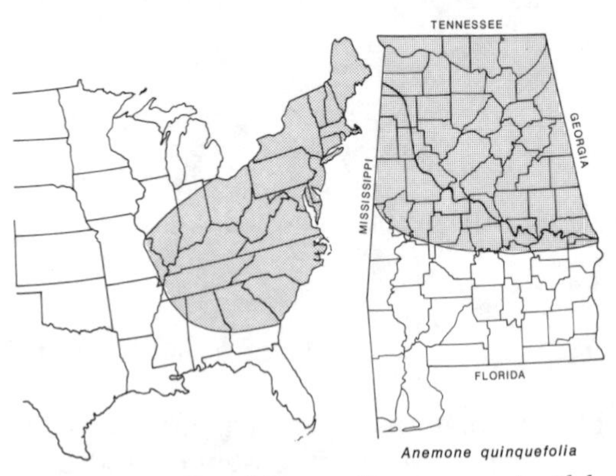

Range in eastern United States Known range in Alabama

Habitat Occurrence: American woodland thimbleweed occurs in rich deciduous woods, chiefly in the mountains but also in the Upper Coastal Plain, Cumberland Plateau, and Piedmont.

Species Recognition: The flowering stem has 2 or more flowers. Plants more than 40 cm tall grow from a thick rootstock or rhizome. Basal leaves and stem leaves are similar in shape and palmately lobed. Seeds are densely hairy.

Geographic Distribution: From Maine to Minnesota, south to Georgia, Alabama, Louisiana, and Kansas. Can be expected anywhere north of the Black Belt in Alabama.

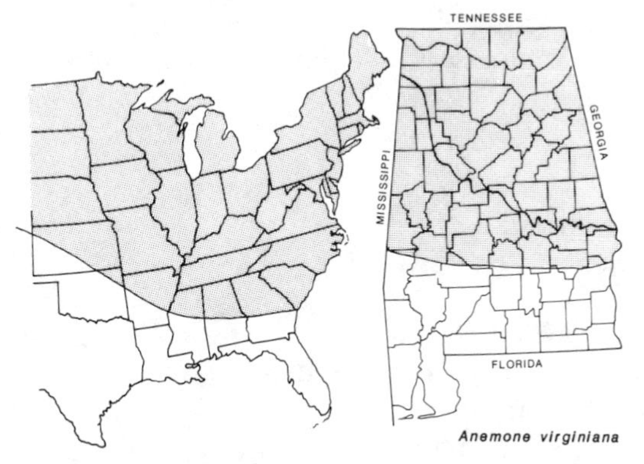

Range in eastern United States   Known range in Alabama

Habitat Occurrence: Tall thimbleweed is most common in rich deciduous woods but also occurs in openings, dry or rocky open woods, thickets, and slopes.

DELPHINIUM LARKSPUR

Delphinium is a genus of herbaceous plants with taproots or tuberous rootstocks. The leaves are alternate, stalked, and divided into few-to-numerous, linear-to-filiform leaflets. The petal-like structures are blue to white, the upper parts of which are elongated spurs. From each flower are produced 1–5 fruits, each having several-to-many seeds and splitting along 1 margin.

Although not all species have been documented to be toxic, all should probably be considered so until proven otherwise.

Toxic Properties: The active ingredients are various diterpenoid alkaloids, the most widely known of which is delphinine. The plants are more toxic when young, and they decrease in toxicity throughout the growing season. The herbaceous part of the plant at flowering time might have as little as 5 percent the amount of toxin as when it was young. Unfortunately, however, the seeds are exceedingly toxic. Symptoms include burning in the mouth, tingling skin, nausea, abdominal pain, weak pulse, labored respiration, and nervous excitement or depression.

Larkspurs are an exceedingly toxic group of plants, despite their unusual beauty when in flower. In fact, in the western United States, larkspur poisoning of cattle is second only to that of the various loco weeds (members of the genus *Astragalus* in the family Fabaceae), and in many states is the main source of cattle poisoning. Mass mortality of livestock attributed to larkspur poisoning originally led to the establishment of the long-continued experimental investigations of poisonous plants by the United States Department of Agriculture. Larkspur should be considered capable of causing death in humans.

Delphinium alabamicum — ALABAMA LARKSPUR

Species Recognition: Plants grow from fascicles of tuberous roots to a height of 1 meter. The leaves are divided into broadly linear segments. The deep-blue-to-purple flowers are 2.5–3 cm broad. Fruits are 3 in a cluster, and seeds have ridges rather than distinct wings.

Geographic Distribution: Endemic to western Alabama; known only from Franklin, Lawrence, Dallas, Perry, and Montgomery counties. May now be absent from the Black Belt localities.

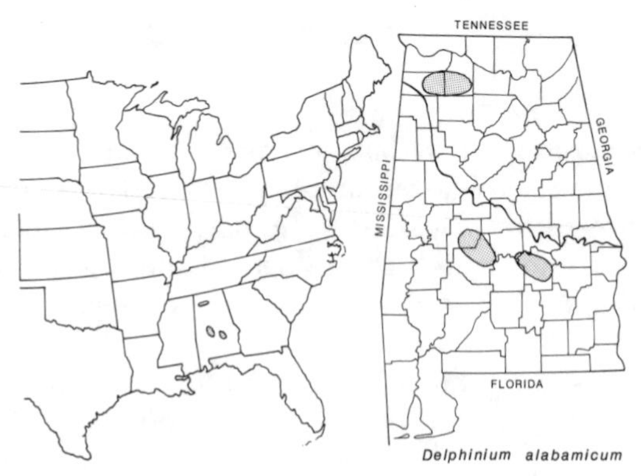

Range in eastern United States — Known range in Alabama

Habitat Occurrence: Alabama larkspur is found exclusively in open limestone glades in heavy clay soil.

Delphinium ambiguum ANNUAL LARKSPUR

Species Recognition: Annual larkspur is an erect herb that grows from a taproot to a height of about 1 meter. The leaves are divided into narrowly linear or filiform segments, especially near the base. The carpels and fruits are solitary in each flower.

Geographic Distribution: Native to Europe but escaped from cultivation in North America from Nova Scotia to Minnesota, and south to Florida and Texas. Can be expected throughout Alabama.

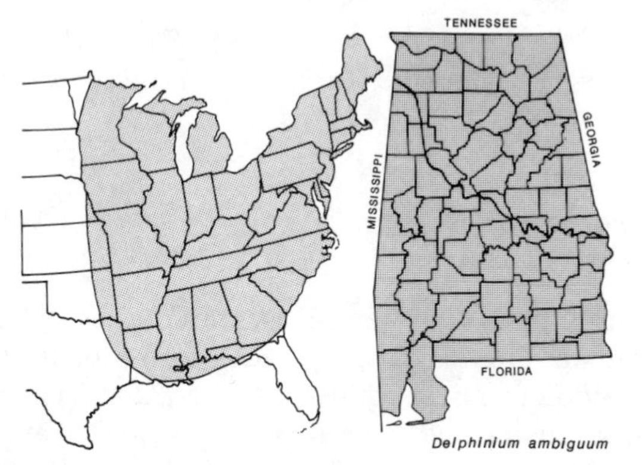

Range in eastern United States Known range in Alabama

Habitat Occurrence: Annual larkspur is found in a variety of disturbed soils, as long as they are dry and sunny. It is especially abundant along roadsides and in fallow fields.

Delphinium carolinianum CAROLINA LARKSPUR

Species Recognition: Plants grow from thickened roots or fusiform tubers to a height of 1.5 meters. The leaves are divided into linear segments. The light blue flowers are mostly less than 2 cm broad. Fruits are 3 in a cluster with winged seeds.

Geographic Distribution: Virginia to Missouri and Oklahoma, south to Florida and Texas. Locally abundant throughout the Black Belt, and occasionally elsewhere in Alabama.

Habitat Occurrence: Carolina larkspur occurs in moist-to-dryish, usually calcareous, soil and is found in clearings, limestone outcrops, glades, and prairies. It is especially common along roadsides.

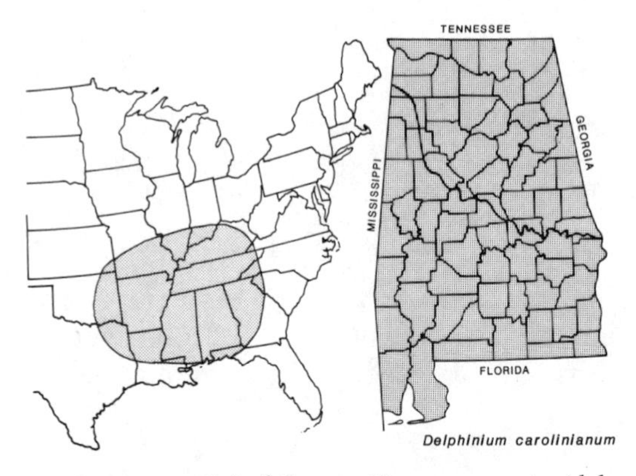

Range in eastern United States Known range in Alabama

***Delphinium tricorne* (Plate 3)** DWARF LARKSPUR

Species Recognition: Plants grow from fascicles of tuberous roots to a height of 0.5 meter. The leaves are divided into broadly linear segments. The blue-violet to near white flowers are 2–2.5 cm broad. The fruits are 3 in a cluster and have seeds with ridges rather than distinct wings.

Geographic Distribution: Pennsylvania to Minnesota and Nebraska, south to Georgia, Mississippi, Arkansas, and Oklahoma. Can be expected north of a line from Jackson County to Greene County in Alabama.

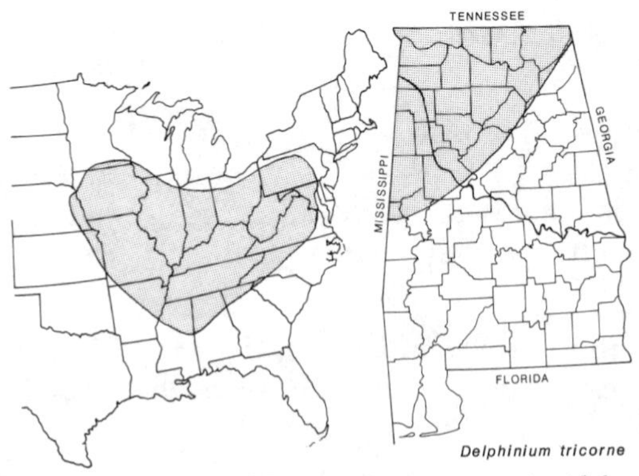

Range in eastern United States Known range in Alabama

Habitat Occurrence: Dwarf larkspur occurs in rich, moist, deciduous woodlands, in mostly neutral acidic or slightly basic loamy soils.

Species Recognition: Plants grow from thickened roots or a fascicle of fusiform tuberous roots to a height of 1 meter. The leaves are divided into linear segments. The flowers are white to whitish green with a tinge of blue and usually less than 2 cm broad. The fruits are 3 in a cluster with winged seeds.

Geographic Distribution: Wisconsin to Manitoba, south to Texas; rare east of the Mississippi River in Illinois and Kentucky; fairly common in central Tennessee. One disjunct population known in northern Alabama.

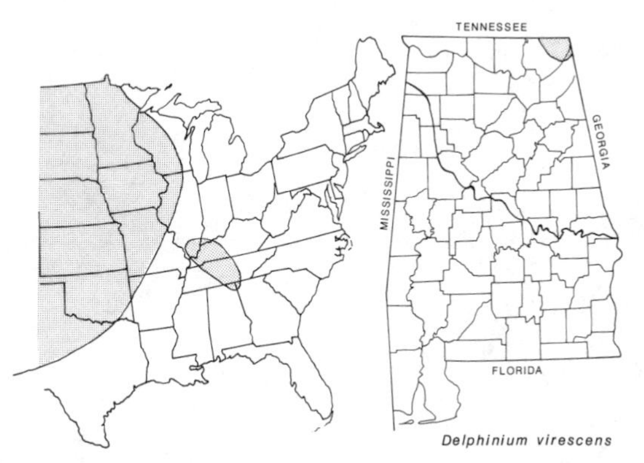

Range in eastern United States Known range in Alabama

Habitat Occurrence: The species occurs in heavy calcareous soils of cedar barrens, prairies, and rocky clearings.

RANUNCULUS BUTTERCUP

Ranunculus is a genus of annual or perennial herbs with palmately compound leaves that are either basal or alternate. The flowers are solitary or clustered into small groups of 2 or 3 with 5 sepals, 5 yellow or, rarely, white petals, many stamens, and many separate carpels. The fruit is an achene.

Toxic Properties: All parts of the plant contain a glycoside, ranunculin. This glycoside is a precursor to a volatile oil, protoanemonin, which degrades into an alcohol anemonol. Apparently, only protoanemonin is toxic or an irritant. On contact with the skin it may cause dermatitis. Respiratory or eye irritations can also occur. If ingested, it will cause burning and redness of the mouth and throat, gastroenteritis, vomiting, dizziness, fainting, urinary trouble, blister eruptions, and, rarely, convulsions and death. High concentrations of the volatile oil in contact with the skin can cause ul-

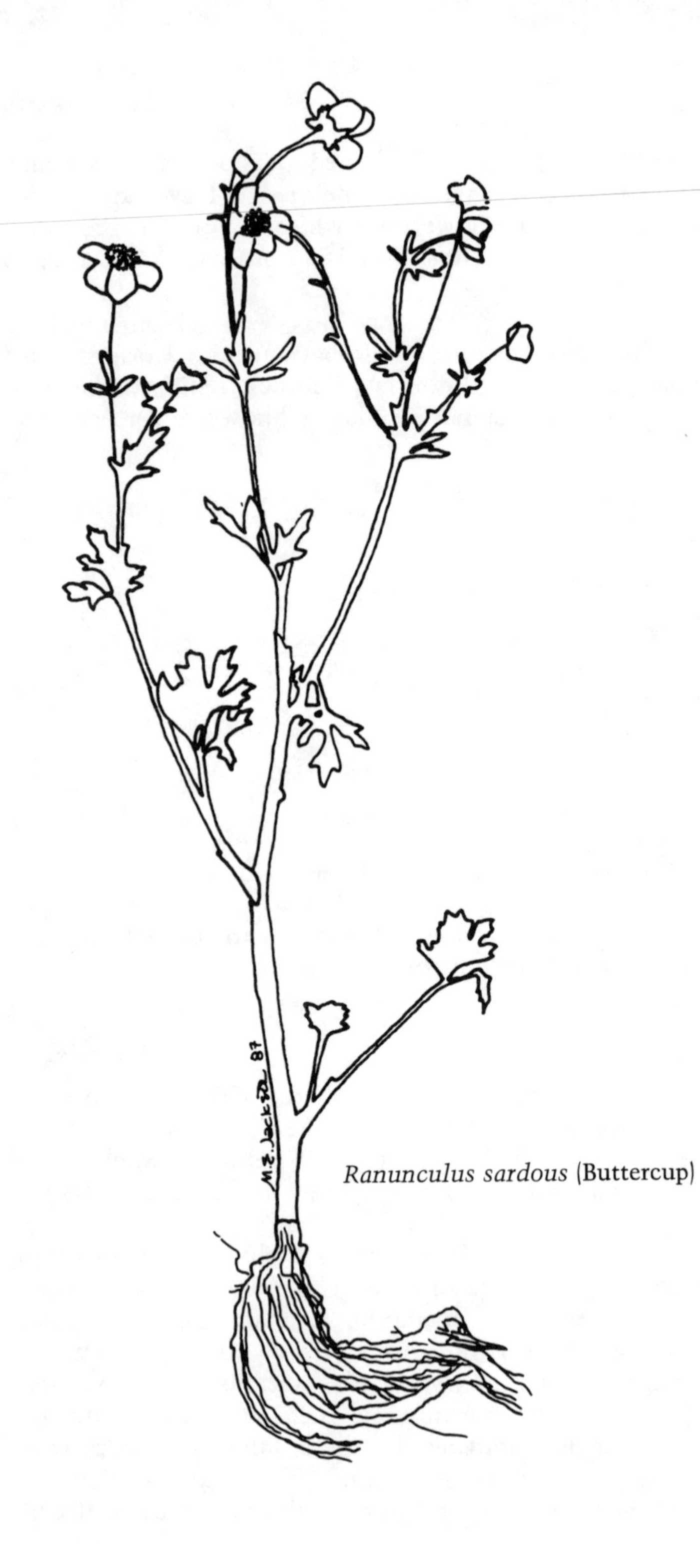

Ranunculus sardous (Buttercup)

cerations. The plant is apparently distasteful and only rarely is enough ingested by an animal to cause any problems other than irritation. When the plant is cooked or dried, the volatile oil degrades into the alcohol; therefore, dried or cooked plants are edible. Apparently, all species of the genus contain these oils. At least 19 species are known to occur in Alabama.

Alabama species: The species of *Ranunculus* known from Alabama include the following: *Ranunculus abortivus* (Kidney-Leaf Buttercup); *R. bulbosus* (Bulbous Buttercup); *R. carolinianus* (Carolina Buttercup); *R. fascicularis* (Early Buttercup); *R. flabellaris* (Yellow Water Buttercup); *R. harveyi* (Harvey's Buttercup); *R. hispidus* (Bristly Buttercup); *R. laxicaulis* (Loose-Stemmed Buttercup); *R. longirostris* (White Water Crowfoot); *R. micranthus* (Small-Flowered Crowfoot); *R. muricatus* (Wrinkled Buttercup); *R. parviflorus* (Small Flower Buttercup); *R. pusillus* (Small Buttercup); *R. recurvatus* (Hooked Buttercup); *R. repens* (Creeping Buttercup); *R. sardous* (Buttercup); *R. sceleratus* (Blister Buttercup); and *R. trilobus* (Three-Lobed Buttercup).

Family BERBERIDACEAE

PODOPHYLLUM MAYAPPLE

Podophyllum is a genus of herbs with an underground rhizome and an erect stem that has 1–2 leaves at its apex. The leaves are palmately lobed and about 15 cm in diameter. Plants with 1 leaf do not flower, but plants with 2 leaves have a single flower in the axis between the 2 leaves. The flower is on a recurved peduncle and has 6 white sepals and 6–9 white petals. The fruit is a yellow-to-red berry that is 3–5 cm in diameter. Only 1 species occurs in Alabama.

Podophyllum peltatum (Plate 3) MAYAPPLE

Toxic Properties: The leaves, stems, flowers, rootstalks, and young fruits of mayapple contain podophyllin, which is a resinous substance. The plant is sometimes sold in herb shops as podophyllin and is used medicinally as a purgative and applied externally to venereal warts. However, even small amounts, either raw or cooked, may cause poisoning with severe diarrhea, vomiting, gastroenteric irritation, and abdominal pain. Larger amounts may cause dizziness, headache, fever, increased breathing, rapid pulse, and coma. A severe irritation or dermatitis can occur with repeated exposure to the plant, especially the root. The mature fruit, when it is yellow or red, contains little of the podophyllin and is edible, although large quantities may cause a mild case of diarrhea.

Geographic Distribution: Western Quebec to southern Ontario and Minnesota, south to Florida and Texas. Can be expected in suitable habitats throughout Alabama.

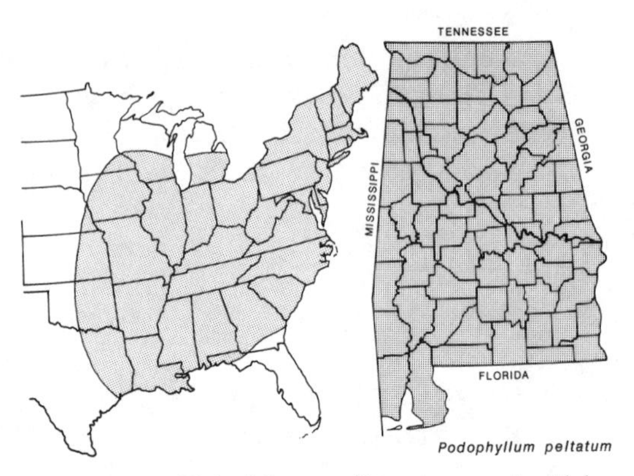

Range in eastern United States   Known range in Alabama

Habitat Occurrence: Mayapple occurs in mixed deciduous woods, alluvial woodlands, meadows, moist roadbanks, and, occasionally, pastures. It is most common in the mountains and Piedmont.

Family MENISPERMACEAE

COCCULUS CORAL BEADS

Only one species occurs in Alabama.

***Cocculus carolinus* (Plate 4)** CORAL BEADS; CAROLINA MOONSEED; TENNESSEE IVY

Species Recognition: This woody vine climbs over vegetation. It has alternate leaves that are cordate at the base, and unlobed and acute at the apex. The flowers have stamens or carpels, but not both, on a single plant. The flowers are in clusters that are about 15 cm long; the fruit is a drupe, a fleshy, 1-seeded fruit with a stone or a pit. The seed is red, spherical, 5–8 mm in diameter. One problem with this species is that the beautiful scarlet seed with the ebony spot belies its danger as a capsule of poison. Beware of the danger and do not display coral beads where children can reach them.

Toxic Properties: All parts of *Cocculus* contain an unknown alkaloid, probably cocculidine or coclifoline. These are exceedingly toxic alkaloids and have been used by natives in Southeast Asia for arrow poison. Symptoms of *Cocculus* ingestion have not been published; however, there is an unpublished record of a dog having died after eating one leaf of *Cocculus*. Dr. Fred Gabrielson, University of Alabama, reports in a personal communication that immature fruits of coral beads fed to mice resulted in death, whereas mature fruits fed to other mice did not.

Geographic Distribution: Southeastern Virginia to southern Illinois and southeastern Kansas. Can be expected throughout Alabama.

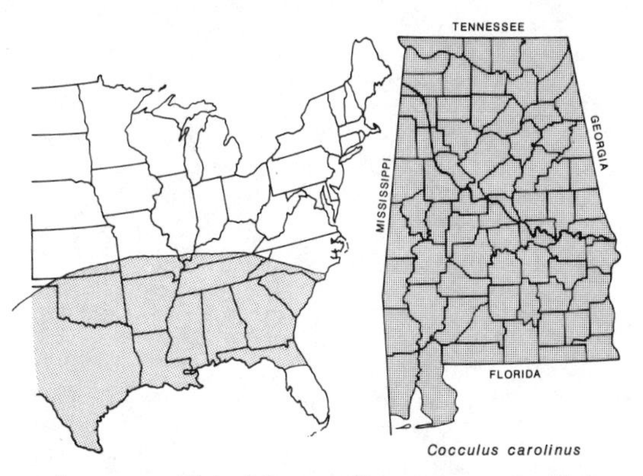

Range in eastern United States Known range in Alabama

Habitat Occurrence: Coral beads grow in rich deciduous woods and thickets and along the edges of fields. Chiefly Coastal Plain and Piedmont.

Family PAPAVERACEAE

ARGEMONE PRICKLY POPPY

Argemone is a genus of coarse annual herbs that grow 0.3–1.2 meters tall with spreading prickly branches and either yellow or white sap. The leaves are alternate, whitish, prickly, pinnate, and divided. The showy flowers have white or yellow petals, are terminal on the stems, and are 3–8 cm across. The fruit is a prickly capsule opening by the reflection of the walls at the apex. Although *Argemone* is not a hallucinogen, it does belong to the same family as the Old World opium poppy.

Toxic Properties: All parts of the plant, especially the seeds, contain the alkaloid isoquinoline as well as the alkaloids sanguinarine, and dihydrosanguinarine. Ordinarily neither humans nor animals are likely to ingest enough of this distasteful plant to cause problems, but were sufficient quantities consumed, the alkaloids could prove toxic. Ingestion of wheat contaminated with large amounts of prickly poppy seed has caused epidemic dropsy in humans in India.

Argemone albiflora WHITE PRICKLY POPPY

Species Recognition: White prickly poppy has white petals and clear-to-white sap.

Geographic Distribution: Introduced from the tropics. White prickly poppy has spread as a weed from Kentucky to Florida, and west to Mississippi. In Alabama most common along the coast.

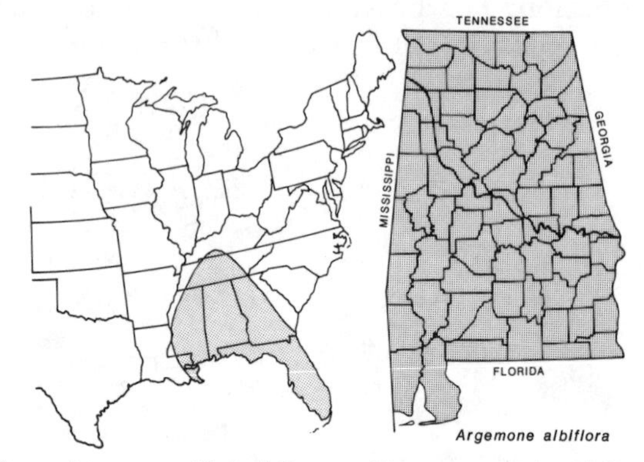

Range in eastern United States Known range in Alabama

Habitat Occurrence: White prickly poppy occurs along roadsides and in waste places.

Argemone mexicana MEXICAN PRICKLY POPPY

Species Recognition: Mexican poppy has yellow petals and bright yellow sap.

Geographic Distribution: Virginia to Tennessee, south to Florida and Texas. In Alabama most common along the coast.

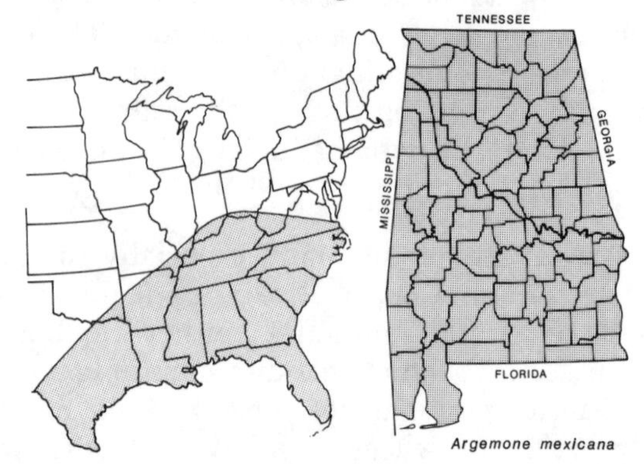

Range in eastern United States Known range in Alabama

Habitat Occurrence: Mexican poppy occurs on sandy roadsides and in waste places.

SANGUINARIA BLOODROOT

One species occurs in Alabama.

***Sanguinaria canadensis* (Plate 4)** BLOODROOT

Species Recognition: Bloodroot consists of 1 lobed leaf and 1 terminal white flower arising from an underground rhizome. The underground rhizome has a red sap. The fruit is a several-seeded, elongate, pointed capsule.

Toxic Properties: The active ingredients are various alkaloids including sanguinarine. This is especially dense in the root. It is used as a medicinal herb in small quantity, but in overdoses is an irritant poison causing nausea, vomiting, and burning of the mucous membranes of the mouth, throat, stomach, and intestines. Symptoms include reduced nerve, heart, and respiratory strength, difficulty in breathing, dilated pupils, faintness, muscular failure, and even death due to cardiac paralysis. The unpleasant taste makes the plant unlikely to be consumed in quantity. The juice irritates the skin and eyes. An additional negative feature was experienced by early white settlers of the eastern United States when they saw the red sap being used as war paint by American Indians.

Geographic Distribution: Eastern Quebec, south to northern Florida and eastern Texas. Can be expected throughout Alabama north of Mobile.

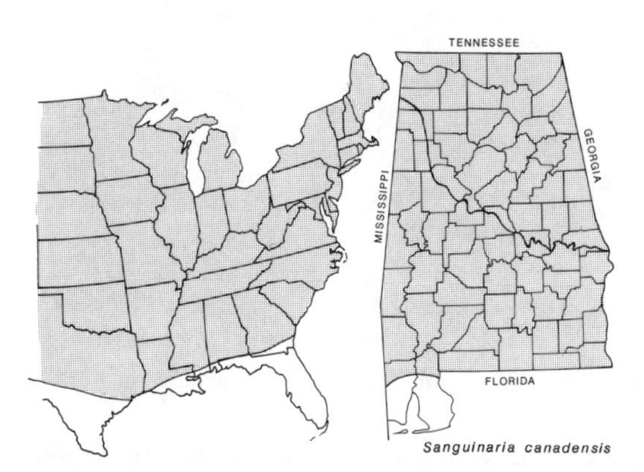

Range in eastern United States Known range in Alabama

Habitat Occurrence: Bloodroot occurs in rich deciduous woods, appearing early in the year before the leaves of deciduous trees appear.

Family FUMARIACEAE

CORYDALIS FUMATORY

Corydalis is a genus of biennial or weak perennial plants with succulent stems that have palmately lobed, pale green leaves. The flowers are in terminal racemes and are axillary to unlobed bracts. They are 0.7–1.5 cm long, yellow to white, and have 1 petal that has a basal spur on it. The fruit is a capsule, which may be either erect or pendulant.

Toxic Properties: Fumatory contains various alkaloids in all parts of the plant. These alkaloids have been known to cause death to cattle, sheep, and horses in quantities of 2–5 percent of the animal's weight. Symptoms include weak breathing, weak heartbeat, and staggering convulsions.

Corydalis flavula FUMATORY

Species Recognition: This species is a weak annual with erect or sprawling leaves and a spurred petal 0.7–0.9 cm long. The fruits are pendulant to spreading. The seeds are 0.2–0.25 cm wide and have a narrow marginal ring around them.

Geographic Distribution: Connecticut to southern Ontario and Minnesota, south to Georgia and Louisiana. Can be expected throughout Alabama from Tuscaloosa northward.

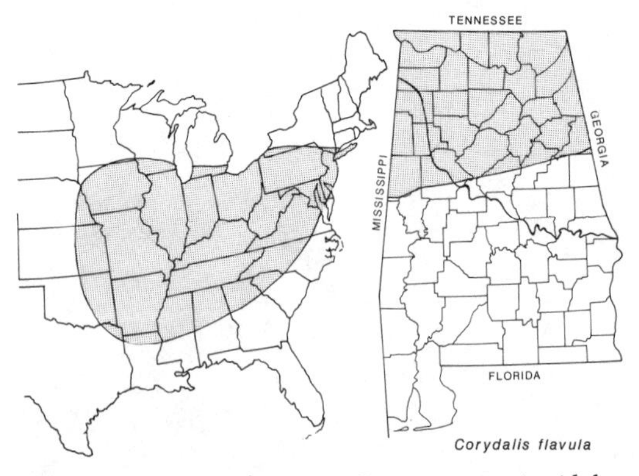

Range in eastern United States Known range in Alabama

Habitat Occurrence: Fumatory occurs in deciduous woods with rich soil and especially in alluvial woods.

Corydalis micrantha SLENDER FUMEWORT

Species Recognition: Slender fumewort is a sprawling annual with petals 0.9–1.5 cm long, erect fruits, and seeds 0.15–0.2 cm wide that do not have a marginal ring.

Geographic Distribution: Illinois to Minnesota and Nebraska, south to the Carolinas, Florida, and Louisiana. Occurs throughout the southern part of Alabama.

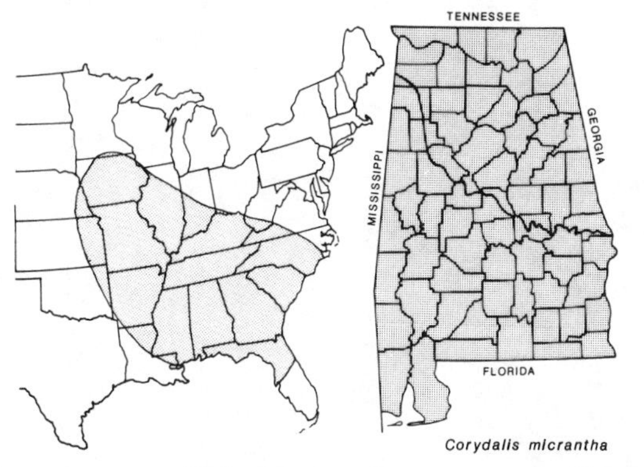

Range in eastern United States Known range in Alabama

Habitat Occurrence: Slender fumewort occurs in sandy or gravelly fields, rocky slopes, rarely in woodlands, but most commonly on roadsides and waste places, especially of the Coastal Plain.

DICENTRA DUTCHMAN'S BRITCHES; STAGGERWEED; SQUIRREL-CORN

One species occurs in Alabama.

Dicentra cucullaria (Plate 4) DUTCHMAN'S BRITCHES

Species Recognition: Dutchman's britches is a perennial from a cluster of underground stems. The plant has basal, palmately divided leaves and a single inflorescence that arises from the bulb. The inflorescence contains 5–10 stalked flowers. The flowers have 2 whitish-to-purplish basal spurs; the fruit is an elongate capsule with black, lustrous, kidney-shaped seeds.

Toxic Properties: Dutchman's britches contains several different alkaloids, in particular aporphine, protoberberine, and protopine, that are found throughout the plant, both above- and belowground. In feeding experiments, cattle that consumed 2 percent of their weight from any part of the plant showed toxic but nonfatal symptoms within a day. Symptoms include trembling, running back and forth with head held high, and salivation. The trembling becomes convulsive, and animals fall down within a few minutes after the onset of symptoms. In the feeding experiments, some animals that ate 4 percent of their weight died.

Geographic Distribution: From Quebec to Ontario and southern Saskatchewan, south to Alabama, Georgia, and Kansas; also in Washington and Oregon. Can be found in the northern part of Alabama.

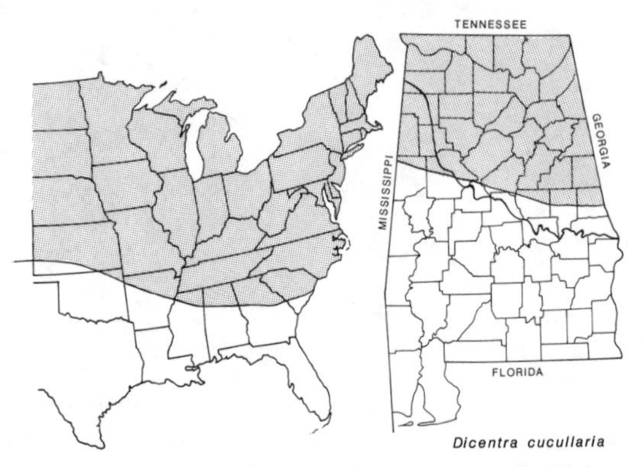

Range in eastern United States   Known range in Alabama

Habitat Occurrence: Dutchman's britches occurs in rich deciduous woods. It appears early in the year before the leaves of deciduous species.

Family MORACEAE

MACLURA — OSAGE ORANGE

Only one species occurs in Alabama.

Maclura pomifera — OSAGE ORANGE

Species Recognition: Osage orange is a medium-sized deciduous dioecious tree with glabrous entire leaves that are acuminate at the apex. Its many small stems are thorny, and the flowers are imperfect. The fruit is the most distinctive part of the tree. It is orange-sized (10–13 cm in diameter), yellowish green, spherical, and warty. Some Alabamians claim that placing an osage orange in a kitchen cabinet will keep bugs away.

Toxic Properties: The milky sap in the stems, leaves, and fruit contains unknown compounds that cause contact dermatitis in a few persons.

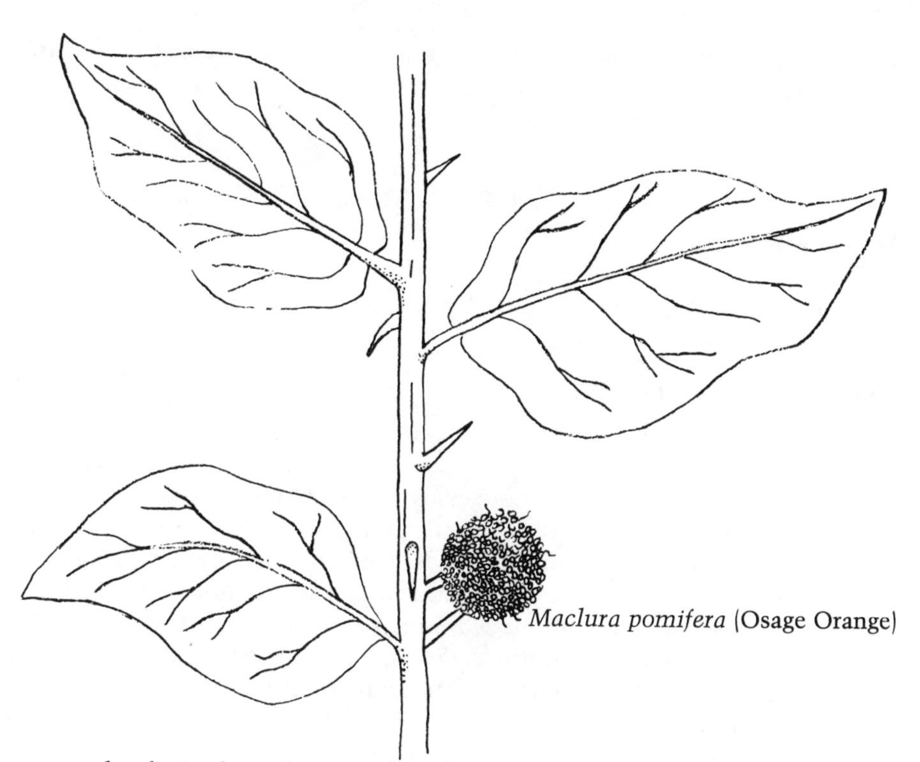
Maclura pomifera (Osage Orange)

The fruits have been said to be toxic when ingested, but toxicity studies and tests in Australia have all been negative.

Geographic Distribution: Native to Arkansas, Texas, and Oklahoma. Osage orange has been planted as a hedge throughout much of the United States and has become established in the wild in the Southeast and Midwest. Particularly common in the Tennessee Valley area and the Black Belt counties of Alabama. Can be found sporadically elsewhere.

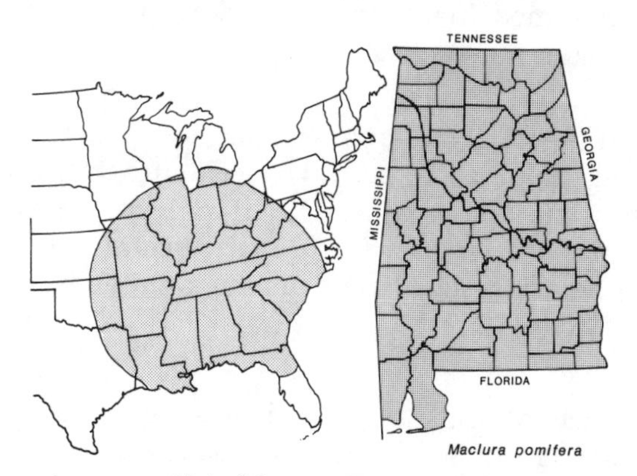

Range in eastern United States   Known range in Alabama

Habitat Occurrence: Osage orange is found most commonly in disturbed sites along roads and fences.

Family CANNABACEAE

CANNABIS MARIJUANA

Only one species occurs in Alabama.

***Cannabis sativa* (*Plate 5*)** MARIJUANA

Species Recognition: Marijuana is a coarse annual herb that reaches a height of 3–5 meters and reproduces by seeds. The leaves are opposite on the lower part of the plant and alternate on the upper parts. They are palmately compound with 3–7 linear, coarsely dentate leaflets. The flowers are imperfect, and the plants are dioecious. The male inflorescences are produced mostly at the tip of the branches, whereas the female inflorescences are produced along the length of the branch. The female flowers mature into a conspicuous achene called hempseed.

Toxic Properties: The toxic chemicals present are tetrahydrocannabinols. These are resinous compounds found in the mature plant, especially in the leaves and the female inflorescences. The effects of smoking or other forms of ingestion of this drug may vary greatly among individuals from a feeling of euphoria and elation to heightened sensitivity to stimulation. Eventually, hallucinations and mental confusion become evident. Large amounts have been known to be toxic to animals, although the plants are bitter and unpleasant tasting and are rarely consumed. Marijuana plants are now unlikely to be accessible for grazing in most areas.

Geographic Distribution: Native to Asia but now widely spread throughout the world, including all of the United States. In alluvial bottomlands of Mississippi and Missouri River valleys, marijuana has formed populations covering several to many hectares often with the assistance of enterprising citizens in a region. Can be expected throughout Alabama.

Habitat Occurrence: Marijuana is found growing naturally in forests, especially those in river bottoms. It is also cultivated indoors in pots, in backyard gardens, and in deep, hidden woods.

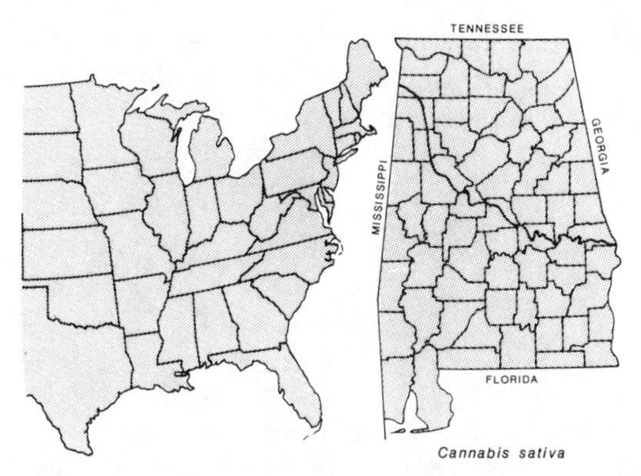

Range in eastern United States   Known range in Alabama

Family PHYTOLACCACEAE

PHYTOLACCA POKEWEED

One species occurs in Alabama.

***Phytolacca americana* (Plate 5)** COMMON POKEWEED

Species Recognition: A large perennial herb that overwinters by means of a large rootstock. The leaves are alternate, pale green, entire to finely serrate along the margins, and acute at both apex and base. The flowers are produced in an elongate raceme at the apex of the stem and branches. They contain 5 white or, more commonly, pinkish sepals, 5–30 stamens, and 5–12 carpels that are united into a ring. The fruit is a purple berry with purple juice and 8–10 large seeds.

Toxic Properties: Pokeweed is a common cause of human and livestock poisoning in the United States. This is especially true for humans, because the young plant is used in making poke salad. The plant apparently becomes much more toxic as it matures. Even with young plants, before being eaten the leaves should be boiled and the water discarded and the leaves reboiled. The toxic properties are phytolaccotoxin (a resin), phytolaccine (an alkaloid), and a saponin. All parts of the plant, especially the roots and purple stem, contain these toxins. Ingestion of small quantities of berries have resulted in no symptoms, but large quantities have caused burning in the mouth, severe abdominal cramps, nausea, vomiting, diarrhea, sweating, salivation, visual disturbance, weakness, spasms, and convulsions. Death in humans has occurred within 24 hours due to respiratory failure.

Geographic Distribution: New England and southern Quebec to southern Ontario and Michigan, and south to Florida and Texas. Occurs throughout Alabama.

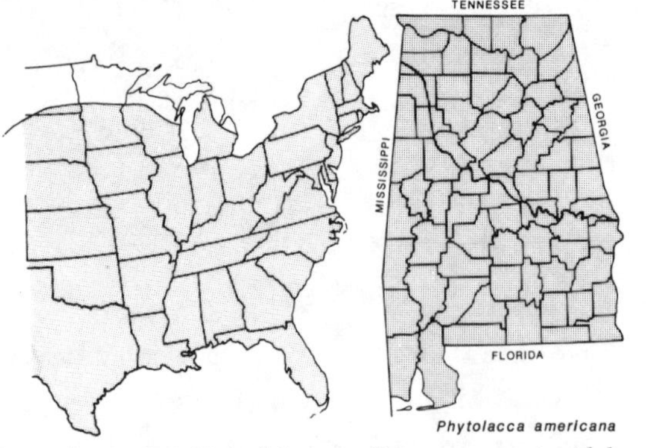

Range in eastern United States Known range in Alabama

Habitat Occurrence: Common pokeweed occurs in rich low grounds, most commonly in recent clearings, fields, and roadsides.

Family NYCTAGINACEAE

MIRABILIS FOUR-O'CLOCK

Mirabilis is a genus of erect coarse perennial herbs with fleshy, tuberous, thickened roots; it frequently has numerous stems arising from the crown. The leaves are opposite and linear to ovate. The flowers are surrounded by a 5-lobed, calyx-like involucre. The calyx is 5-lobed and united almost to the apex. Two species occur in Alabama.

Toxic Properties: The seeds and root contain different alkaloids, in particular oxymethylanthraquinone and tribonelline. These 2 alkaloids are effective as laxatives and possibly cause stomach pain, vomiting, and diarrhea.

Mirabilis albida WHITE FOUR-O'CLOCK

Species Recognition: The white four-o'clock has leaves that are lance-elliptic, and the main leaves are more than 2 cm wide.

Geographic Distribution: South Carolina to Kansas and Iowa, south to Alabama and Texas. Can be expected in the Black Belt counties of Alabama.

Habitat Occurrence: The white four-o'clock occurs in dry soils, meadows, and hillsides.

Mirabilis nyctaginea (Wild Four-O'clock)

Mirabilis nyctaginea WILD FOUR-O'CLOCK

Species Recognition: The wild four-o'clock has leaves that are deltoid-ovate and are more than 2 cm wide.

Geographic Distribution: Wisconsin to Montana, south to Tennessee, Alabama, Texas, and Colorado. Can be expected in the northern counties of Alabama.

Habitat Occurrence: The wild four-o'clock occurs in weedy places, dry soil, limestone, gravel, and sandy post oak areas.

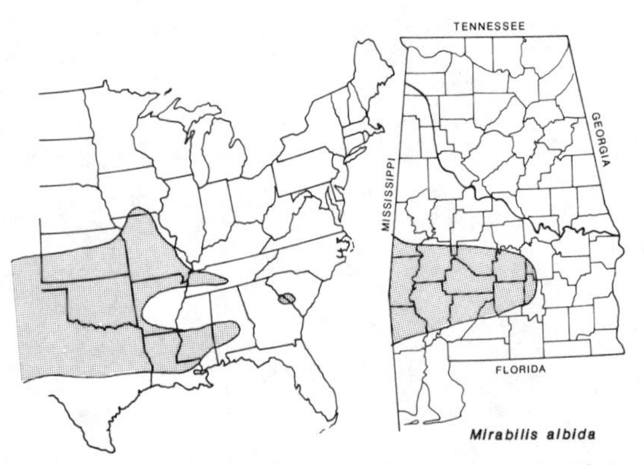

Range in eastern United States Known range in Alabama

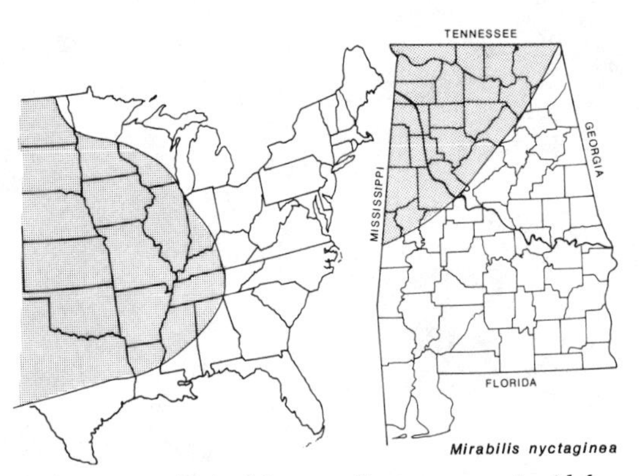

Range in eastern United States Known range in Alabama

Family CHENOPODIACEAE

ATRIPLEX SALT BUSHES

Atriplex is a genus of annual or perennial herbs with opposite or alternate leaves. The leaves are either entire, toothed, or lobed. The fruits are enclosed by 2 valvelike bracts. Three species occur in Alabama.

Toxic Properties: Salt bushes contain soluble oxalates, possibly nitrates, and saponins in all parts of the plant. One species concentrates the chemical element selenium. High concentrations of any of these chemicals are dangerous when ingested. However, many species are eaten cooked as a potherb after the first boiling water is poured off and new water is brought to a boil.

Atriplex patula ATRIPLEX

Species Recognition: *Atriplex patula* can be recognized by its greenish color, the opposite leaves on the bottom part of the plant, and the hastate blades.

Geographic Distribution: Newfoundland to Ohio and British Columbia, south to California, Florida, and Texas. In Alabama most common in Mobile and Baldwin counties.

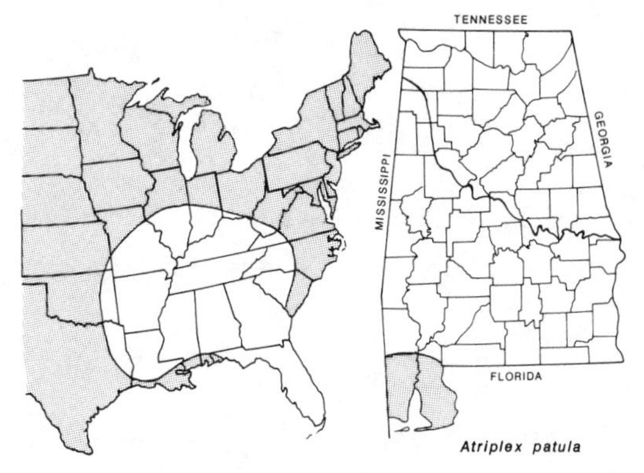

Range in eastern United States   Known range in Alabama

Habitat Occurrence: *Atriplex patula* occurs in saline soils, salt marshes, and along sand dunes.

Atriplex pentandra SEABEACH ORACH

Species Recognition: Seabeach orach is a gray-green plant with alternate leaves throughout that are dentate along the margins.

Geographic Distribution: Florida to Texas, and south to the West Indies, Venezuela, Colombia, and Peru. In Alabama can be found only along the Gulf Coast.

Habitat Occurrence: Seabeach orach occurs on sandy seashores.

Atriplex rosea ATRIPLEX

Species Recognition: *Atriplex rosea* is an erect green herb with all leaves alternate, most of them dentate.

Geographic Distribution: Wyoming to southern Washington, south to California and Mexico. *Atriplex rosea* has been introduced from New York to Florida. In Alabama known only along the Gulf Coast.

Habitat Occurrence: *Atriplex rosea* occurs along roadsides, waste places, cultivated areas, and alkaline soils.

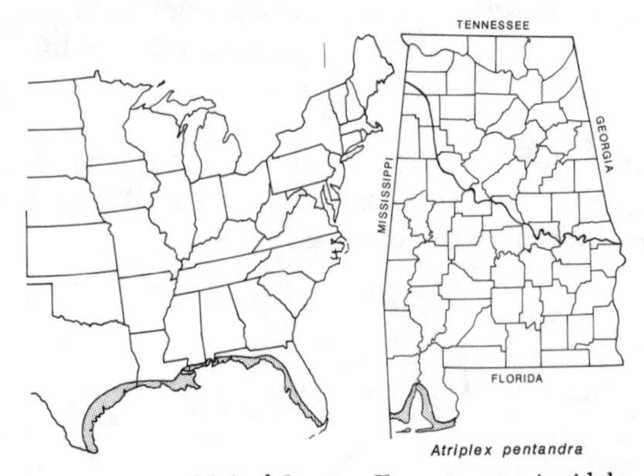

Range in eastern United States   Known range in Alabama

Range in eastern United States   Known range in Alabama

CHENOPODIUM — GOOSEFOOT; PIGWEED

Chenopodium is a genus of erect annual, or occasionally perennial, monoecious herbs. The leaves are alternate, pinnately veined, often 3-nerved, petiolate, and occasionally appear in the general shape of a goose's foot. The flowers are mostly sessile in small clusters that are collected into panicles or spikes. Eight species have been reported from Alabama, but only 2 are known to be toxic.

Toxic Properties: All parts of the plant contain soluble oxalates, nitrates, and the terpene ascaridol. Overdoses of the plant have caused headache, nausea, hallucinations, and gastroenteritis. Convulsions, paralysis, coma, and fatalities have been reported in humans and livestock. Nonetheless, both species have been used as a potherb or as a tea. They can apparently be ingested safely after having been cooked and the water poured off.

Chenopodium album LAMB'S-QUARTERS

Species Recognition: Plants are erect and freely branching without being covered by glandular resin dots. The leaves are thick and fleshy rhombic-lanceolate, 3-nerved, irregularly sinuate to entire. The inflorescence is without leafy bracts. The seeds are glossy black and 1–1.5 mm broad.

Geographic Distribution: Throughout the eastern United States. Occurs in every county of Alabama.

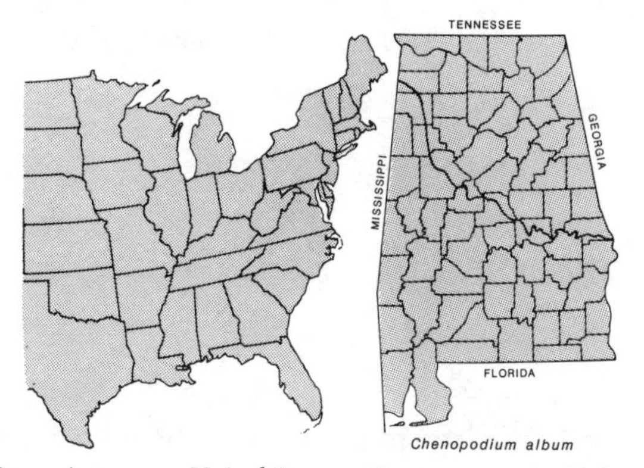

Range in eastern United States   Known range in Alabama

Habitat Occurrence: Lamb's-quarters is a weed of cultivated waste ground and barnyards.

***Chenopodium ambrosioides* (Plate 5)** MEXICAN TEA; AMERICAN WORMSEED

Species Recognition: Mexican tea is an erect annual or weak perennial that branches only sparingly. The leaves are covered by short pubescence and completely covered on the lower surface with glandular resin dots. The leaves are lanceolate, pinnately veined with more than 3 veins, nearly entire to coarsely serrate or dentate. The inflorescence is usually leafy and bracted; the seeds are glossy black or dark brown and less than 1 mm wide.

Geographic Distribution: Northern New England, New York, southern Ontario, Wisconsin, and Iowa, south to Florida and Texas. Occurs throughout Alabama.

Habitat Occurrence: Mexican tea occurs in waste places, cultivated grounds, and pastures.

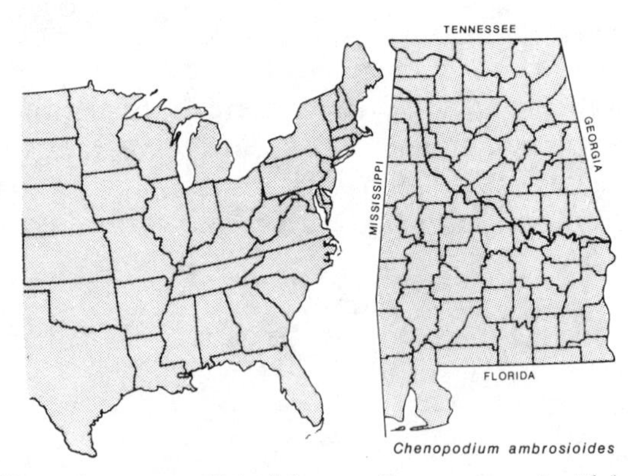

Range in eastern United States   Known range in Alabama

SALSOLA TUMBLEWEED

One species occurs in Alabama.

Salsola kali TUMBLEWEED

Species Recognition: Tumbleweed is an annual herb with a much-branched stem that has alternate linear-to-filiform leaves that are almost terete, giving the appearance of spines. When green, the plant is more or less succulent. The flowers are in the axis of upper leaves that more or less enclose the flower and fruit. Perianth parts become winged and the fruit is 1-seeded.

Toxic Properties: Tumbleweed concentrates soluble oxalates. Large amounts of oxalates or oxalic acid may cause calcium deficiency. In sufficient quantity, these compounds will rapidly cause electrolyte imbalance, nervous symptoms, reduced blood coagulation, and the formation of oxalate crystals in the kidney and in the urinary tract.

Geographic Distribution: Minnesota to Saskatchewan, Washington, California, Texas, and along the Gulf Coast and Atlantic Coast, north to the Carolinas. In Alabama known only from Mobile County.

Habitat Occurrence: Tumbleweed occurs in sandy soil, valleys, on dry plains, along roadsides, and disturbed areas.

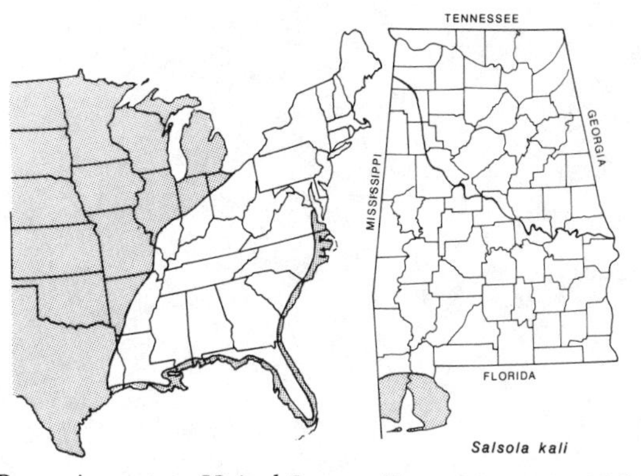

Range in eastern United States Known range in Alabama

Family AMARANTHACEAE

AMARANTHUS PIGWEED

Amaranthus is a genus of erect herbs with alternate entire leaves; the flowers are imperfect, and the male and female flowers are usually on the same plant in dense terminal or axillary spikes. Each is subtended by 3 conspicuous red, green, or purple bracts. The fruit is a 1-seeded indehiscent fruit.

Toxic Properties: All parts of the plants are known to contain oxalates, nitrates, and other compounds. Ingestion of the leaves by livestock causes vitamin deficiency, bloating, and gastroenteric irritation. Death can occur but is rare because the concentration of toxins is low.

Amaranthus albus TUMBLEWEED

Species Recognition: An erect herb with small axillary spines, floral bracts that are not widened, and flowers in axillary clusters. Petals and sepals are present. The fruit is smooth.

Geographic Distribution: Widely distributed throughout all of North America and adventive in Europe, Asia, Africa, and South America. In Alabama reported only from Mobile and Baldwin counties.

Habitat Occurrence: Tumbleweed is restricted to waste places and cultivated areas.

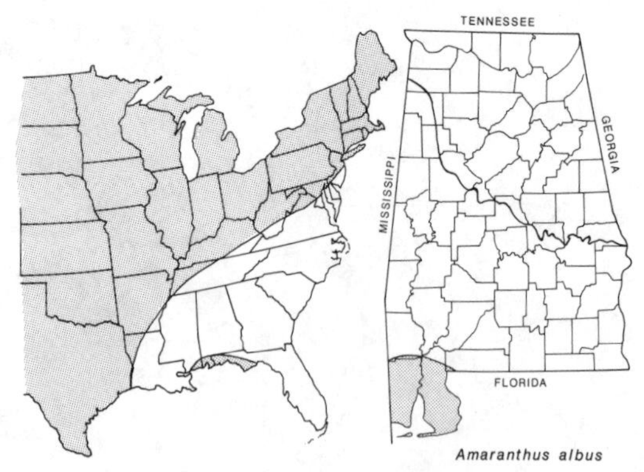

Range in eastern United States Known range in Alabama

Amaranthus australis SOUTHERN WATER HEMP

Species Recognition: An erect herb without spines, with floral bracts that are not widened. The flowers are in axillary spikes; sepals and petals are absent; and the fruit is smooth.

Geographic Distribution: Florida to Texas, and south to Venezuela. In Alabama known along the Mobile River.

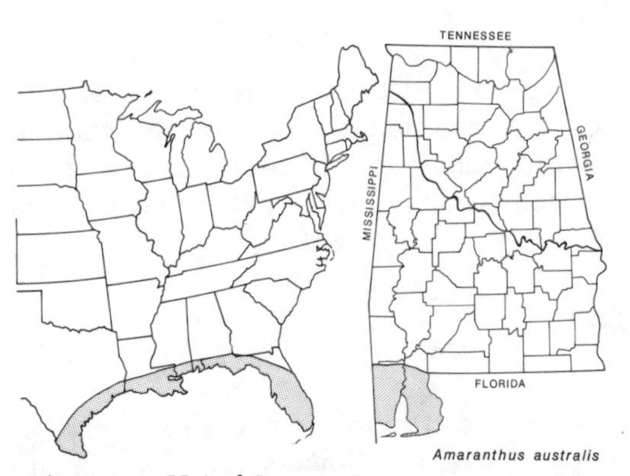

Range in eastern United States Known range in Alabama

Habitat Occurrence: Southern water hemp is a plant of salty and marshy places, occasionally growing in shallow water.

Amaranthus spinosus SPINY AMARANTH

Species Recognition: *Amaranthus spinosus* is an erect herb with axillary spines and without widened floral bracts. The spikes are axillary, and sepals and petals are present. The fruit is smooth.

Amaranthus spinosus (Spiny Amaranth)

Geographic Distribution: Throughout North America from New England to Minnesota, and south to the tropics. Throughout Alabama.

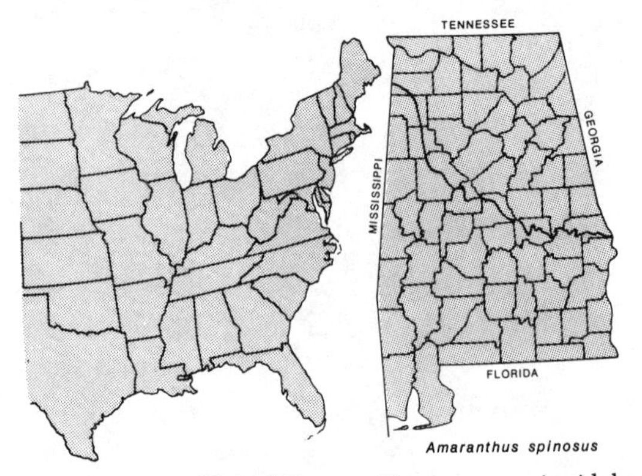

Range in eastern United States   Known range in Alabama

Habitat Occurrence: Spiny amaranth is a common weed of cultivated areas and waste places.

Additional Species in Alabama: Other species with similar appearances and properties as those mentioned above also occur in this region.

Amaranthus crassipes AMARANTHUS

Range in eastern United States   Known range in Alabama

Amaranthus hybridus GREEN PIGWEED

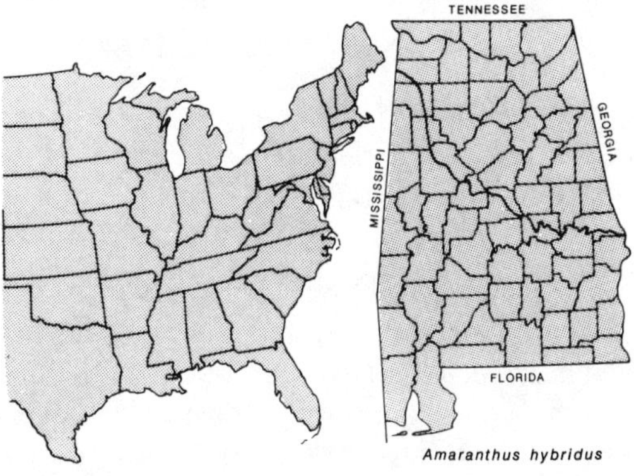

Range in eastern United States Known range in Alabama

Amaranthus retroflexus PIGWEED

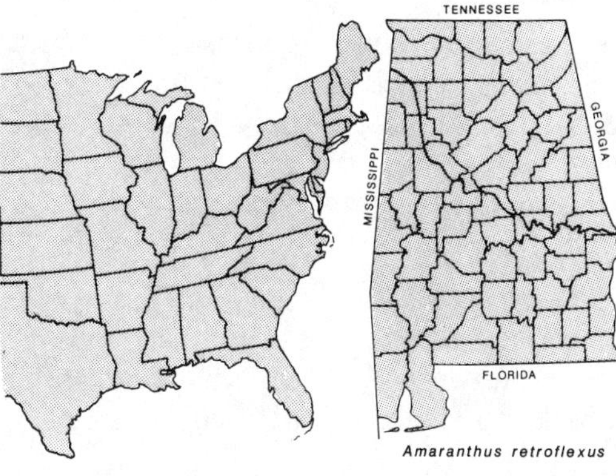

Range in eastern United States Known range in Alabama

Amaranthus viridis AMARANTHUS

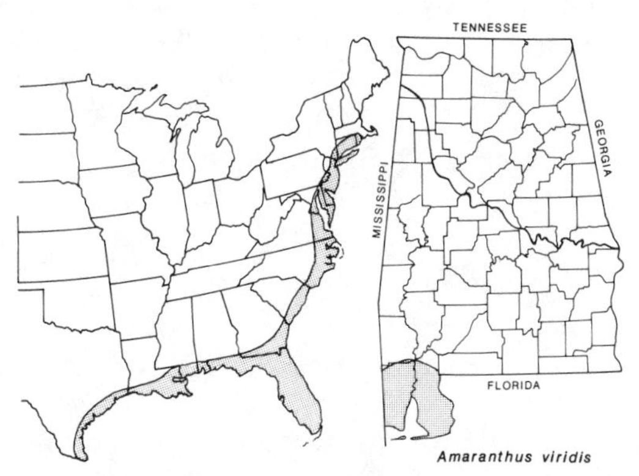

Range in eastern United States Known range in Alabama

Family BATACEAE

BATIS SALTWORT

Batis is a genus with only one species.

Batis maritima (_Plate 6_) SALTWORT

Species Recognition: Saltwort is a succulent subshrub with trailing stems that root at the nodes and form large patches from which erect flowering branches arise. The leaves are opposite, fleshy, cylindrical, and linear, with a small downward projection. The flowers are in conelike clusters and have either stamens or carpels. The fruits are formed from several different flowers making a multiple of 2–8 fruits in 1 cluster. As the name suggests, saltwort is a halophyte, one of the plant species noted for thriving in salty habitats.

Toxic Properties: All parts of the plant contain some unknown toxin, which is thought possibly to be accumulated nitrates or oxalates. Nonetheless, the plant can be eaten safely in small quantities and is best when cooked. Large amounts of saltwort have been suspected of causing poisoning in livestock, but there are no experimental studies to substantiate this.

Geographic Distribution: Atlantic Coast and Gulf Coast from South Carolina to Florida and west Texas. In Alabama can be expected only in extreme southern Mobile and Baldwin counties.

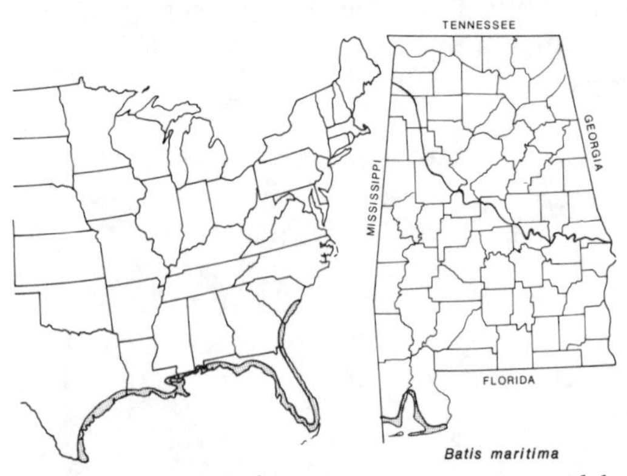

Range in eastern United States Known range in Alabama

Habitat Occurrence: Saltwort is an inhabitant of salt marshes or open, brackish marshes.

Family POLYGONACEAE

POLYGONUM — SMARTWEED OR KNOTWEED

Polygonum is a genus of perennial herbs with erect to decumbent stems that have alternate linear-to-lanceolate leaves, swollen stems at the nodes and tubular stipules that surround the stem and grow upward. The flowers are pink to white in terminal or, rarely, axillary spikes. There are 6 perianth parts with 3 sepals and 3 petals. The fruit is a 3-angled nutlet. There are 17 species in Alabama, and none of them is considered to be seriously toxic; therefore, the genus will be considered as a whole.

Toxic Properties: The sap of most species contains an acrid compound whose composition has not been determined. Prolonged exposure to this compound may cause dermatitis in some people. Grazing animals possibly suffer gastroenteritis when *Polygonum* is taken internally. Various species are used medicinally and are eaten as a cooked potherb. Some plants are possibly photosensitizers when ingested. They cause livestock or humans to be sensitive to the sun and develop white spots on the skin.

Geographic Distribution: Throughout the United States. Can be expected throughout Alabama.

Habitat Occurrence: *Polygonum* may be a weed in lawns, but occurs more commonly in open areas along shorelines and in wet ditches, rivers, and lakes.

Alabama species: The following species of *Polygonum* occur in Alabama: *P. aviculare*; *P. cespitosum*; *P. convolvulus*; *P. densiflorum*; *P. hirsutum*; *P. hydropiper* (Marshpepper Smartweed); *P. hydropiperoides* (Swamp Smartweed); *P. lapathifolium*; *P. orientale*; *P. pensylvanicum* (Pennsylvania Smartweed); *P. persicaria*; *P. punctatum*; *P. sagittatum*; *P. scandens*; *P. setaceum*; and *P. tenue*.

RUMEX — DOCK SORRELL

Rumex is a genus of erect herbs with alternate leaves, swollen nodes, and sheathing stipules. The fruit and the flowers are in terminal or axillary spikes. There are 3 sepals and 3 petals. The fruit is a 3-angled nut or achene, but the sepals remain as 3 wings on a mature fruit. Three species are known to be toxic, and all occur in Alabama.

Toxic Properties: The leaves contain soluble oxalates. Dangerous levels of oxalates are sometimes accumulated. Some species have been suspected causes of livestock poisoning in Europe, Australia, and New Zealand. The juice of the leaves may cause dermatitis to sensitive persons. For symptoms of soluble oxalates see the genus *Oxalis*.

***Rumex acetosella* (Plate 6)** SHEEP SORRELL

Species Recognition: Sheep sorrell is an erect herb that is perennial by means of rhizomes. The leaves are hastate or sagittate. The plants are dioecious, and the sepals, when fully mature, are less than 3 mm long.

Geographic Distribution: Introduced as a weed from Europe and has spread throughout the eastern United States and westward. Can be expected throughout Alabama.

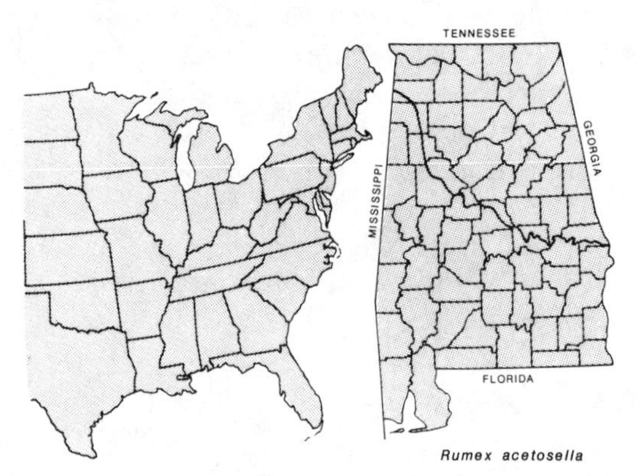

Range in eastern United States Known range in Alabama

Habitat Occurrence: Sheep sorrell occurs in open sandy or rocky acid soil.

Rumex crispus CURLY DOCK

Species Recognition: Curly dock is an erect, monoecious perennial herb. The leaves are without basal lobes; the margins of the sepals are entire to slightly toothed; the leaves are finely crispid along the margin. The mature fruiting calyx is less than 6 mm long.

Geographic Distribution: Throughout the eastern United States, and farther west, as an introduced weed from Europe. Occurs throughout Alabama.

Habitat Occurrence: Curly dock is a weed of cultivated ground, waste soils, old fields, and various waste places.

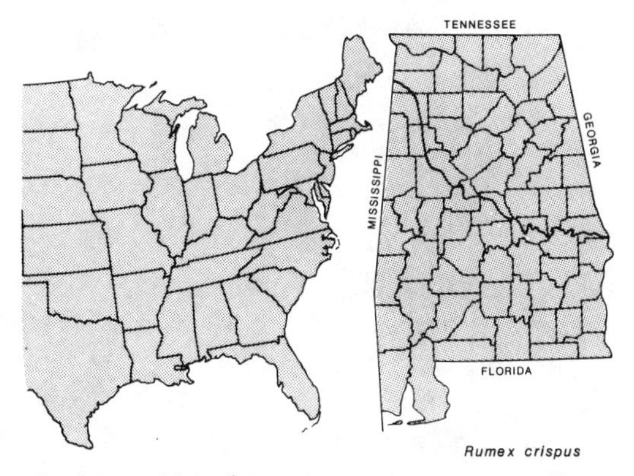

Range in eastern United States Known range in Alabama

Rumex hastatulus HALBERD SORRELL

Species Recognition: Halberd sorrell is an erect perennial herb without rhizomes. The plants are dioecious, the leaves are hastate, and the calyx, when fully developed, is more than 3 mm long.

Geographic Distribution: Coastal from Massachusetts to Long Island, south to Florida, and inland west to Texas. In Alabama occurs from Tuscaloosa County southward.

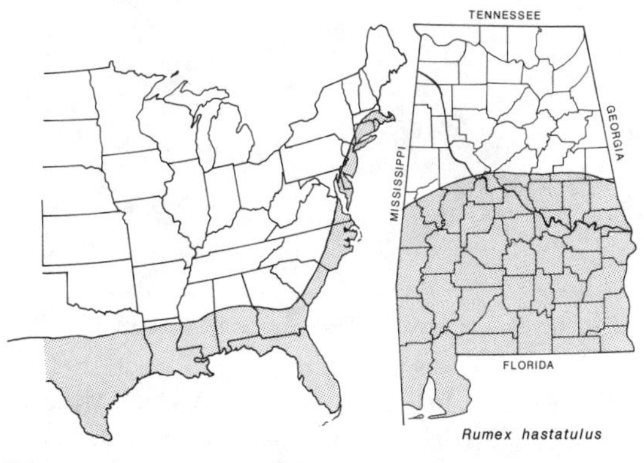

Range in eastern United States Known range in Alabama

Habitat Occurrence: Halberd sorrell occurs in sandy soil, open pastures, and on lawns.

Family MALVACEAE

MODIOLA GROUND IVY

Only one species occurs in Alabama.

Modiola caroliniana ***(Plate 6)*** CAROLINA BRISTLE MALLOW

Species Recognition: Carolina bristle mallow is a spreading or prostrate herb with stiff stems that grow to a length of about 0.6 meters. The leaves are palmately lobed and veined with blades about 6 cm across. The flowers are in the axils of the leaves, solitary, and about 0.5 cm across. The petals are orangish, and the flowers have many stamens that surround 4–5 stigmas. The fruit is a many-seeded capsule.

Toxic Properties: All parts of the plant contain an unknown compound. Ingestion is suspected to cause incoordination and prostration in goats, sheep, and cattle and posterior paralysis in goats. Sheep and cattle display nervous disturbances with convulsions preceding death.

Geographic Distribution: From northern Virginia to Texas and Florida, California, and Mexico. Can be found throughout Alabama.

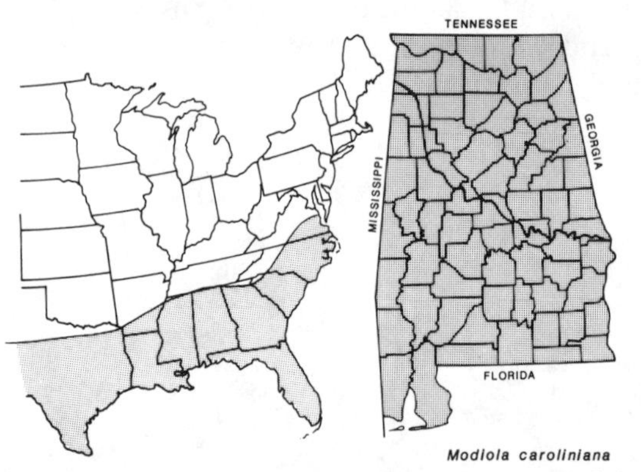

Range in eastern United States Known range in Alabama

Habitat Occurrence: Carolina bristle mallow is a weed that occurs on lawns, roadsides, and in disturbed areas where there is full sunlight.

Family ERICACEAE

Toxic Properties: The active ingredient of species in this family is a resinous substance, andromedotoxin, that is present in all parts of the plant, including the pollen. Even honey made from the nectar is bitter and toxic. When sufficient quantities of the leaves, pollen, or honey are ingested, humans begin watering in the mouth, eyes, and nose. Within 6 hours, nausea and vomiting begin. Sweating, abdominal pain, headache, weakness, tingling of the skin, slow pulse, uncoordination, low blood pressure, and respiratory difficulty may also occur. Symptoms often culminate in convulsions, progressive paralysis, coma, and death, the latter due to respiratory failure.

KALMIA — WICKY; MOUNTAIN LAUREL

Kalmia is a genus of evergreen shrubs or small trees that occurs in soils of poor quality. The genus is distinctive in that the flowers have small pockets in which the anthers are lodged before pollination. Once the pollen is mature, the anthers are released abruptly when an insect visiting the flower contacts the filament of an anther. The anther quickly springs forward out of the pocket and deposits pollen on the insect which transports it to the next flower it visits.

Kalmia hirsuta — WICKY; HAIRY LAUREL

Species Recognition: Wicky is a low shrub that grows less than 1 meter tall, and has bark that breaks into thin strips. The leaves are evergreen, less than 2 cm long, simple, entire, leathery, and scattered along the stem, rather than being clustered at the apex of the stem. The entire stem is covered with stiff hairs. The flowers are pink and marked with red near the stamen pockets. The fruit is a capsule covered with stipitate glands.

Geographic Distribution: Coastal South Carolina to Florida and coastal Alabama. In Alabama wicky is restricted to the southernmost counties.

Range in eastern United States   Known range in Alabama

Habitat Occurrence: Wicky is characteristic of seasonally wet pine savannas in the Lower Coastal Plain. It is especially common in pine savannas that burn periodically and have sandy soil.

Kalmia latifolia MOUNTAIN LAUREL

Species Recognition: Mountain laurel is a shrub that grows more than 3 meters tall and has bark that breaks into thin strips. The leaves are 5–12 cm long, simple, entire, leathery, dark green, and shiny. All are produced at the apex of the stem. The flowers are pink to pinkish white and often have purple spots near the stamen pockets. The fruit is a capsule and is mostly covered with stipitate glands.

Geographic Distribution: New England to Indiana, and south to western Florida and Louisiana. Can be expected throughout Alabama, although rare in the Lower Coastal Plain.

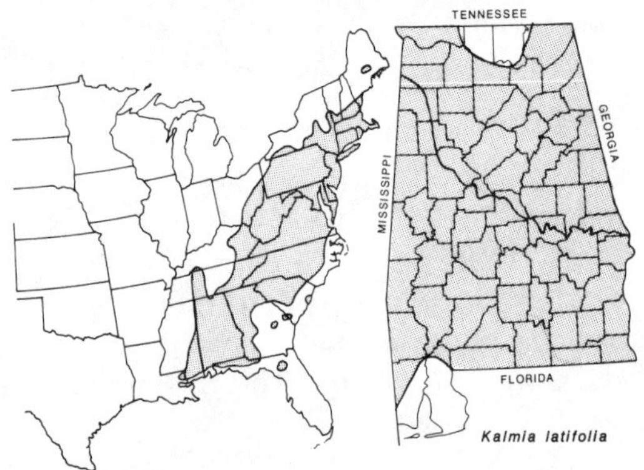

Range in eastern United States Known range in Alabama

Habitat Occurrence: Mountain laurel occurs most commonly in acid conditions, especially in sandy soils or along the edges of sandstone cliffs, usually on north-facing slopes. Rarely, the species occurs in swampy habitats.

LEUCOTHOE FETTERBUSH

Leucothoe is a genus of erect shrubs with leathery serrate leaves. Flowers are produced either in the axils of the leaves or at the ends of the stems. The petals are fused into a tube that is cylindrical. The fruit is a capsule, dark brown to almost black, that splits open to release many seeds.

Leucothoe axillaris LEUCOTHOE

Species Recognition: *Leucothoe axillaris* is an evergreen shrub with axillary flowers.

Geographic Distribution: Southeast Virginia to Florida and Mississippi. Can be expected in the southern one-third of Alabama.

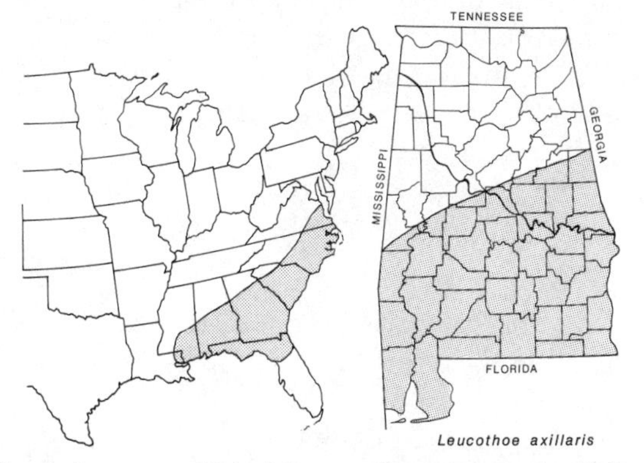

Range in eastern United States   Known range in Alabama

Habitat Occurrence: *Leucothoe axillaris* occurs in low woods, especially those with acid soils.

Leucothoe racemosa LEUCOTHOE

Species Recognition: *Leucothoe racemosa* is a deciduous shrub with flowers produced at the tips of the stems.

Geographic Distribution: Massachusetts to eastern Pennsylvania, south to Florida and Louisiana. Can be expected throughout Alabama.

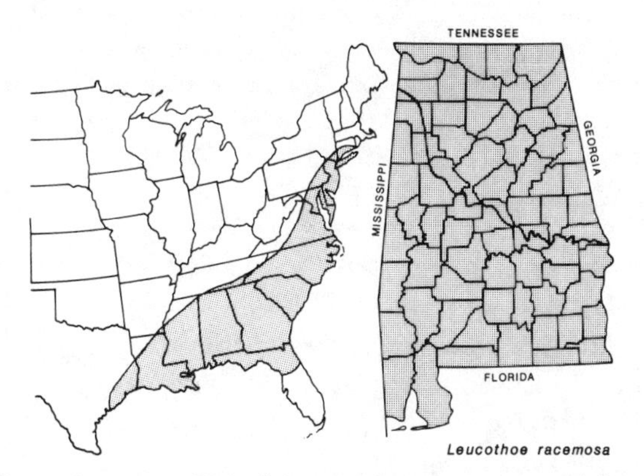

Range in eastern United States   Known range in Alabama

Habitat Occurrence: *Leucothoe racemosa* occurs in moist thickets and pond shores, especially in sandy acid soil.

Lyonia is an erect shrub that grows 2–3 meters tall with entire, alternate, leathery leaves that often are enrolled. The flowers are produced either in the axils of the leaves or at the tip of the stem. They have petals united into a tube that is shaped more or less like an urn. The fruit is a brown capsule that dehisces to release many seeds.

Lyonia ligustrina MALEBERRY

Species Recognition: *Lyonia ligustrina* is a shrub with deciduous leaves. It has a nearly globous corolla and a capsule less than 0.5 cm long.

Geographic Distribution: New England to West Virginia and Kentucky, south to South Carolina, Georgia, and Alabama. Can be expected throughout the eastern half of Alabama and rarely in the western half.

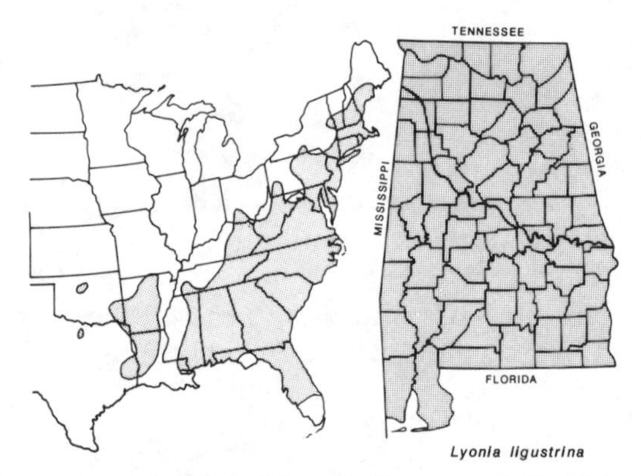

Range in eastern United States Known range in Alabama

Habitat Occurrence: Maleberry occurs in wet or dry thickets, swamp woods, and low pinelands.

Lyonia lucida TETTERBUSH

Species Recognition: Tetterbush is a low shrub with evergreen leaves, a cylindrical corolla, and a capsule more than 0.45 cm long. This species is said to be nontoxic, but is included because of the high amount of toxicity of related species.

Geographic Distribution: Virginia to Florida and Louisiana. Occurs in the southern half of Alabama.

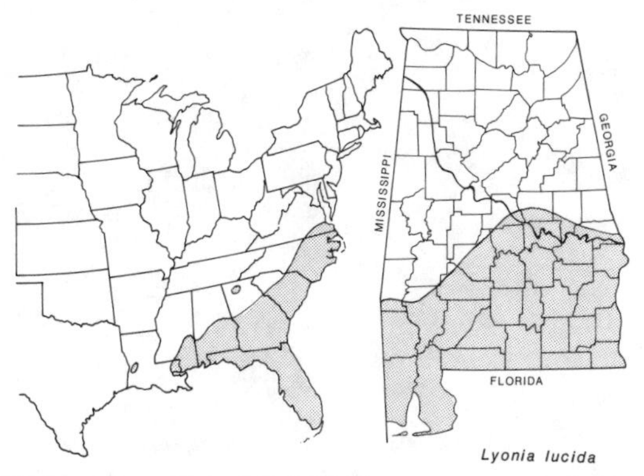

Range in eastern United States Known range in Alabama

Habitat Occurrence: Tetterbush occurs in low woods and thickets.

RHODODENDRON RHODODENDRON; AZALEA

Rhododendron is a genus of shrubs with evergreen or deciduous leaves that are entire and may or may not be covered by hair. The flowers are white to pink, rarely yellow or orange. The fruit is a capsule. The petals are united into a tube and are spreading at the apex. There are 10 stamens, and the corolla is more than 2 cm broad.

Rhododendron alabamense ALABAMA AZALEA

Species Recognition: Alabama azalea is an erect deciduous shrub with pubescent twigs that have no glands on them, a corolla tube that is glandular, and a pubescent capsule. The leaves are pubescent beneath. The pedicels and the calyces are glandular. The mature flower buds are pubescent, and the corolla is white with an orange spot.

Geographic Distribution: Restricted to Alabama, ranging from Morgan County, south to Pike County.

Habitat Occurrence: Alabama azalea occurs in dry woods, especially deciduous woods that are mixed with pine.

Rhododendron canescens SWEET AZALEA

Species Recognition: Sweet azalea has deciduous serrate leaves, pubescent twigs that are without glands, a corolla tube that contains glands with pubescence on the capsule. The leaves are pubescent beneath;

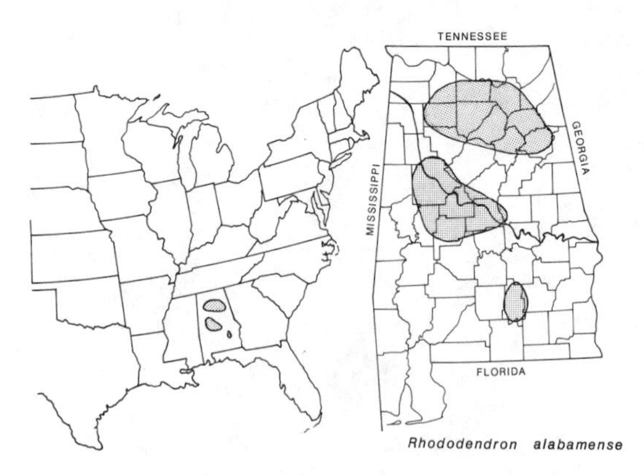

Range in eastern United States Known range in Alabama

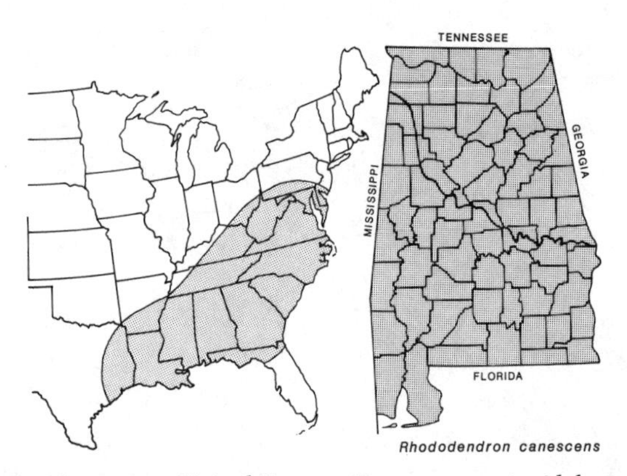

Range in eastern United States Known range in Alabama

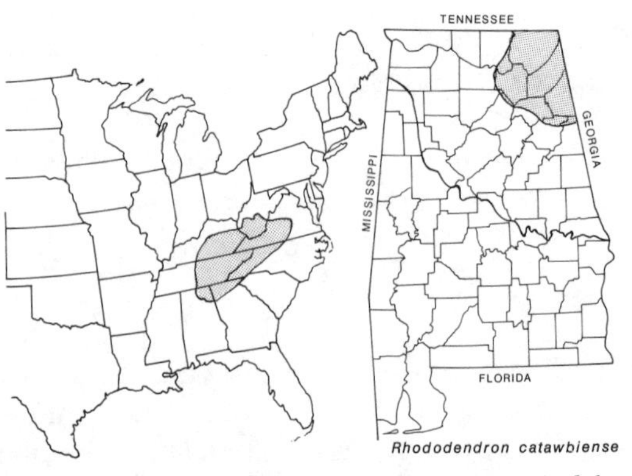

Range in eastern United States Known range in Alabama

the pedicels and calyces do not have glands. The flower is pink or white, and there are no strigose hairs along the midveins of the leaves.

Geographic Distribution: North Carolina to Tennessee and Arkansas, south to Florida and Texas. Occurs throughout Alabama.

Habitat Occurrence: Sweet azalea occurs along stream banks in rather acid, sandy soil.

***Rhododendron catawbiense* (*Plate 6*)** PINK RHODODENDRON

Species Recognition: Pink rhododendron has evergreen entire leaves with 10 stamens; the leaves have no glandular punctae beneath. The flowers are pink, very showy, and are in large terminal clusters.

Geographic Distribution: West Virginia to Georgia and Alabama. Can be expected in the northeastern corner of Alabama.

Habitat Occurrence: Pink rhododendron occurs in acid soils, on mountain slopes, summits, bluffs, and cliffs.

Additional Species in Alabama: Other species with similar appearances and properties as those mentioned above also occur in this region.

Rhododendron arborescens SMOOTH AZALEA

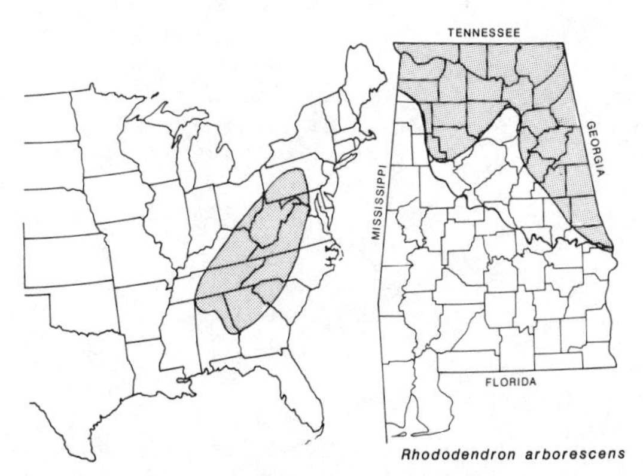

Range in eastern United States Known range in Alabama

Rhododendron austrinum ORANGE AZALEA

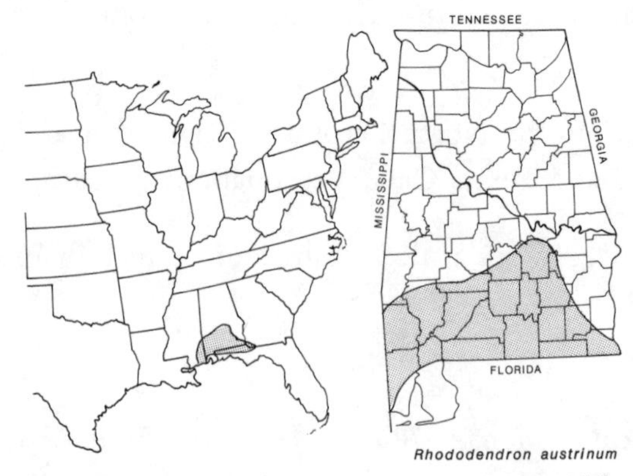

Range in eastern United States Known range in Alabama

Rhododendron calendulaceum FLAME AZALEA

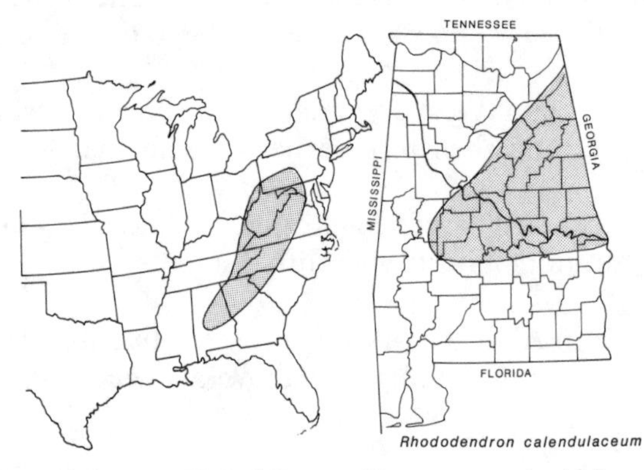

Range in eastern United States Known range in Alabama

Rhododendron flammeum (Plate 7) FLAME AZALEA

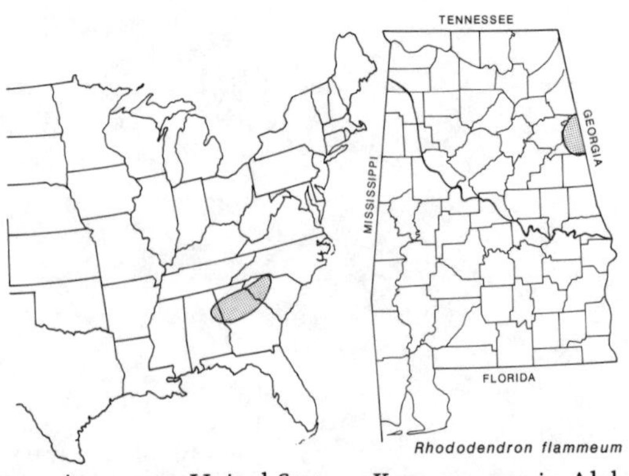

Range in eastern United States Known range in Alabama

Rhododendron minus SPOTTED RHODODENDRON

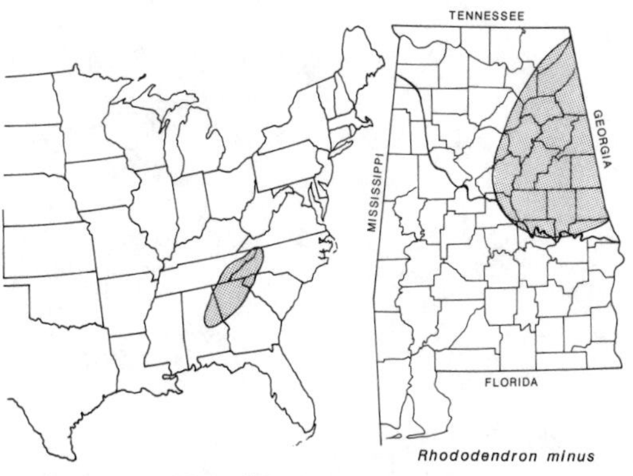

Range in eastern United States Known range in Alabama

Rhododendron periclymenoides EARLY AZALEA; HONEYSUCKLE AZALEA

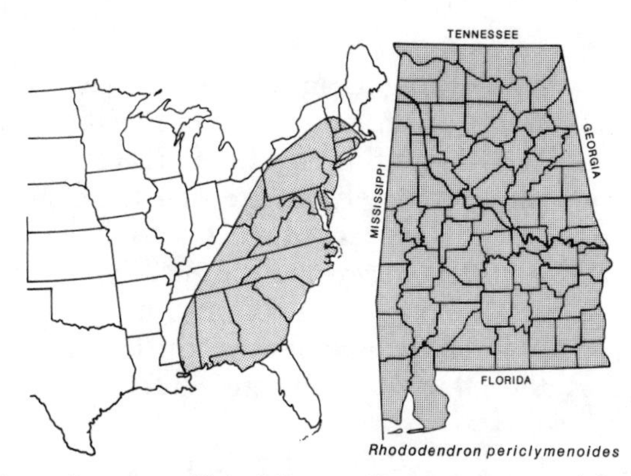

Range in eastern United States Known range in Alabama

Rhododendron prunifolium PLUM LEAF

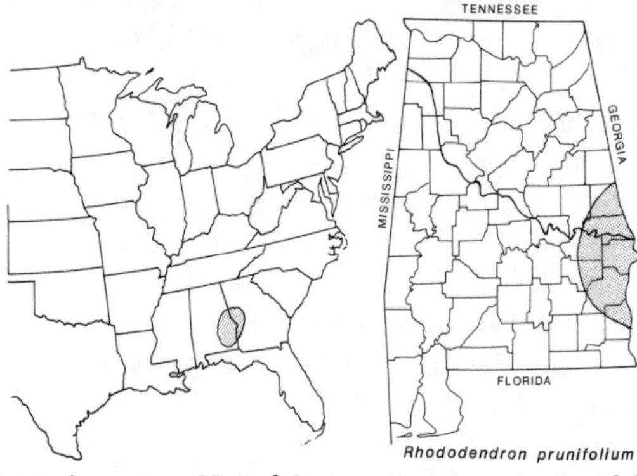

Range in eastern United States Known range in Alabama

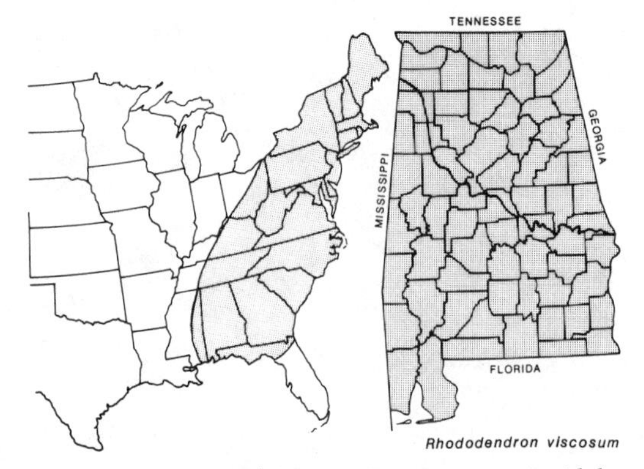

Range in eastern United States Known range in Alabama

Family SAXIFRAGACEAE

HYDRANGEA HYDRANGEA

These shrubs grow to 2–3 meters in height and have brown exfoliating bark on old stems. The leaves are opposite and lobed or unlobed, and are densely tomentose beneath. The flowers are produced in compact clusters. Some of the flowers are completely sterile and are represented only by showy bracts. The fertile flowers lack these bracts, have white petals, and produce brown capsules that are strongly ribbed. Cultivated hydrangeas have a distinctive feature; they react opposite to litmus paper to soil type. Those growing in acidic soil have blue flowers; those in basic soils have pink flowers. If the soil is neutral, the flowers are white.

Toxic Properties: Hydrangea leaves and buds contain a cyanogenic glycoside. Apparently there are other toxins, because the symptoms produced from ingestion of hydrangea are not necessarily those caused by cyanogenic glycosides. Symptoms include abdominal pain, bloody diarrhea, nausea, and vomiting.

Hydrangea arborescens SEVENBARK; WILD HYDRANGEA

Species Recognition: Sevenbark can be recognized by its serrate leaves and the flowers produced in flat-to-roundtop clusters.

Geographic Distribution: Southern New York to Ohio, Indiana, Illinois, and Missouri, south to Georgia and Oklahoma. Occurs from Lee County northward in Alabama.

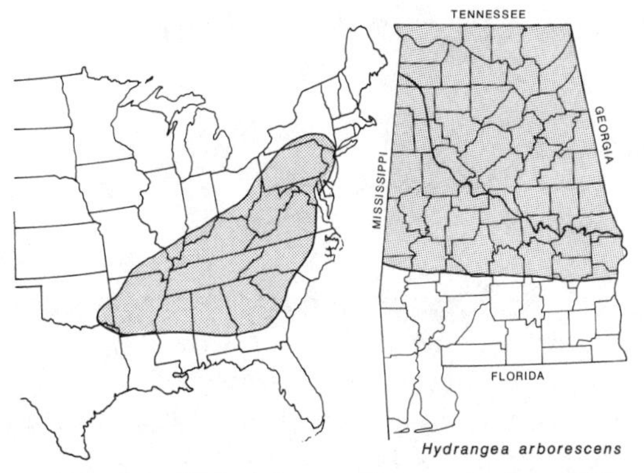

Range in eastern United States   Known range in Alabama

Habitat Occurrence: Sevenbark occurs in rich woods, on calcareous rocky slopes, and banks of streams.

***Hydrangea quercifolia* (*Plate 7*)** OAK-LEAF HYDRANGEA

Species Recognition: The oak-leaf hydrangea can be recognized by the lobed leaves and by flowers that grow in elongate clusters.

Geographic Distribution: Gulf Coast states from Louisiana to Florida. Occurs throughout Alabama.

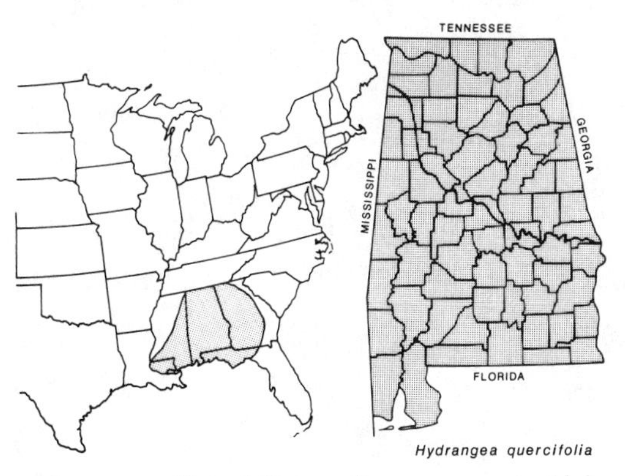

Range in eastern United States   Known range in Alabama

Habitat Occurrence: Oak-leaf hydrangea occurs in rich woods, especially with calcareous rocky slopes, and along stream banks.

Family ROSACEAE

ERIOBOTRYA LOQUAT

One species occurs in Alabama.

Eriobotrya japonica LOQUAT

Species Recognition: Loquat is an evergreen tree that grows about 7 meters tall. The young branches and lower surfaces of the leaves are covered with rusty-colored hairs. The flowers are white, about 1 cm wide, and clustered in terminal panicels. The fruit is about 3 cm long, yellow, and contains a few large seeds.

Toxic Properties: The seeds of loquat contain the cyanogenic glycoside, amygdalin, which forms hydrogen cyanide. Ingestion of large quantities of seeds containing hydrogen cyanide can cause death. See the treatment under *Prunus* for possible symptoms. The fruit flesh often is eaten raw or used in preparation of jams and jellies.

Geographic Distribution: Cultivated from Florida to Texas and Mexico. In Alabama can be expected anywhere from Tuscaloosa south.

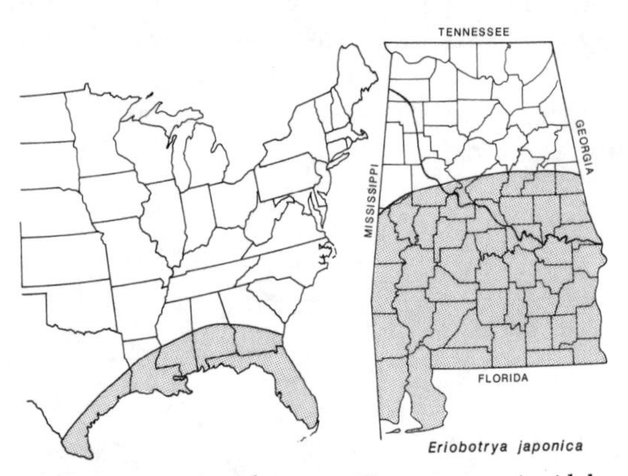

Range in eastern United States Known range in Alabama

Habitat Occurrence: Loquat is cultivated, and it persists at old homesites and along roadsides.

PRUNUS PLUM; CHERRY

Prunus is a genus of shrubs or small trees that reach about 7 meters, with alternate simple leaves that are usually deciduous. A few species are tardily deciduous or evergreen. The inflorescences are either umbels or terminal racemes. There are 5 fused sepals and 5 petals that appear to grow from the top of the ring formed by the sepals. There are many stamens and 1 ovary. The fruit is a fleshy drupe.

Toxic Properties: The leaves, bark, and especially the seeds contain cyanogenic glycosides as well as the proteins amygdalin and mandelonitrile. These are acted on by enzymes to produce hydrogen cyanide. Cyanic poisoning symptoms occur within one hour. Lethal doses produce spasms, and death is due to respiratory failure. The fruit has caused numerous deaths of humans and livestock. The cyanogenic glycosides are found in the stones of the fruits. The seeds of some species, such as the almond, are eaten after being roasted, which deactivates the toxin. Large quantities of green seeds can be harmful.

Prunus caroliniana CAROLINA LAUREL CHERRY

Species Recognition: The leaves are evergreen, the petioles do not have any glands near the summit, and the flowers and fruits are in racemes.

Geographic Distribution: Coastal Plain from North Carolina, south to Florida, west to Mississippi and Tennessee. Can be expected throughout Alabama, especially in the southern half.

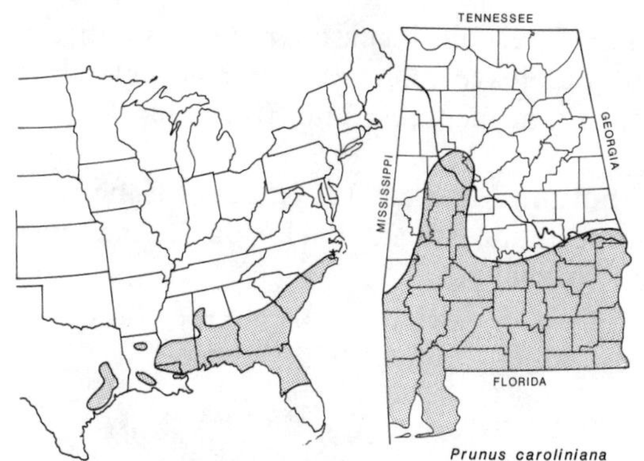

Range in eastern United States Known range in Alabama

Habitat Occurrence: Carolina laurel cherry occurs on fencerows, in thickets, low woods, and forests. It is apparently rare as a native plant, but more abundant as it escapes from cultivation.

Prunus persica PEACH

Species Recognition: The peach is a small tree with flowers solitary or in small clusters, and pedicels that are less than 0.5 cm long. The fruit covering is velvety.

Geographic Distribution: New England to Michigan, and south to the Gulf Coast. Can be expected throughout Alabama, but especially in the sandy soil areas of the central part of the state.

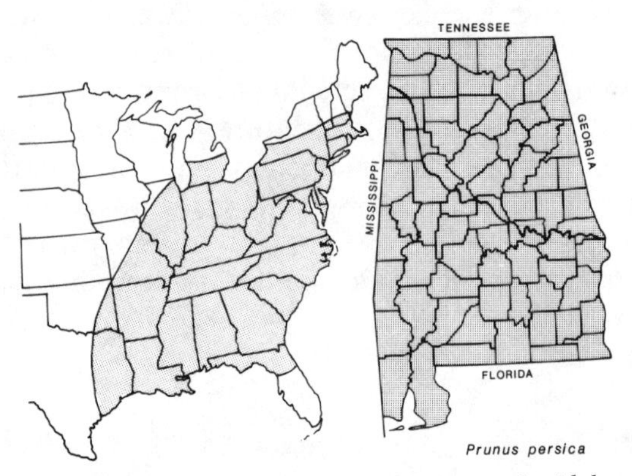

Range in eastern United States   Known range in Alabama

Habitat Occurrence: The peach occurs in roadside thickets and waste places and persists after cultivation around homesites.

***Prunus serotina* (Plate 7)** WILD BLACK CHERRY

Species Recognition: The wild black cherry is a small tree 7–8 meters tall with deciduous leaves. The petiole has 2 glands near its apex, and the flowers are in racemes. The fruit is black.

Geographic Distribution: Maritime Provinces of Canada, west to Minnesota and North Dakota, south to Florida, Texas, and Mexico. Occurs throughout Alabama.

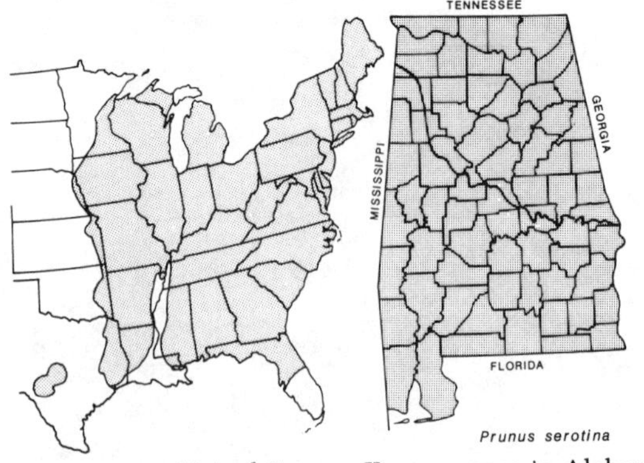

Range in eastern United States   Known range in Alabama

Habitat Occurrence: Wild black cherry is found especially in dry woodlands, along fencerows, and waste places.

Prunus serotina (Wild Black Cherry)

Family FABACEAE

BAPTISIA WILD INDIGO

Baptisia is a genus of glabrous herbs that often have a glaucous covering The leaves are usually 3-foliolate, and the flowers are in terminal clusters of 5–10. The petals usually are creamy yellow to white, and the fruit is an inflated legume. The entire plant will often blacken as it dries. Four species occur in Alabama.

Toxic Properties: All parts of the plant contain certain alkaloids and glyco-sides. The alkaloids contained are quinolizidines. These alkaloids cause severe diarrhea and anorexia. In certain cases, cattle have died after consuming a single plant.

Baptisia bracteata WILD INDIGO

Species Recognition: *Baptisia bracteata* has pale yellow or cream flower that are in a terminal raceme. Each flower has, at its base, an obviou leaf or bract (5 mm or more wide). The leaflets of the leaves are 2– cm long and about 8–14 mm wide. The fruit body is 3–4 cm lon and 1.5–2 cm in diameter.

Geographic Distribution: North Carolina to Georgia and Alabama. Occurs only in the eastern half of Alabama.

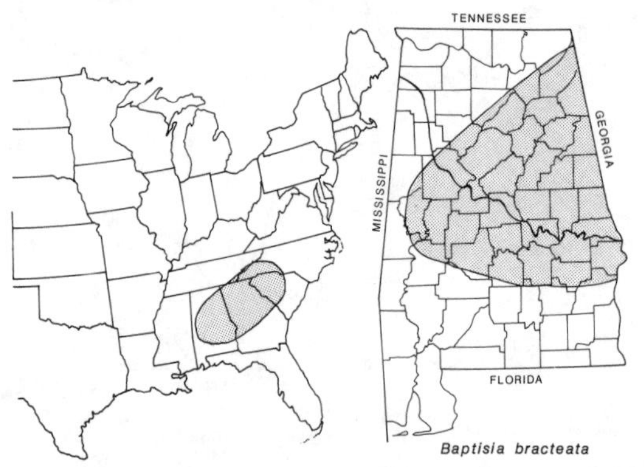

Range in eastern United States Known range in Alabama

Habitat Occurrence: *Baptisia bracteata* occurs in oak woods, sandhills, and open woods. It can be found in the Piedmont and Coastal Plain.

Baptisia lactea WILD INDIGO

Species Recognition: *Baptisia lactea* has white flowers that are in axillar or lateral racemes. The leaves are 2.5–6 cm long and 1.5–2 cm wide The fruit is 2.5–4 cm long and about 1.5 cm in diameter.

Geographic Distribution: Ontario and Minnesota, south to North Carolina, Florida, and Texas. Can be expected throughout Alabama.

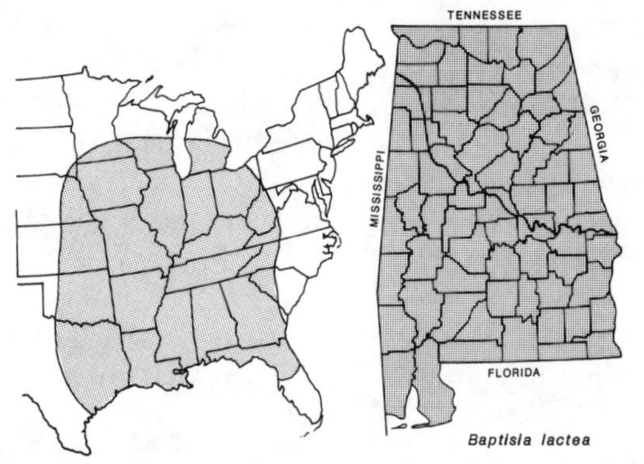

Range in eastern United States   Known range in Alabama

Habitat Occurrence: *Baptisia lactea* occurs on riverbanks and in oak-hickory woods, secondary woods, and flatwoods.

Baptisia lanceolata WILD INDIGO

Species Recognition: *Baptisia lanceolata* has 1–6 yellow-to-cream-colored flowers that are produced mostly in the axis of the leaves. The leaves have leaflets that are widest at the middle and are 5–8 cm long and 1–1.5 cm wide. The legume is 1.5–2.5 cm long and 1–1.5 cm in diameter.

Geographic Distribution: Coastal Plain, from North Carolina to Alabama. Occurs in Mobile, Baldwin, and possibly other counties along the southern tier of Alabama.

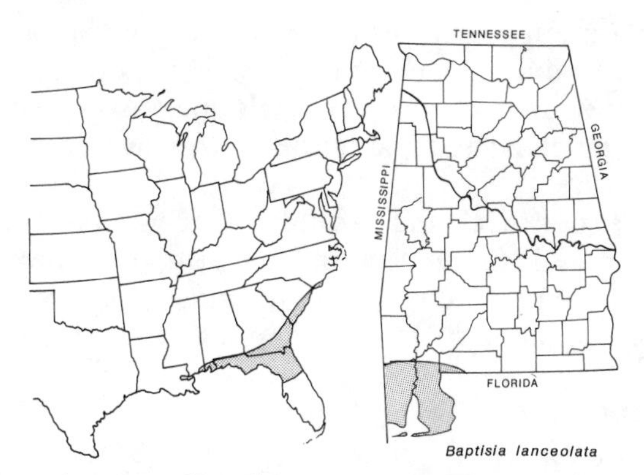

Range in eastern United States   Known range in Alabama

Habitat Occurrence: *Baptisia lanceolata* occurs in dry pinelands, sandhills, flatwoods, scrub pine, oak-hickory woods, and occasionally in bogs.

Baptisia megacarpa APALACHICOLA WILD INDIGO

Species Recognition: Apalachicola wild indigo has yellow-to-cream-colored flowers that are all borne in racemes, never singly, and are usually found at the apex of the stem. The leaves are, with leaflets, 4–8 cm long and about 2 cm wide. The fruits are 3.5–5 cm long and about 2 cm in diameter.

Geographic Distribution: Middle Florida through southern Georgia and Alabama. Occurs in central to southern Alabama.

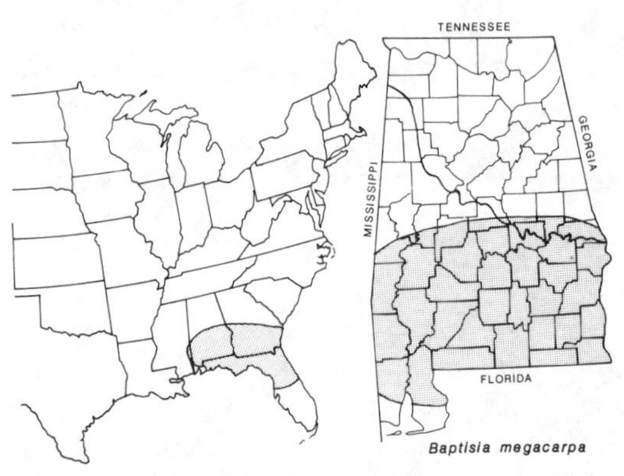

Range in eastern United States Known range in Alabama

Habitat Occurrence: Apalachicola wild indigo occurs in rich soil, mostly along riverbanks, bluffs, and floodplains.

CASSIA SENNA

All the North American species of *Cassia* are low herbs with pinnately compound leaves having 7 or more leaflets. The leaves are alternate. The flowers are yellow with 5 petals and 5 sepals and are only slightly zygomorphic. The fruit is an elongated pod with few-to-many seeds.

Toxic Properties: The leaves and possibly other parts of the plant contain anthraquinones. Extracts with this compound are used as purgatives. If the compound is ingested in large quantity it may cause distress and occasionally death in animals.

Cassia fasciculata PARTRIDGE PEA

Species Recognition: Partridge pea is an erect herb with pinnately compound leaves that have more than 11 leaflets per leaf, and the leaflets are shorter than 0.5 cm. The flower is 1 cm or more wide. The fruit is less than 5 cm long.

Geographic Distribution: Northern Massachusetts, New York, and southern Ontario, west to Minnesota and South Dakota, south to Florida and Texas. Occurs throughout Alabama.

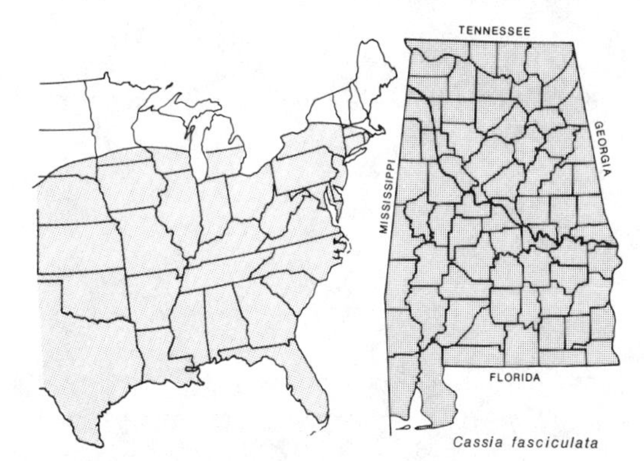

Range in eastern United States   Known range in Alabama

Habitat Occurrence: Partridge pea occurs in sandy open soil, occasionally in tidal marshes and sandy coastal dunes.

Cassia marilandica AMERICAN SENNA

Species Recognition: American senna is an erect herb with 10–15 leaflets that are more than 2 cm long. The flowers are about 0.5–1 cm wide and the pod is 5–9 cm long.

Geographic Distribution: Pennsylvania to Iowa and Kansas, south to Florida and Texas. Occurs throughout Alabama.

Habitat Occurrence: American senna is found in waste places, dry roadsides, and thickets.

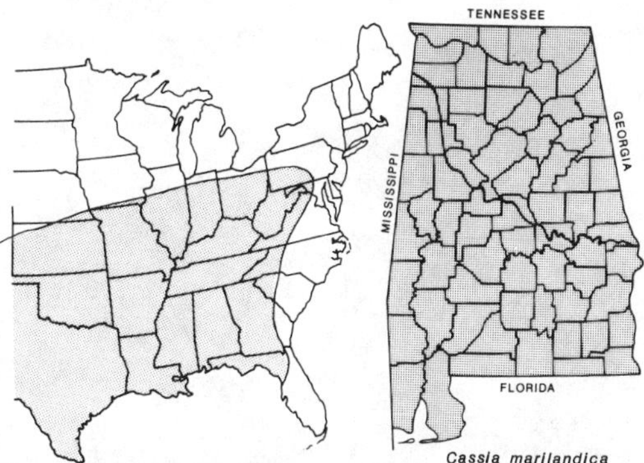

Range in eastern United States   Known range in Alabama

Cassia obtusifolia COFFEE POD; SICKLEPOD

Species Recognition: Coffee pod is an erect herb with fewer than 10 leaflets per plant. The leaflets are 2 cm or more long. The flowers are 0.5–1 cm in diameter, and the fruit is more than 10 cm long. The pod is terete in cross section.

Geographic Distribution: Virginia to Indiana, Illinois, Iowa, and eastern Kansas, south to Florida and Texas. Can be expected throughout Alabama.

Habitat Occurrence: Coffee pod occurs in waste places, cultivated lands, and shores.

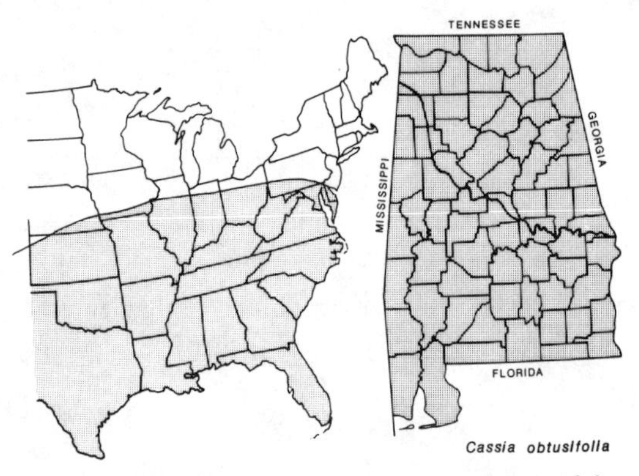

Range in eastern United States Known range in Alabama

Additional Species in Alabama: Other species with similar appearances and properties as those mentioned above also occur in this region.

Cassia nictitans WILD SENSITIVE PLANT

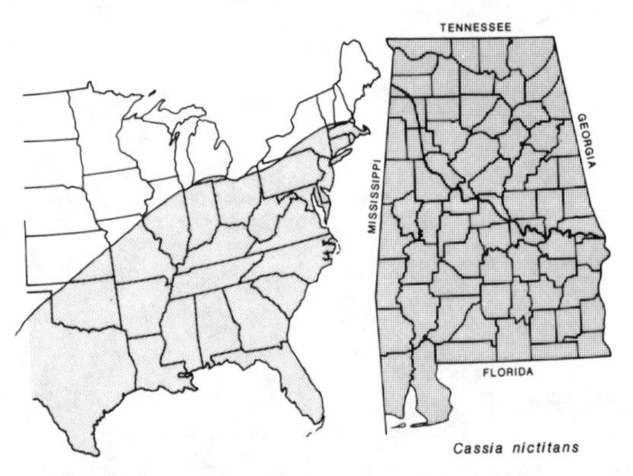

Range in eastern United States Known range in Alabama

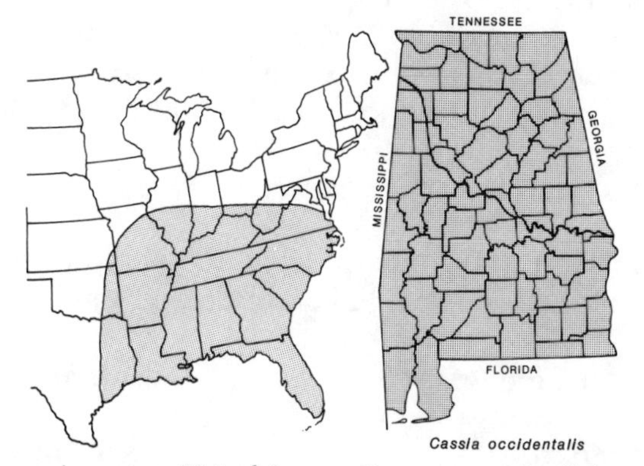

Range in eastern United States Known range in Alabama

CROTALARIA RATTLEBOX

Crotalaria is a genus of herbaceous plants with leaves that are composed of 1 blade or divided into 3 bladelike structures. The yellow flowers are shaped like those of the pea, and are produced at the top of the stem, either 1 or 2 together or 15 or more scattered along a common stalk. The most distinctive part of the plant is the fruit that appears as a large inflated pea pod and has many dark hard seeds that separate from the fruit wall when it is mature and rattle as the fruit is shaken. The shaking fruit is startlingly similar to the sound made by the large rattlesnakes of the genus *Crotalus*. Five species of *Crotalaria* are known to occur in Alabama.

Toxic Properties: The active ingredients are a group of alkaloids of the pyrrolizidine group. These include monocrotaline, retusamine, and other alkaloids. These alkaloids are in all parts of the plant including the leaves, stem, seeds, and roots. They are toxic to all livestock and fowl. In cases of acute poisoning, congestion of the liver, gastroenteritis, bloody feces, and yellowish mucous membranes occur within one to two weeks. Other symptoms include listlessness, loss of appetite, and diarrhea; death can occur, probably because of cardiac failure. Little can be done to bring about recovery after acute poisoning. Only 2 species have been documented to be toxic, but the other 3 species should be considered so until proven otherwise.

Crotalaria retusa YELLOW RATTLEBOX

Species Recognition: Yellow rattlebox is an erect herb that grows to a height of about 0.8 meters from a woody taproot. The leaves have 1 blade and are widest near their apex. They are 5–10 cm long. The stipules are very narrow and pointed and less than 2 mm long. The stems bear a very fine pubescence. The flowers are yellow or occasionally purple and tend to hang downward.

Geographic Distribution: Introduced from Europe. Known in North Carolina and Florida and should be expected in southern Alabama.

Habitat Occurrence: Yellow rattlebox occurs in fields and waste places.

Crotalaria sagittalis ARROWHEAD RATTLEBOX

Species Recognition: Arrowhead rattlebox is an annual herb that grows to a height of about 0.5 meters. It usually has many branches. The stems and leaves bear conspicuous, spreading whitish-to-brownish hairs. The leaves have 1 blade and are linear to oval. The flowers are yellow and less than 6 mm long. The fruits are 18–25 mm long.

Geographic Distribution: Northern Massachusetts and southern Vermont, west to Minnesota and South Dakota, and south to Florida, Texas, and Mexico. Can be expected throughout the northern two-thirds of Alabama.

Habitat Occurrence: Arrowhead rattlebox occurs in woodland borders, openings, and fields. It is primarily in the Piedmont, the mountains, and the Upper Coastal Plain.

Crotalaria spectabilis SHOWY RATTLEBOX

Species Recognition: Showy rattlebox is an erect herb growing as tall as 2 meters. The stems are without hairs; the leaves are composed of 1 blade and are markedly whiter near the apex. The longer leaves grow up to 18 cm. The flowers are 2.5 cm or more long, bright yellow and numerous at the apex of the plant. The fruits are about 5 cm long.

Geographic Distribution: Introduced from the American tropics. Showy rattlebox occurs from eastern Virginia to southeastern Missouri south to Texas and Florida. This species should be expected throughout Alabama.

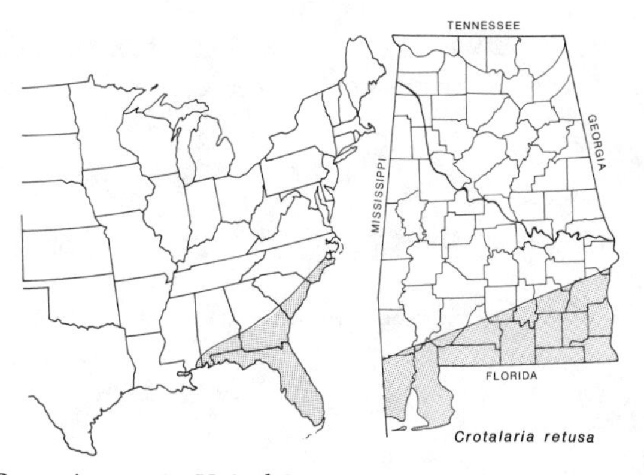

Range in eastern United States Known range in Alabama

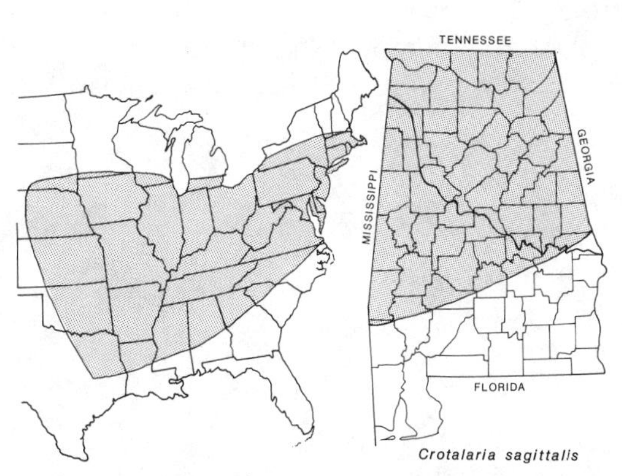

Range in eastern United States Known range in Alabama

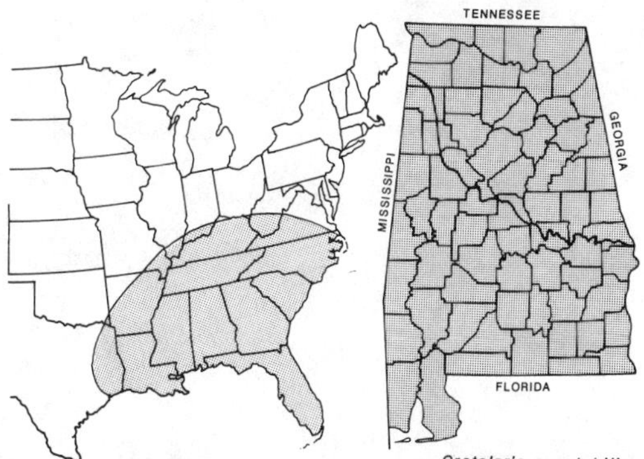

Range in eastern United States Known range in Alabama

Habitat Occurrence: Showy rattlebox occurs in open fields and along open roadsides as well as in other disturbed areas.

ERYTHRINA CORAL BEAN

Only 1 species occurs in Alabama.

Toxic Properties: The raw seeds contain alkaloids, including erysodine, hypaphorine, and other similar ones. The young leaves and flowers are edible when they are cooked. The seeds are used as rat and dog poison in Mexico.

***Erythrina herbacea* (Plate 8)** CORAL BEAN

Species Recognition: Coral bean is an erect shrub with a green stem and alternate, pinnately compound leaves that have 3 diamond-shaped leaflets. The flowers are red and tubular, about 3 cm long. The fruit is a pod.

Geographic Distribution: Coastal Plain from North Carolina to Florida, west to Texas and Mexico. Can be expected in the southern half to two-thirds of Alabama.

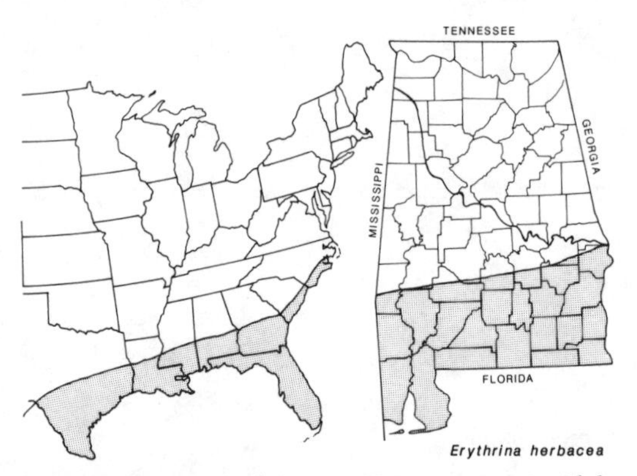

Range in eastern United States Known range in Alabama

Habitat Occurrence: Coral bean occurs in sandy woods, especially deciduous forests.

LATHYRUS WILD PEA

Lathyrus is a genus of viney herbs with alternate, pinnately compound leaves terminating in a tendril. The irregular flowers are in terminal panicles. The fruit is a legume or pod.

Toxic Properties: The seeds of *Lathyrus* contain the protein beta-aminopropionitrile. Ingestion of too many seeds results in lameness, paralysis, and skeletal deformities. Numerous human deaths have been attributed to diets composed partly or entirely of seeds from *Lathyrus* especially in the eighteenth and nineteenth centuries during conditions of poverty or drought. Three species occur in Alabama, 2 of which have been shown to be toxic.

Lathyrus hirsutus SINGLETARY PEA; HAIRY PEA

Species Recognition: The stem is winged and the petioles are without wings. The flowers are 9–13 mm long. The calyx tube is as long as the calyx teeth. The fruits are covered with hairs that have swollen bases.

Geographic Distribution: Virginia to Arkansas, south to Alabama and Mississippi. Known in the northern part of Alabama.

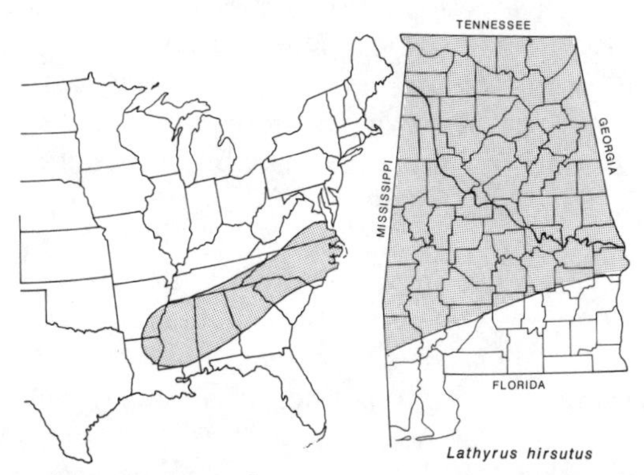

Range in eastern United States Known range in Alabama

Habitat Occurrence: Singletary pea occurs on roadsides and in borders of fields and thickets.

Lathyrus latifolius EVERLASTING PEA; BROADLEAF PEAVINE

Species Recognition: Everlasting pea has a winged stem, winged petioles, and leaflets that are 2.5 cm long or more.

Geographic Distribution: Throughout the eastern half of the United States north of Florida. Occurs in the northern half of Alabama.

Habitat Occurrence: Everlasting pea has escaped from cultivation to roadsides, thickets, and waste places.

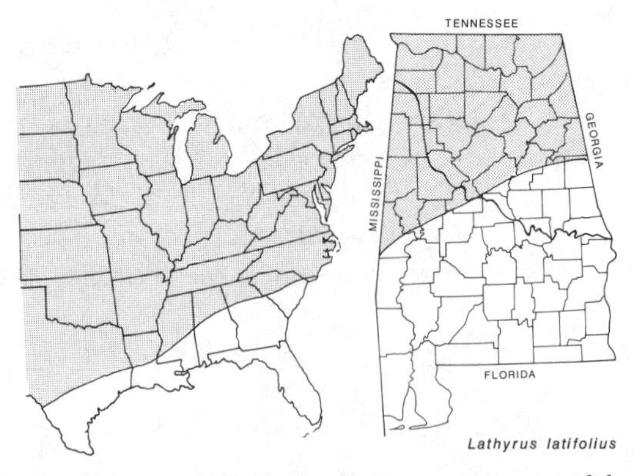

Range in eastern United States   Known range in Alabama

LUPINUS BLUEBONNET; LUPINE

Lupinus is a genus of low perennial shrubs or herbs with alternate, simple, or palmately compound leaves. The leaflets are 5–7 cm long, ranging in shape from linear to obovate. The inflorescence is a terminal raceme, often with showy flowers. The flowers are irregular, blue, white, rarely red, or yellow with 10 stamens united in 1 tubular structure. The fruit is a many-seeded pod.

Toxic Properties: Very little is known about the toxicity of the Alabama species of *Lupinus*. All parts of the plants including the leaves, pods, and especially the seeds contain various alkaloids, including lupinine, sparteine, hydroxylupanine, and others. There have been no toxicity tests on the Alabama species, but many western species have proven to be deadly when ingested by farm animals.

Lupinus diffusus SKY-BLUE LUPINE

Species Recognition: *Lupinus diffusus* has leaves that are unifoliate; that is, the leaves appear to be simple rather than palmately compound. They are covered with appressed pubescence. The flower is blue with a conspicuous white-to-green spot on the big petal.

Geographic Distribution: North Carolina, south to Florida and Mississippi. Can be expected in the southern half of Alabama.

Habitat Occurrence: Sky-blue lupine occurs in Coastal Plain, pinelands, and oak ridges, especially in sandy soil.

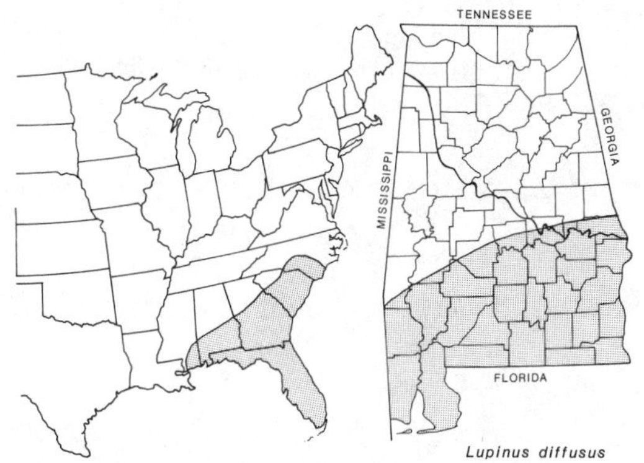

Range in eastern United States   Known range in Alabama

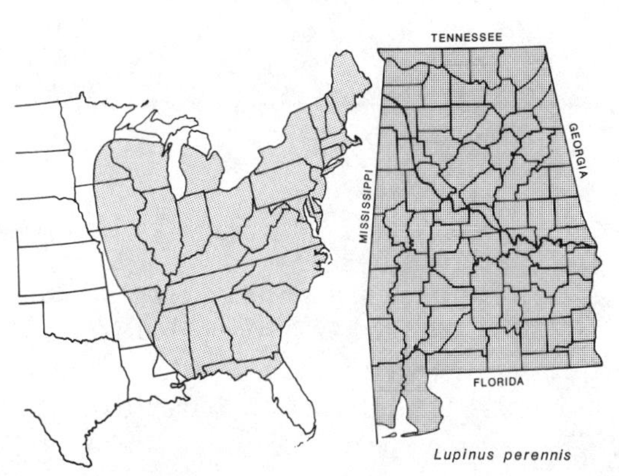

Range in eastern United States   Known range in Alabama

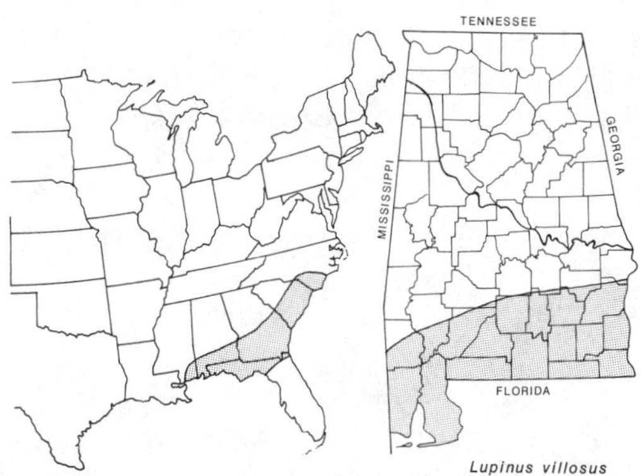

Range in eastern United States   Known range in Alabama

Lupinus perennis SUNDIAL LUPINE

Species Recognition: Sundial lupine has palmately compound leaves with 7–11 leaflets. The flower is pale blue.

Geographic Distribution: Maine to Minnesota, south to Florida and Louisiana. Can be expected throughout Alabama.

Habitat Occurrence: Sundial lupine occurs in open woods, fields, and acid sandy soils.

Lupinus villosus (Plate 8) LADY LUPINE

Species Recognition: The lady lupine leaf has 1 leaflet. The plant is covered by a shaggy pubescence. The large flower is purplish with deep reddish purple spots.

Geographic Distribution: North Carolina to Florida, and west to Mississippi. Can be expected in the southernmost counties of Alabama.

Habitat Occurrence: Lady lupine occurs in dry pinelands and sandy barrens, especially in the Coastal Plain.

MELILOTUS SWEET CLOVER

Melilotus is a genus of annual or biennial herbs that are erect and have leaves with 3 leaflets. The leaflets are serrate; that is, with sawteeth along the margin. The flowers are arranged in slender axillary or terminal racemes. The fruit is a short legume that is longer than the calyx, but normally has only 1 seed. Three species occur in Alabama. All have shown some form of toxicity.

Toxic Properties: Sweet clover contains the glycoside coumarin which is itself not toxic. However, whenever sweet clover is subjected to fungal decay, coumarin is converted to dicoumarin, which is highly toxic. Symptoms of dicoumarin poisoning include internal bleeding with massive hemorrhaging. Dicoumarin, in fact, has been used medicinally to prevent blood clotting in humans. Vitamin K is an antidote for dicoumarin poisoning.

Melilotus alba WHITE SWEET CLOVER

Species Recognition: White sweet clover is distinctive with its white flowers and dark-brown-to-black fruit.

Geographic Distribution: Naturalized from Europe, occurring throughout the eastern United States. Found throughout Alabama.

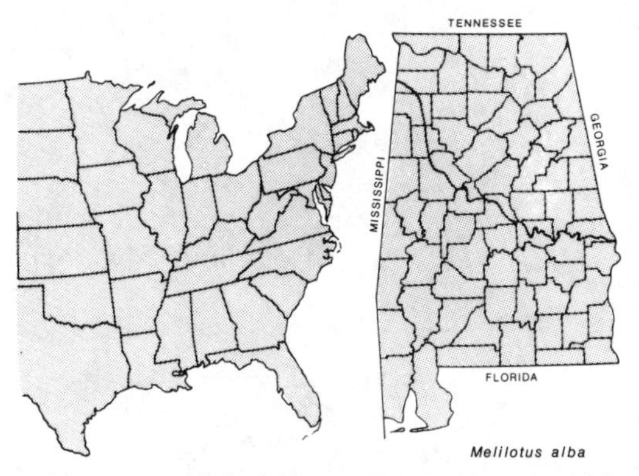

Range in eastern United States Known range in Alabama

Habitat Occurrence: White sweet clover occurs in fields, roadsides, and waste places.

Melilotus indica SOUR CLOVER

Species Recognition: Sour clover has yellow flowers with yellow-to-reddish fruits and petals that are about 2 mm long.

Geographic Distribution: Naturalized from Eurasia, occurring from Nova Scotia to Minnesota, and south to Florida, Alabama, and Mississippi. Can be found throughout Alabama.

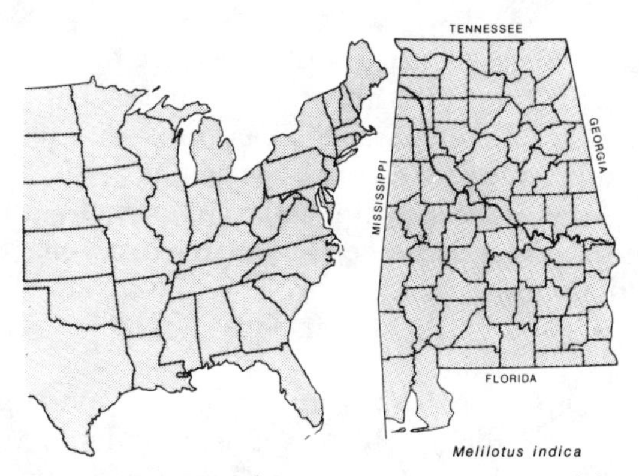

Range in eastern United States Known range in Alabama

Habitat Occurrence: Sour clover occurs in roadsides, grasslands, and waste places.

Melilotus officinalis YELLOW SWEET CLOVER

Species Recognition: Yellow sweet clover has yellow flowers with petals that are more than 3 mm long and fruits that are light brown.

Geographic Distribution: Throughout the eastern United States and southern Canada. Can be expected throughout Alabama.

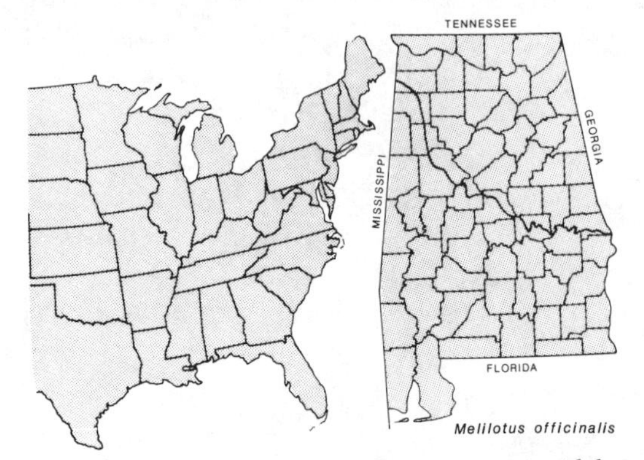

Range in eastern United States Known range in Alabama

Habitat Occurrence: Yellow sweet clover occurs in waste areas, cultivated grounds, and on roadsides.

ROBINIA BLACK LOCUST

Only one of the Alabama species will be treated.

Robinia pseudoacacia BLACK LOCUST

Species Recognition: Black locust is a tree growing to about 25 meters tall and with a trunk up to 0.3 meters in diameter. The trunk is straight, slender, and the smaller stems are covered with unbranched spines that resemble large rose thorns. The spines persist for several years. The leaves are alternate, pinnately compound with entire leaflets that are elliptical and number 3–10 pairs. Inflorescences are druping racemes that are 10–20 cm long. The flowers are white and showy. The fruit is a straight, flat, many-seeded pod.

Toxic Properties: The inner bark, sprouts, leaves, and seeds contain the toxic compound robin, which is a heat labile phytotoxin. In addition, there is a glycoside, robitin. These compounds occasionally cause poisoning when ingested by animals or, rarely, by humans. Symptoms are loss of appetite, fatigue, nausea, cold extremities, dilated pupils, weak and irregular pulse, labored respiration, and bloody diarrhea. Recovery normally comes in several days, and death rarely occurs.

Geographic Distribution: Pennsylvania and West Virginia to south Indiana, Iowa, and Oklahoma, south to northern Florida, Louisiana, and Texas. Occurs throughout Alabama.

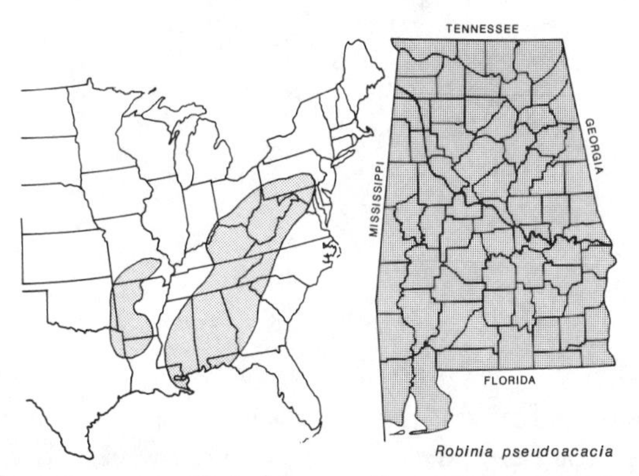

Range in eastern United States   Known range in Alabama

Habitat Occurrence: Black locust occurs in woods and thickets, especially in the mountains, but has been planted over much of the eastern United States for timber and fence posts.

SESBANIA COFFEE BEAN

Sesbania is a genus of herbaceous shrubs that grow to about 5 meters. They have alternate pinnately compound leaves with more than 10 pairs of leaflets. The leaflets are entire, about 1.5–2 cm long and about 0.5 cm wide. The flowers are yellow to orange, irregular, and have 10 stamens. The fruit is a pod.

Toxic Properties: The seeds and possibly other parts of the plant contain various saponins that are poisonous to livestock, fowl, and man. Symptoms observed in livestock have been weakness, diarrhea, rapid and irregular pulse, labored breathing, and coma, followed by death. Symptoms appear approximately 1 day after ingestion. One to 2 ounces have been fatal to livestock.

***Sesbania drummondii* (Plate 8)** DRUMMOND'S RATTLEBOX; COFFEE BEAN

Species Recognition: Drummond's rattlebox is an erect shrub with orange flowers. The fruit has 2–4 seeds per pod and is 4-winged.

Geographic Distribution: From Florida, to Texas and Mexico. In Alabama, can be expected from Wilcox County south.

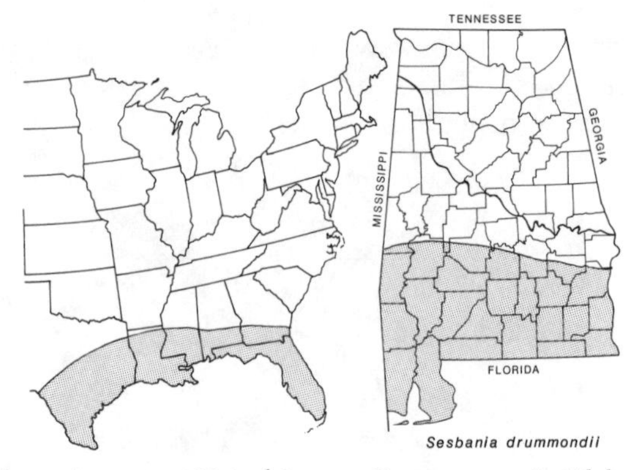

Range in eastern United States   Known range in Alabama

Habitat Occurrence: Drummond's rattlebox occurs in edges of low woods, along banks of bayous, marshes, roadsides, pond margins, edges of ditches, railroad embankments, in mixed woods, on sandy beaches, and prairies.

Sesbania macrocarpa COFFEEWEED

Species Recognition: Coffeeweed is an erect shrub with yellow flowers and pods that are about 0.5 cm in diameter, terete, and about 15–20 cm long.

Geographic Distribution: South Carolina to Florida and west to Texas. In Alabama can be expected from Montgomery County south.

Habitat Occurrence: Coffeeweed occurs on railroad embankments, swamp margins, roadsides, banks of bayous, fallow fields, edges of woods, cultivated fields, edges of ditches, and in pastures.

Sesbania punicea PURPLE RATTLEBOX

Species Recognition: Purple rattlebox has purple flowers and 4-winged pods with 2–4 seeds. The pods are 5–6 cm long by 1.5–2.5 cm in diameter.

Geographic Distribution: North Carolina to Florida, and west to Texas. In Alabama can be expected from Tuscaloosa County south.

Habitat Occurrence: Purple rattlebox occurs along swamp edges, roadsides, edges of pine woods, and ditches.

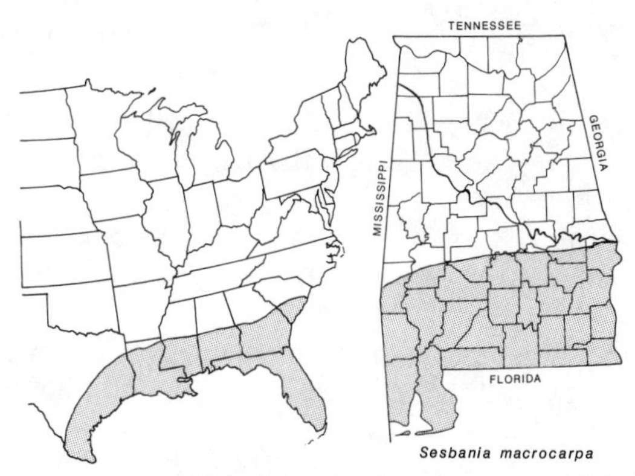

Range in eastern United States   Known range in Alabama

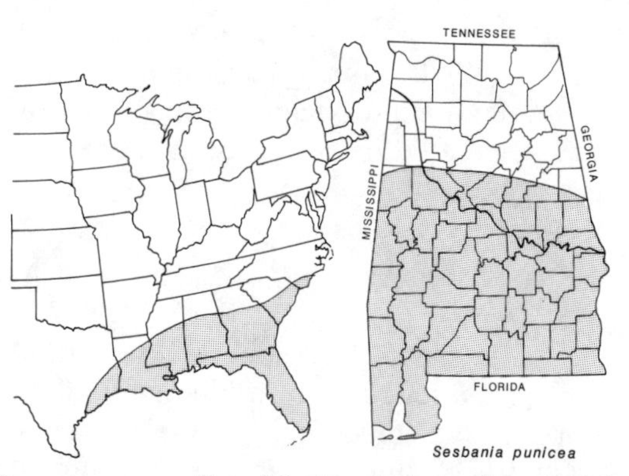

Range in eastern United States   Known range in Alabama

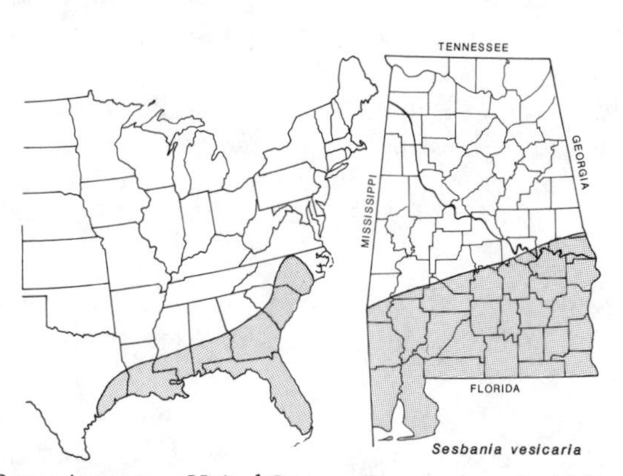

Range in eastern United States   Known range in Alabama

Sesbania vesicaria BLADDERPOD RATTLEBOX

Species Recognition: Bladderpod rattlebox has yellow flowers and pods 2–6 cm long with 2–4 seeds and without any wings.

Geographic Distribution: North Carolina to Florida, west to Texas. In Alabama occurs from Wilcox County southward.

Habitat Occurrence: Bladderpod rattlebox occurs on banks of bayous, sandbars, roadsides, and in pastures, swamps, and pine woods.

TRIFOLIUM CLOVER

Trifolium is a genus of decumbent or, rarely, erect herbs with alternate palmately compound leaves, having 3, very rarely, 4 leaflets. The leaflets are minutely serrulate. The flowers are white to crimson and are produced in a cluster of sessile or near sessile flowers arising from 1 receptacle. The fruit is a small pod with a few seeds. Four species that occur in Alabama have been known to have some type of toxic properties. The toxicity is different for each of the species; therefore toxic properties will be covered at the species level rather than at the generic level.

Trifolium hybridum ALSIKE CLOVER

Species Recognition: Alsike clover has globous flower heads that are about as wide as long, less than 2.5 cm in diameter, and produced on peduncles 1 cm or more long. The flowers are white, and the heads are short racemes. The stems are erect or ascending, never creeping. The peduncles are terminal or axillary from the upper stem leaves.

Toxic Properties: All parts of the plant contain some unknown chemical that causes an allergenic dermatitis to some persons. Horses, cows, sheep, and pigs that have eaten very large quantities have had dermatitis due to photosensitization. In addition, liver, digestive, and nervous disorders have been reported. When this syndrome or reaction begins in animals, it is fatal unless controlled in the early stages.

Geographic Distribution: Cultivated over much of the United States and Canada. Can be expected throughout Alabama.

Habitat Occurrence: Alsike clover is cultivated in fields and pastures but spreads to roadsides, clearings, and alongside woods.

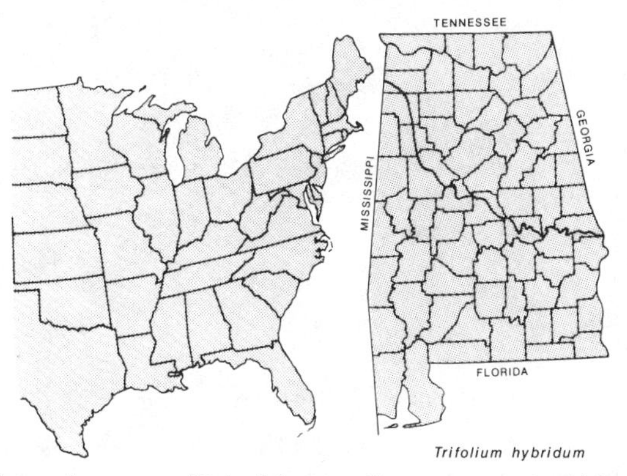

Range in eastern United States Known range in Alabama

***Trifolium incarnatum* (Plate 9)** CRIMSON CLOVER

Species Recognition: Crimson clover has flower heads on long peduncles, 1 cm or more long and longer than wide. The petals are bright red, about 1 cm long, and longer than the calyx lobes. The calyx lobes and pedicels have short barbed hairs. As the flower sets seeds, the flowers become stiff and wiry.

Toxic Properties: The stiff and wiry hairs of overripe crimson clover can be dangerous to horses. Death from impaction has followed ingestion in Delaware, Virginia, and North Carolina. Dense, feltlike bulbs with a diameter of 8–11 cm composed almost entirely of crimson clover hairs have been removed from their intestines.

Geographic Distribution: Planted as a winter annual, particularly in the east-central states. Crimson clover has escaped to much of the United States, and can be expected throughout Alabama.

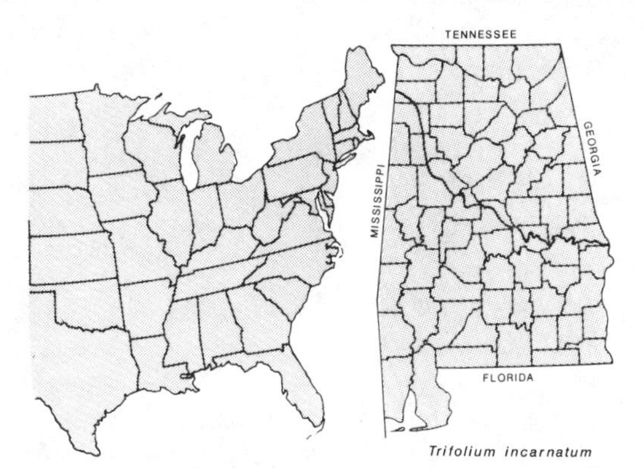

Range in eastern United States Known range in Alabama

Trifolium incarnatum (Crimson Clover)

Habitat Occurrence: Crimson clover occurs on roadsides, empty lots, edges of ditches, edges of woods, and in open fields.

Trifolium pratense RED CLOVER

Species Recognition: Red clover has sessile flower heads that are nearly spherical. The flowers are reddish. The stems are decumbent to ascending.

Toxic Properties: All parts of the plant contain some unknown chemical that, especially when ingested late in the year by cattle, horses, and sheep, can cause dermatitis due to photosensitization. Visual disturbances, stiffness of walk, slobbering, diarrhea, anemia, and abortion also have been reported in livestock. *Trifolium* can be planted as a forage crop without risk to livestock in circumstances where the primary foods of the animals are other plants, such as grass.

Geographic Distribution: Cultivated as a forage crop throughout much of the United States. Can be expected throughout Alabama.

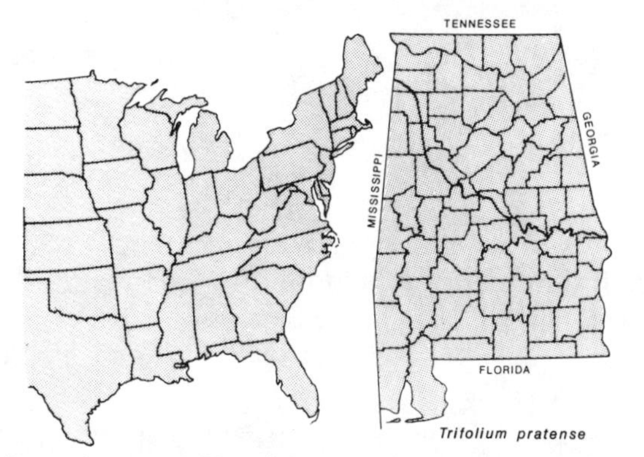

Range in eastern United States Known range in Alabama

Habitat Occurrence: Red clover occurs on roadsides, open fields, and railroad embankments.

Trifolium repens WHITE CLOVER

Species Recognition: White clover has nearly spherical flower heads, about as wide as long, that are produced on peduncles 1 cm or more long. The flowers are white to yellow-white. Mature flowering heads are less than 2.5 cm across. The stem is prostrate or creeping.

Toxic Properties: All parts of the plant contain a cyanogenic glycoside that has not been thoroughly analyzed. Small amounts are safely eaten and used medicinally. Ingestion of a very large amount may cause bloat in cattle, slobbering in horses, or dermatitis due to photosensitization. Some strains cause cyanide poisoning.

Geographic Distribution: Planted as a forage crop throughout the United States. Occurs throughout Alabama.

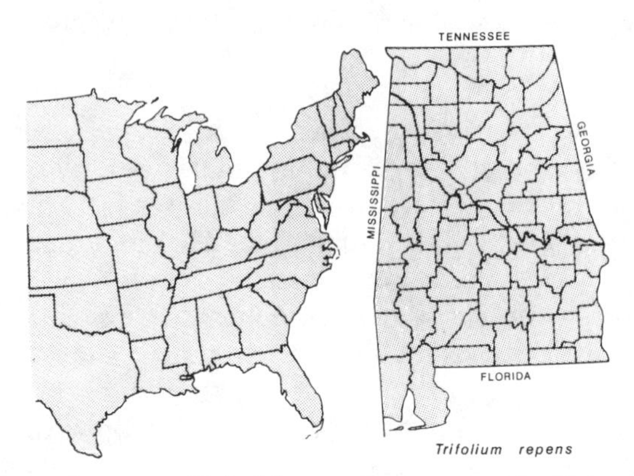

Range in eastern United States   Known range in Alabama

Habitat Occurrence: White clover occurs in open fields, along the edges of woods, roadsides, and in pastures.

WISTERIA WISTERIA

Wisteria is a genus of woody vines with alternate pinnately compound leaves and a large inflorescence of pendulous, showy purple, blue, or white flowers that are fragrant. The fruit is an elongate hairy pod with several seeds.

Toxic Properties: Wisteria's pods, seeds, and bark contain a glycoside, wisterin, and a resin of unknown origin. Human poisoning from eating the seeds of wisteria has occurred in the United States. Symptoms include mild-to-severe gastroenteritis, nausea, vomiting, abdominal pain, and diarrhea. In severe cases, dehydration and collapse may occur. Two seeds have caused severe poisoning. Recovery is normally within twenty-four hours. The flowers are reputed to be edible.

Wisteria frutescens AMERICAN WISTERIA

Species Recognition: Ovary and pod are glabrous, the pedicels are 6–10 cm long, and the raceme is 4–12 cm long.

Geographic Distribution: Florida to Louisiana, north along the Coastal

Plain to southeastern Virginia. Occurs throughout the southern half of Alabama.

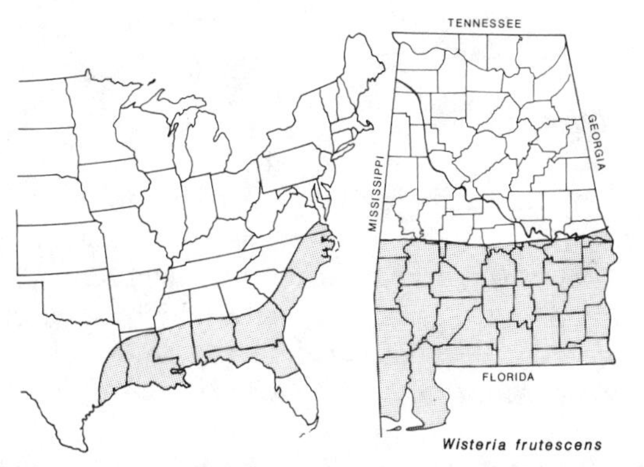

Range in eastern United States   Known range in Alabama

Habitat Occurrence: *Wisteria* occurs along the banks of rivers, wooded areas, railroad embankments, and near old homesites, usually climbing on trees or old buildings.

Wisteria sinensis CHINESE WISTERIA

Species Recognition: The pods and ovary are velvety. The pedicels are 1–2.5 cm long, the leaflets are usually 7–13 cm long, and the 10–50-cm raceme is thick and cylindrical.

Geographic Distribution: Introduced into the United States from China and naturalized in the Southeast. Can be found throughout Alabama.

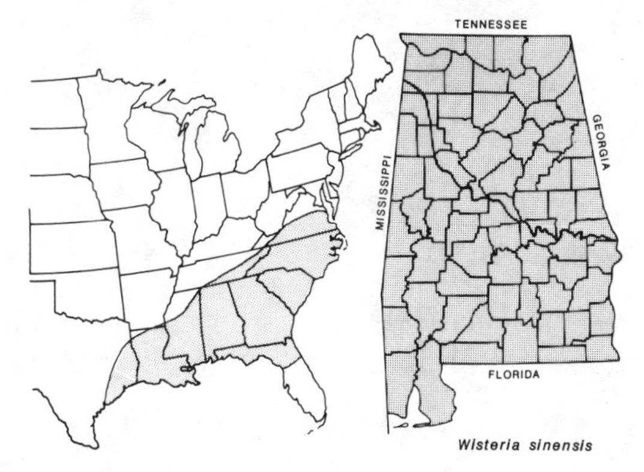

Range in eastern United States   Known range in Alabama

Habitat Occurrence: Chinese wisteria occurs in mixed woods, on banks of creeks, and edges of woods.

Family THYMELAEACEAE

DIRCA LEATHERWOOD

One species occurs in Alabama.

Dirca palustris ***(Plate 9)*** LEATHERWOOD

Species Recognition: Leatherwood is a freely branched shrub with jointed branchlets and grows to a height of 3.5 meters. The bark is fibrous and remarkably tough. The leaves are deciduous, alternate, widely elliptic, and 4–9 cm long. Flowers are inconspicuous, yellowish, in clusters of 2–3, and appear before the young leaves. The fruit is a red drupe 6–8 mm long.

Toxic Properties: The active ingredient is apparently a bitter resin in the bark. The resin is an irritant to the gastrointestinal tract and may cause vomiting and diarrhea. There also is an unknown property in the fruit that reportedly is narcotic and poisonous.

Geographic Distribution: New Brunswick to Ontario and Minnesota, south to northern Florida and Louisiana. Extremely rare in Alabama and occurs from Winston County south to Clark and Monroe counties.

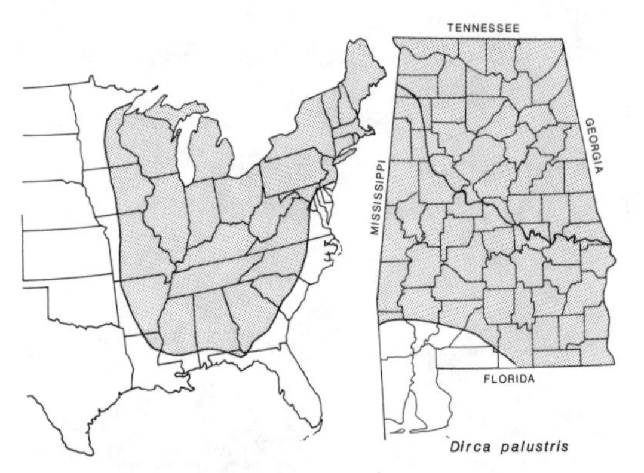

Range in eastern United States Known range in Alabama

Habitat Occurrence: Leatherwood occurs in neutral or slightly acid soils along margins of streams in rich, mixed, or deciduous woods.

Family LORANTHACEAE

PHORADENDRON MISTLETOE

One species occurs in Alabama.

Species Recognition: American mistletoe is an erect herb that grows parasitically in the crown of deciduous tree species, especially oak. The stems are green; the leaves are opposite and green and are nearly round to slightly ovate. The flowers are imperfect and are produced in axillary racemes. The fruit is a white berry.

Toxic Properties: All parts, especially the berries, contain the alkaloid betaphenylethylamine and tyramine. Berries and tea from berries have caused poisoning and death to humans and livestock. Symptoms appear in 1–2 hours and include nausea, vomiting, diarrhea, sweating, dilated pupils, labored respiration. Death due to cardiovascular collapse can occur in about 10 hours. Deaths have resulted from using the berries to produce abortion. There have been no reports of the foliage of *P. serotinum* causing poisoning, but the foliage of other species of *Phoradendron* has been responsible for livestock loss. Although the berries may poison humans and livestock, mistletoe is apparently harmless to birds and is dependent upon them for dispersal. Migrating birds eat the berries and leave the seeds in their droppings on another tree, perhaps many kilometers away from the original plant.

Geographic Distribution: New Jersey, West Virginia, Ohio, southern Missouri, southeastern Kansas, south to Florida and Texas. Occurs throughout Alabama.

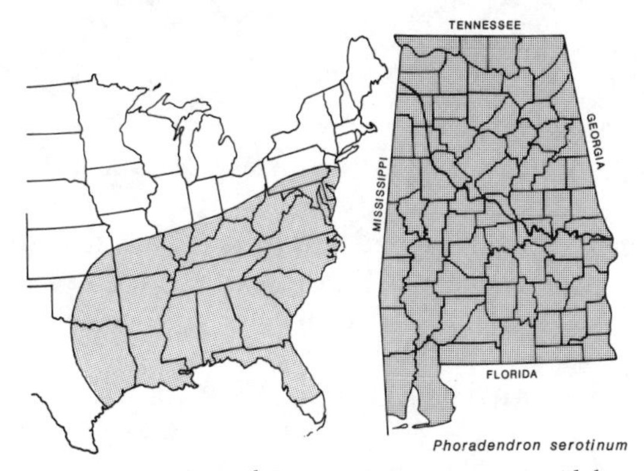

Range in eastern United States Known range in Alabama

Habitat Occurrence: American mistletoe is found in the canopy of deciduous species wherever they might occur.

Family CELASTRACEAE

EUONYMUS SPINDLE TREE

Only one species occurs in Alabama.

Toxic Properties: The leaves, bark, seeds, and roots contain some unknown toxin that has been used medicinally as a purgative. When taken in excess, the toxins cause vomiting, diarrhea, weakness, chills, convulsions, and sometimes coma.

***Euonymus americanus* (*Plate 10*)** STRAWBERRY BUSH; HEARTS-A-BURSTING

Species Recognition: Strawberry bush is an erect shrub that grows to about 1 meter with green stems and opposite serrate leaves. The fruit is a capsule with a red warty coating. When mature, the fruit breaks open and exposes 5–6 red seeds.

Geographic Distribution: Southeastern New York to Pennsylvania and West Virginia, west to southern Illinois and Missouri, south to Florida and Texas. Occurs throughout Alabama.

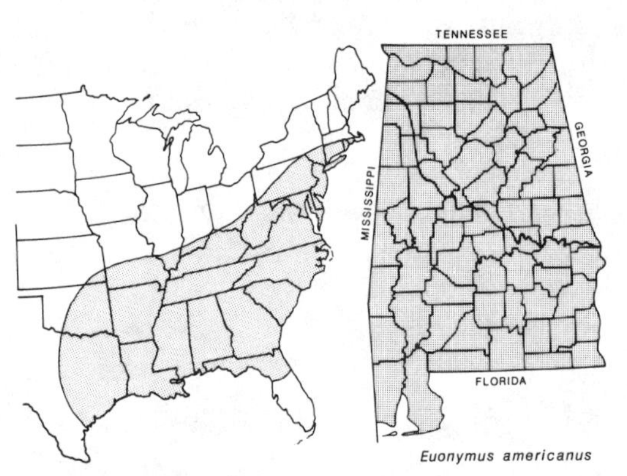

Range in eastern United States Known range in Alabama

Habitat Occurrence: Strawberry bush occurs in rich deciduous woods with sandy soil, and especially on the sides of ravines.

Family EUPHORBIACEAE

ACALYPHA THREE-SEEDED MERCURY

Acalypha is a genus of annual herbs with alternate lance-shaped leaves. The flowers are imperfect and are produced in the axils of the leaves. The large bracts are foliaceous at the base of the female flowers. These bracts are variously lobed, having 5–15 lobes. The fruit is a 3-lobed capsule with 1 seed in each lobe.

Toxic Properties: *Acalypha* has never been shown definitely to be toxic, but there is suspicion that several species have poisoned livestock. The toxic compounds are unknown.

Acalypha gracilens ACALYPHA

Species Recognition: *Acalypha gracilens* is a small weedy herb with petioles less than a quarter the length of the leaf blades. The bracts at the base of the female flowers are shallowly lobed.

Geographic Distribution: Maine and New York to Minnesota, south to southern Florida and southern Texas. Can be expected throughout Alabama.

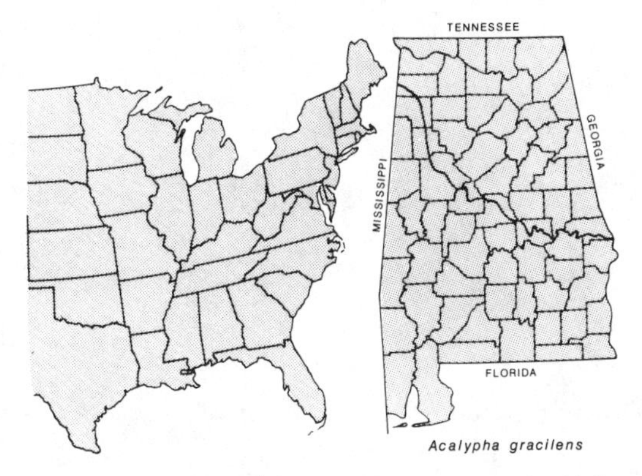

Range in eastern United States   Known range in Alabama

Habitat Occurrence: This species occurs in dry sandy soils, often in disturbed areas with full sun.

Acalypha ostryaefolia ACALYPHA

Species Recognition: *Acalypha ostryaefolia* is an erect herb, up to about 1 meter tall, with sharply serrate leaves that have petioles at least half as long as the leaf blade. The bracts of the female flowers are deeply and finely lobed with 10–15 segments.

Geographic Distribution: Southern Virginia to southwestern Iowa and Kansas, south to Florida, Texas, and Mexico. Can be expected throughout Alabama.

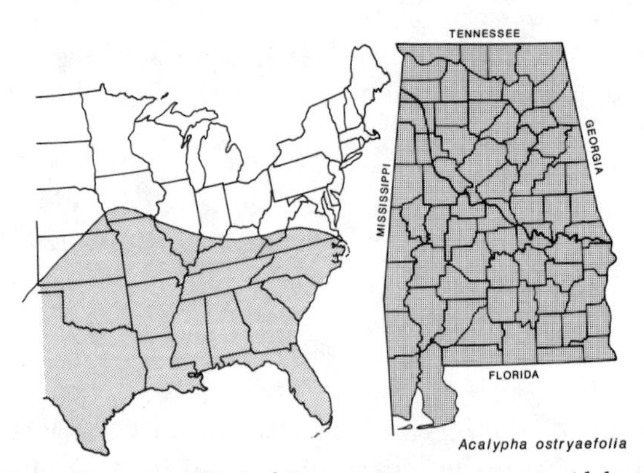

Range in eastern United States Known range in Alabama

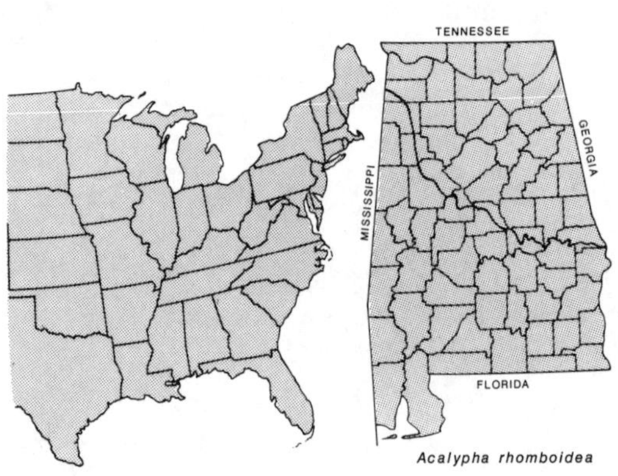

Range in eastern United States Known range in Alabama

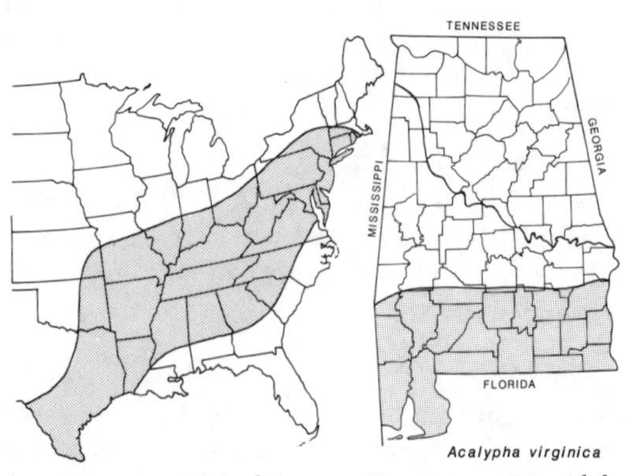

Range in eastern United States Known range in Alabama

Habitat Occurrence: *Acalypha ostryaefolia* is most common in disturbed thickets and along roadsides, waste places, and in cultivated fields, especially in sandy soil.

Acalypha rhomboidea ACALYPHA

Species Recognition: *Acalypha rhomboidea* is an herb, growing up to about 1 meter, with coarsely toothed leaves and petioles about half as long as the leaf blade. The bracts of the female flowers usually have 8–9 deep lobes.

Geographic Distribution: Southern Nova Scotia to western Ontario, Minnesota, and Nebraska, south to Florida, Alabama, Arkansas, and Oklahoma. Can be expected throughout Alabama.

Habitat Occurrence: *Acalypha rhomboidea* is most common along roadsides, fields, and borders of woods, especially in waste places.

Acalypha virginica ACALYPHA

Species Recognition: *Acalypha virginica* is an erect herb, up to about 1.5 meters in height. Its coarse lanceolate leaves have petioles about half as long as the blades, with very large bracts subtending the female flowers. The bracts are shallowly toothed with 8–10 teeth.

Geographic Distribution: Northern Massachusetts to Missouri and Kansas, south to Georgia and Texas. Can be expected in the southern half of Alabama.

Habitat Occurrence: *Acalypha virginica* is most common in open dry soil, especially recently disturbed areas.

ALEURITES TUNG OIL TREE

Aleurites fordii TUNG OIL TREE

Species Recognition: Tung oil trees grow to about 10 meters. The leaves, heart-shaped at the base, are fairly large and have 5 main veins that arise at the base. The large and showy flowers, 2.5–3 cm across, are white and borne in clusters. Each flower has only 1 sex, containing either stamens or pistils.

Toxic Properties: All parts of the tung oil tree contain the toxic substances toxalbumin and saponin, which when ingested cause severe reactions. Symptoms include gastroenteritis, nausea, vomiting, abdominal cramps, diarrhea, dizziness, poor reflexes, and dehydration. The onset of symptoms occurs within 30 minutes, and recovery is usually within 24 hours, but too much toxic material can cause death. Dermatitis has been known to develop from contact with the 2 toxic substances.

Geographic Distribution: Native to China, and once cultivated widely in northern Florida, southern Alabama, and southern Georgia as a source of tung oil, which is used in paint, varnish, and wood treatment. Cultivation has decreased, and plantations are disappearing. However, many trees persist, either as cultivated individuals or as escaped specimens in the coastal sections of Alabama, Mississippi, and Georgia. In Alabama the species can be found in the southern tier of counties. There is a tree, however, in Lee County, at the agricultural experiment station associated with Auburn University.

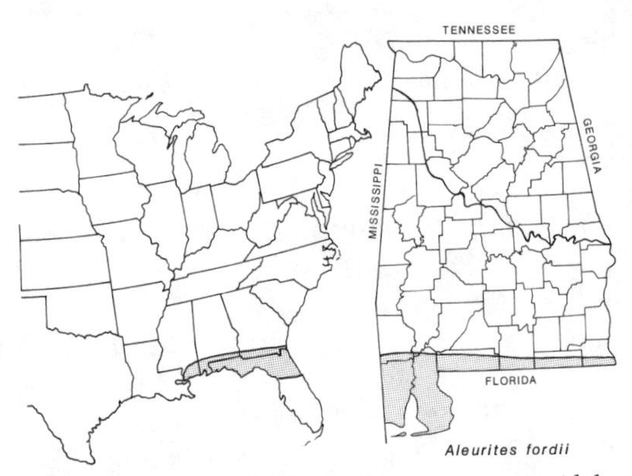

Range in eastern United States   Known range in Alabama

Habitat Occurrence: Tung oil trees are most common in sandy soils receiving abundant sunlight.

CROTONOPSIS RUSHFOIL

One species occurs in Alabama.

Toxic Properties: *Crotonopsis* is closely related to *Croton*, superficially resembles it in many ways, and should be considered to have toxicity similar to *Croton*. There are no known records of poisoning by *Crotonopsis*, but because it is so closely related, and so similar to *Croton*, it should be considered toxic.

Crotonopsis elliptica RUSHFOIL

Species Recognition: *Crotonopsis elliptica* is an erect annual, freely branching, with entire grayish leaves that are covered with reddish glandular dots. The leaves are linear to elliptic and less than 0.5 cm wide. The fruits contain 1 seed per capsule. The genus is very similar to *Croton* but differs by having only 1 fruit per capsule where *Croton* has 3.

Geographic Distribution: Southern Connecticut to eastern Pennsylvania, southern Indiana, and southern Illinois, south to northern Florida, eastern Texas, and eastern Kansas. Occurs throughout Alabama.

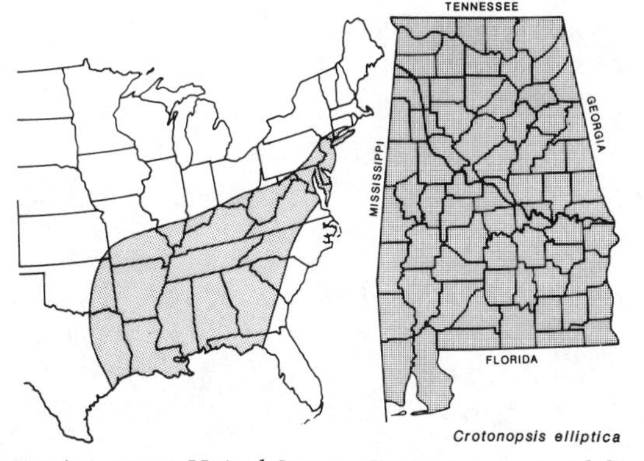

Range in eastern United States Known range in Alabama

Habitat Occurrence: *Crotonopsis elliptica* occurs in dry sandy soil and rocky barrens.

EUPHORBIA SPURGE

The spurges are all weedy annual or perennial herbs with simple entire or dentate leaves that are opposite or alternate. All have a special type of inflorescence called a cyathium. This is a cluster of many flowers that appears to be 1. In the center of the cluster is a female flower with 3 fused carpels on a long stalk that hangs outside and bends over the edge of the cyathium. Along the rim of the cyathium are usually 5 petal-like colored glands, and on the inside surface is a cluster of 10–40 male flowers having 1 stamen each. The fruit is a capsule. *Euphorbia* is the only genus that contains this cyathium.

Toxic Properties: All species of *Euphorbia* contain a milky latex sap with the substance euphorbon as well as various resins and glycosides. This sap is a severe external irritant to many persons and animals, causing blistering, inflammation, and burning. It is an eye irritant and may cause temporary blindness. Internally, it can cause inflammation and burning of the mouth and throat. Some species cause vomiting, diarrhea, and even death. The seeds are also poisonous.

Euphorbia chamaesyce SPURGE

Species Recognition: *Euphorbia chamaesyce* is a decumbent herb with leaves that are rounded at the base and toothed on the margins. The leaves are less than twice as long as wide. The fruit is hairy.

Geographic Distribution: Southeastern Virginia to Florida, west to Texas and Mexico. Can be expected throughout south Alabama.

Habitat Occurrence: *Euphorbia chamaesyce* occurs in open gravelly or sandy waste places, and on roadsides.

Euphorbia commutatus WOOD SPURGE

Species Recognition: Wood spurge has leaves with symmetrical bases that are arranged alternately on the stem. The leaves are less than twice as long as wide, less than 5 cm long, and entire. The glands of the cyathium have narrow yellowish appendages.

Geographic Distribution: Pennsylvania, south to Florida, Alabama, and Texas. Can be expected throughout Alabama.

Habitat Occurrence: Wood spurge occurs along streams, sandy slopes, or calcareous rocks.

Euphorbia corollata FLOWERING SPURGE

Species Recognition: Flowering spurge is an erect herb with entire oblanceolate leaves that are more than 3 times as long as wide and symmetrical at the base. The glands at the top of the cyathium are white, large, and showy.

Geographic Distribution: New York and southern Ontario, west to Minnesota, Wisconsin, and Nebraska, south to Florida and Texas. Occurs throughout Alabama.

Habitat Occurrence: Flowering spurge occurs in dry open woods, clearings, and roadsides.

Additional Species in Alabama: Other species with appearances and properties similar to those mentioned above also occur in this region.

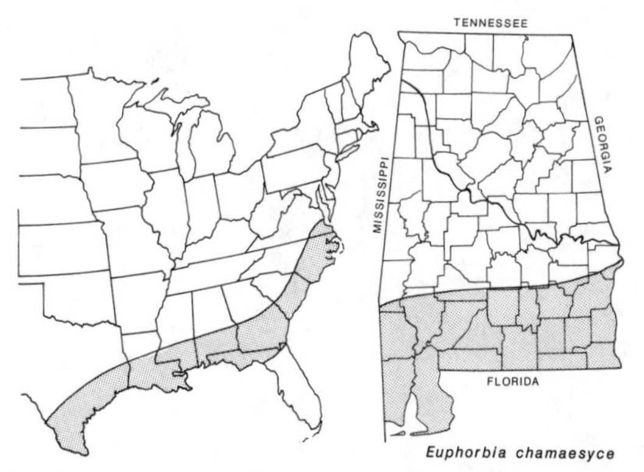

Range in eastern United States   Known range in Alabama

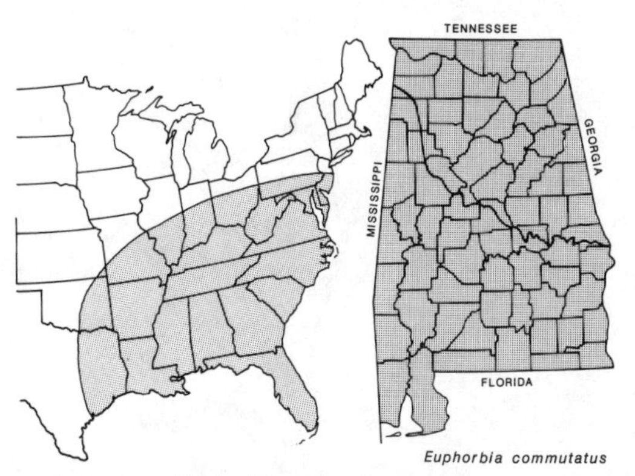

Range in eastern United States   Known range in Alabama

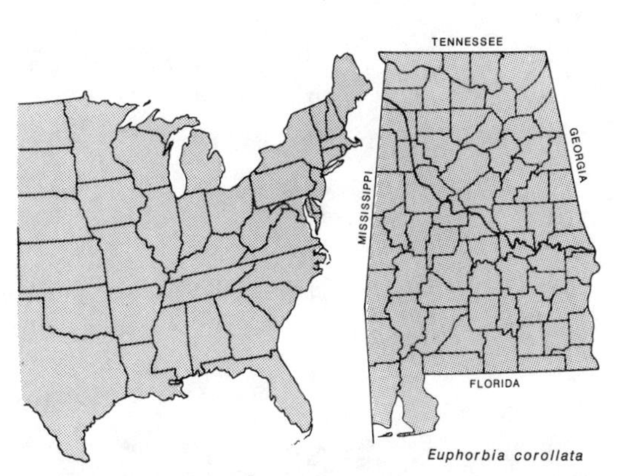

Range in eastern United States   Known range in Alabama

Euphorbia corollata (Flowering Spurge)

Euphorbia ammonnioides EUPHORBIA

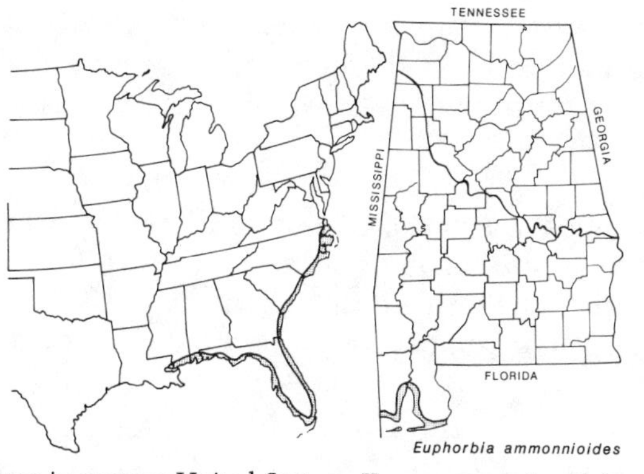

Range in eastern United States   Known range in Alabama

Euphorbia maculata SPOTTED SPURGE

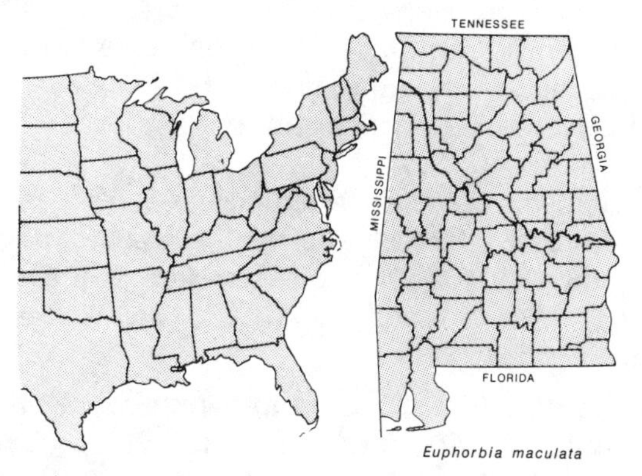

Range in eastern United States   Known range in Alabama

PHYLLANTHUS PHYLLANTHUS

Phyllanthus is a genus with 2-ranked leaves and imperfect flowers produced in the axils of the leaves. The flowers are solitary, or 2–3 are clustered together, and they are without petals. The fruit is a capsule with 3 or 6 seeds. The seeds have a coat with either small punctae or horizontal ridges.

Toxic Properties: The poisonous compound is unknown and there is uncertainty about where it occurs in the plant. Calves have died after grazing on *Phyllanthus* in Texas. Pathological studies have found that 1.5 percent of the animal's weight of the plant is necessary to produce death. Symptoms include listlessness, diarrhea, and exhaustion. Lesions are found on the kidney, and cirrhosis of the liver occurs.

Phyllanthus caroliniensis CAROLINA PHYLLANTHUS

Species Recognition: *Phyllanthus caroliniensis* is an erect annual herb with flowers produced on both the main stem and all lateral branches. The seeds have horizontal rows of punctae, translucent or colored dots.

Geographic Distribution: Pennsylvania and Illinois, south to the Gulf Coast and Mexico. Carolina phyllanthus is unusual for this family, occurring from the northern Arctic region, south through the tropics into the southern temperate latitudes. Can be expected throughout Alabama.

Habitat Occurrence: *Phyllanthus caroliniensis* is most common in disturbed open sites, especially along roadsides and in fields.

Phyllanthus urinaria PHYLLANTHUS

Species Recognition: *Phyllanthus urinaria* is an erect annual herb with flowers produced only along the lateral branches, not on the main stem. The seeds have vertical ridges.

Geographic Distribution: An Old World species introduced into the southeastern United States. *Phyllanthus urinaria* was introduced into Florida in the 1950s and has since spread to Louisiana, Alabama, and Texas. In Alabama found in Mobile, Tuscaloosa, and Covington counties.

Habitat Occurrence: *Phyllanthus urinaria* is found most commonly in disturbed, sunny, open habitats, especially with sandy soil.

STILLINGIA STILLINGIA

Stillingia is a genus of small shrubs with glabrous serrate-to-crenate leaves. The flowers are in terminal spikes that are without petals. The fruits are 3-seeded capsules. The stems are essentially without milky sap.

Toxic Properties: *Stillingia* contains an alkaloid, stillingine. Small amounts of stillingine are used medicinally for laryngitis, bronchitis, and syphilis. The oil of *Stillingia* is a powerful irritant. Overdoses are said to cause dizziness, burning of the mouth, throat, and gastrointestinal tract, nausea, vomiting, and diarrhea.

Stillingia aquatica CORKWOOD

Species Recognition: Corkwood is an erect shrub with 1 stem arising from

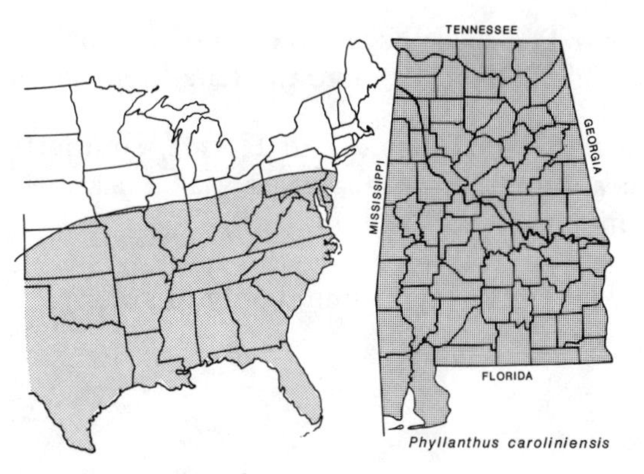

Range in eastern United States Known range in Alabama

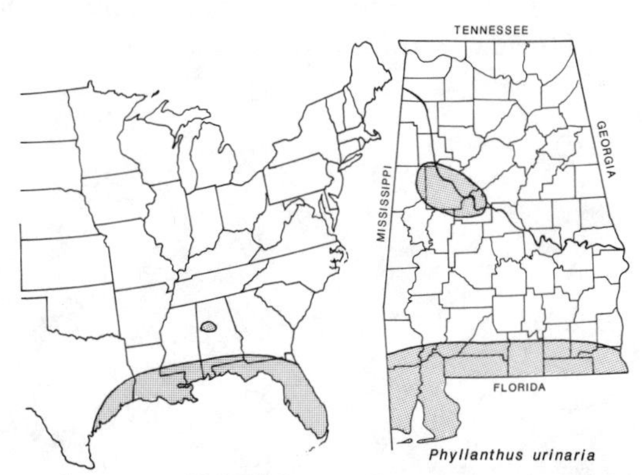

Range in eastern United States Known range in Alabama

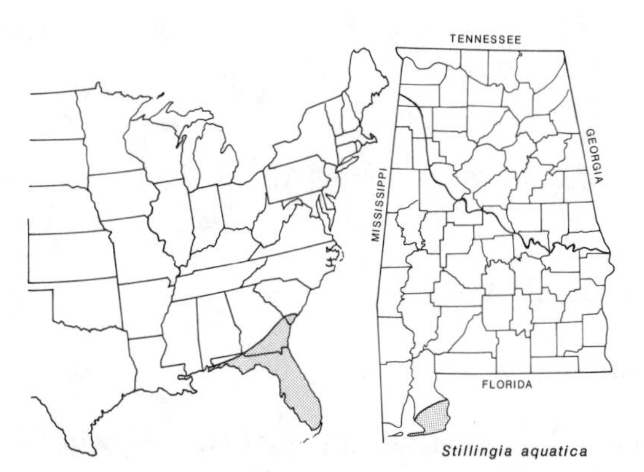

Range in eastern United States Known range in Alabama

the crown. The seeds have a conspicuous, horseshoe-shaped hilum that extends halfway or more the length of the seed.

Geographic Distribution: Restricted to Florida, southern Georgia, South Carolina, and southern Alabama. In Alabama known only from Baldwin County.

Habitat Occurrence: Corkwood is found only in swamps, often in standing water to 0.6 meters deep.

Stillingia sylvatica QUEEN'S DELIGHT

Species Recognition: Queen's delight has several herbaceous or near-herbaceous stems arising from the crown. The seeds lack a conspicuous hilum.

Geographic Distribution: Southeastern Virginia to Kansas and New Mexico, and south to Florida. Found in the southern half of Alabama.

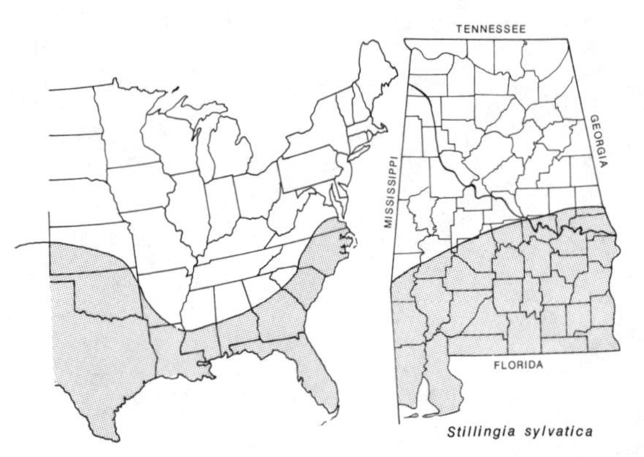

Range in eastern United States Known range in Alabama

Habitat Occurrence: Queen's delight occurs in dry, especially sandy, flats or pine woods along the Coastal Plain.

Family RHAMNACEAE

BERCHEMIA RATTAN

One species occurs in Alabama.

Berchemia scandens RATTAN

Species Recognition: Rattan is a scandent or twining vine with lustrous reddish brown stems and alternate glabrous leaves that are entirely o slightly crenate. The lower surface of the leaf has conspicuous pinnate venation. The fruits are black drupes in small clusters and become dry once they are fully mature.

Toxic Properties: All parts of the plant, especially the fruits, are thought to contain some unknown toxin. The species is included here because so many of the Rhamnaceae are toxic.

Geographic Distribution: Rattan occurs from Virginia to Tennessee and southern Missouri, south to Florida and Texas. Can be expected throughout Alabama.

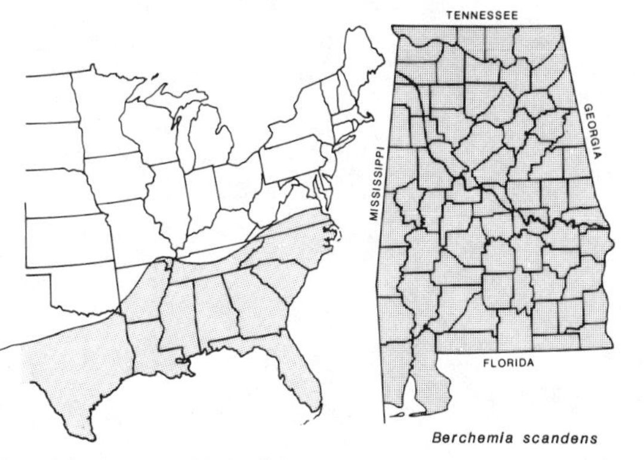

Range in eastern United States   Known range in Alabama

Habitat Occurrence: Rattan occurs in moist sandy woods, swamp forests, and along stream banks. It is found chiefly in the Coastal Plain.

RHAMNUS BUCKTHORN

Only one species occurs in Alabama.

***Rhamnus caroliniana* (Plate 10)** CAROLINA BUCKTHORN

Species Recognition: A large shrub or small tree with alternate crenately serrate leaves that have conspicuous raised veins on the lower surface. The flowers are perfect and solitary in the axils of the leaves. The fruit is black and upon drying produces 2–3 seedlike stones that become free from each other.

Toxic Properties: The leaves, bark, and fruit contain unknown glycosides. These glycosides are used in formation of cascara sagrada, a powerful laxative. It has been known to cause serious gastroenteric irritation, abdominal pain, and severe diarrhea, beyond all expectations of a simple laxative. Although Carolina buckthorn has not been shown experimentally to be poisonous, all other species of *Rhamnus* that have been tested are toxic, and so it may be assumed that this species is also.

Geographic Distribution: Virginia to Nebraska, and south to Florida and Texas. Occurs throughout Alabama.

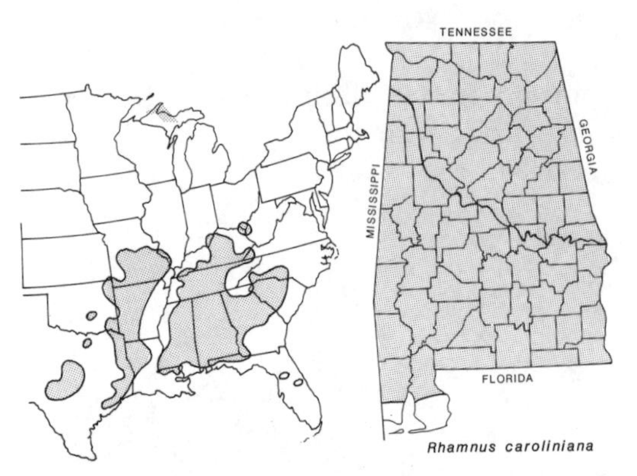

Range in eastern United States   Known range in Alabama

Habitat Occurrence: Carolina buckthorn is most common in deciduous bottomlands and along the edges of forests.

Family VITACEAE

PARTHENOCISSUS VIRGINIA CREEPER

Only one species occurs in Alabama.

***Parthenocissus quinquefolia* (Plate 11)** VIRGINIA CREEPER

Species Recognition: A vine with palmately compound leaves that have 5 coarsely serrate-to-dentate leaflets. The flowers are perfect and are produced in inflorescences opposite the leaves. The fruits are blue berries 0.5–0.7 cm in diameter and contain 1–3 seeds. The vine climbs high by means of adhesive disks at the ends of tendrils. Virginia creeper is sometimes misidentified as poison ivy but is easily distinguished by the presence of 5 leaflets instead of 3.

Toxic Properties: Records show that a few children have died from ingesting Virginia creeper berries. A feeding experiment with fresh berries demonstrated that 10–12 fruits were toxic to a guinea pig. The toxic compounds in the berries are unknown. Symptoms include vomiting, diarrhea, dilated pupils, sweating, and weak pulse. Some victims apparently become very sleepy.

Geographic Distribution: New England to Minnesota, south to Florida and Texas. Occurs throughout Alabama.

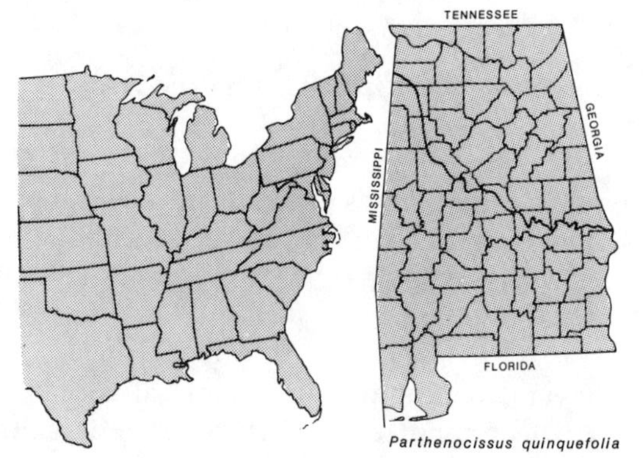

Range in eastern United States Known range in Alabama

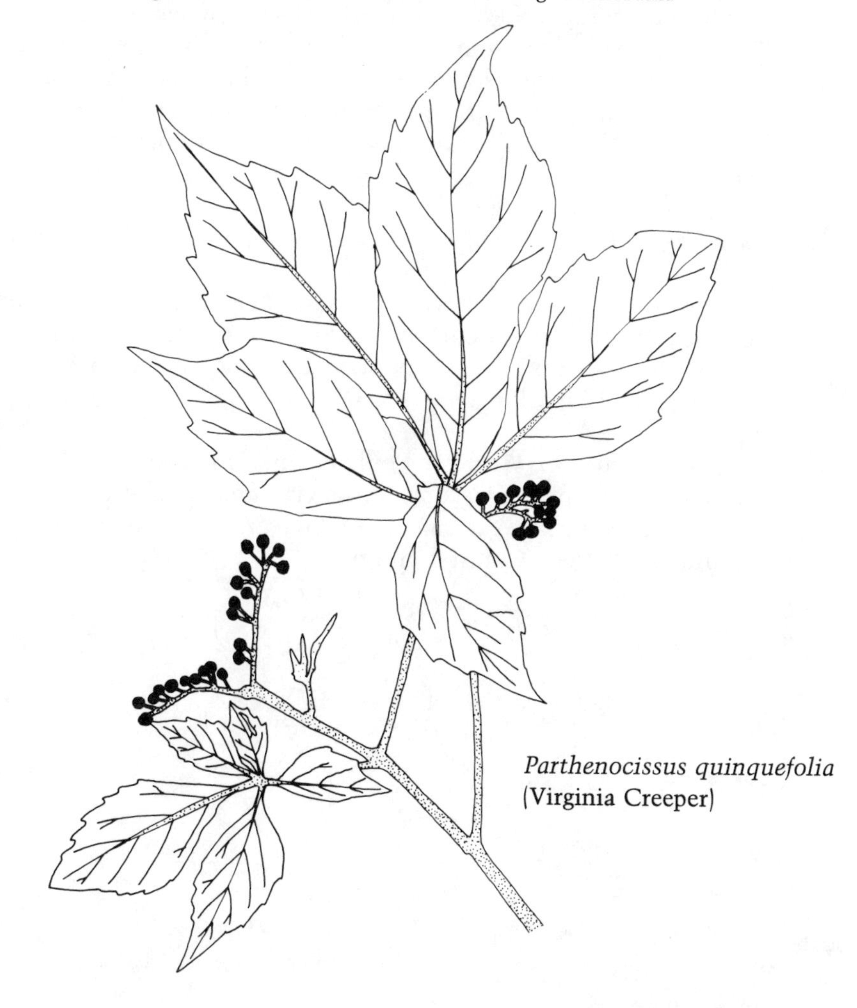

Parthenocissus quinquefolia
(Virginia Creeper)

Habitat Occurrence: Virginia creeper occurs in woods, on rocky banks, in open woodlands, on fences, and at the edges of forests.

Family HIPPOCASTANACEAE

AESCULUS BUCKEYE

Aesculus is a genus of large shrubs or small trees with opposite palmately compound leaves. The flowers are produced in terminal clusters and range in colors from white with red spots, to yellow, to red. The fruits are thick-shelled capsules with 1–3 large brown seeds.

Toxic Properties: Several compounds including alkaloids, glycosides, and saponins have been reported as toxic constituents of the genus. One glycoside, aesculin, however, has been shown to be particularly toxic. The compound occurs in all parts of the plant but especially in young leaves, in stem tips, and in the seeds. There are reports in Europe of young children who died after having eaten seeds of this genus. Although not all species of the genus have been demonstrated to be toxic, all should be so considered until proven otherwise.

Aesculus glabra OHIO BUCKEYE

Species Recognition: Ohio buckeye is a small tree that grows to a height of about 10 meters. The flowers are yellow to greenish yellow, with petals and stamens that are about equal in length. The inflorescence is less than 20 cm long.

Geographic Distribution: Western Pennsylvania to Iowa and Nebraska, south to Alabama, Mississippi, Arkansas, and Oklahoma. Can be expected in the northwest quarter of Alabama on a line from Jackson County to Greene County, and north.

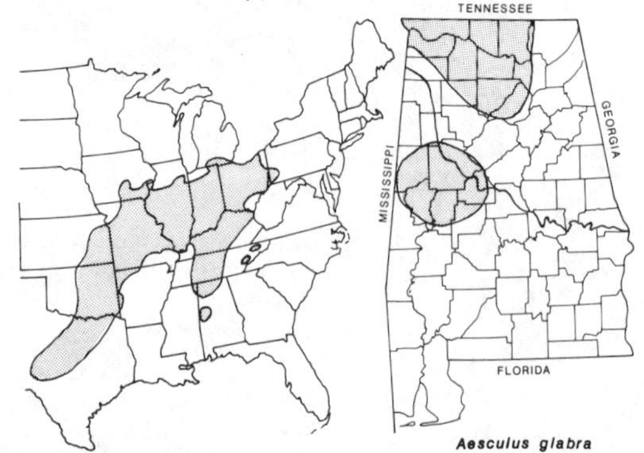

Range in eastern United States Known range in Alabama

Habitat Occurrence: Ohio buckeye occurs in rich deciduous woods, especially on the Cumberland Plateau and Upper Coastal Plain.

***Aesculus octandra* (Plate 11)** YELLOW BUCKEYE

Species Recognition: Yellow buckeye is a large shrub or tree that grows about 30 meters tall. The flowers are yellow, with the stamens up to twice the length of the petals. The inflorescence is usually less than 20 cm long. The petals are of 2 distinct lengths, and the pedicels have hairs along their length.

Geographic Distribution: Western Pennsylvania, Michigan, and Iowa, south to Georgia and Alabama. In Alabama known only from Madison County in the extreme northern part of the state.

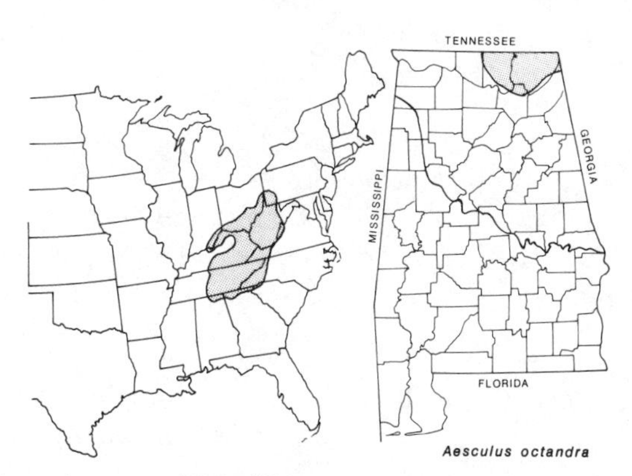

Range in eastern United States Known range in Alabama

Habitat Occurrence: Yellow buckeye is most common in rich deciduous woods, especially along rivers and streams.

Aesculus parviflora SMALL-FLOWER BUCKEYE; BOTTLEBRUSH BUCKEYE

Species Recognition: Small-flower buckeye is a shrub that reaches a height of about 10 meters and has white flowers, stamens 3 or 4 times as long as the petals, and inflorescences that are 20–30 cm long.

Geographic Distribution: Georgia and Florida, west to Mississippi. Occurs throughout Alabama.

Habitat Occurrence: Small-flower buckeye is most common in rich deciduous woods, especially those that are on cliffs overlooking rivers.

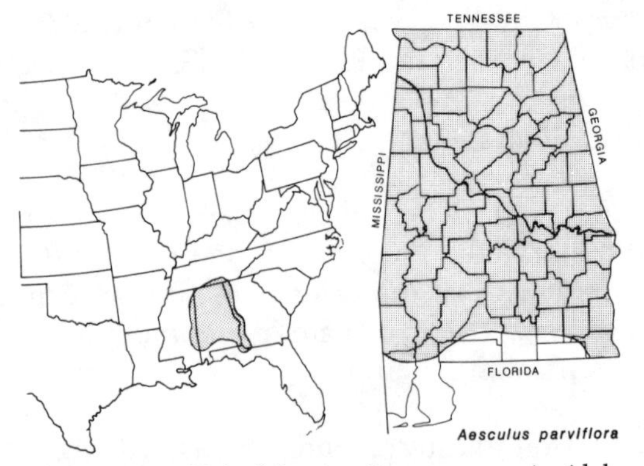

Range in eastern United States   Known range in Alabama

Aesculus pavia RED BUCKEYE

Species Recognition: A small shrub about 3–5 meters tall with red flowers that are occasionally tinged with yellow. The petals are of 2 distinct lengths. The stamens are less than twice as long as the petals, but are exerted beyond the lateral petals. The inflorescence is less than 20 cm long.

Geographic Distribution: Florida to Louisiana, north to Virginia and West Virginia. Can be expected in all parts of Alabama.

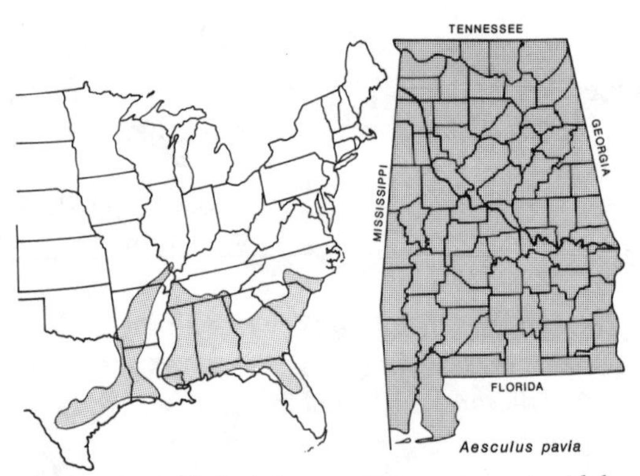

Range in eastern United States   Known range in Alabama

Habitat Occurrence: Red buckeye is most common in deciduous woods, thickets, and often along roadsides. It is an understory plant that does best in partially shaded habitats.

Aesculus sylvatica PAINTED BUCKEYE

Species Recognition: Painted buckeye is a small shrub or tree that grows to a height of about 13 meters. The flowers are yellow with petals of 2 distinct lengths. The stamens are shorter than the petals. The inflorescence is less than 20 cm long.

Geographic Distribution: Southeastern Virginia to Florida and Alabama along the Coastal Plain and outer Piedmont. In Alabama most common in the northeast quarter of the state but also found in the western border of counties.

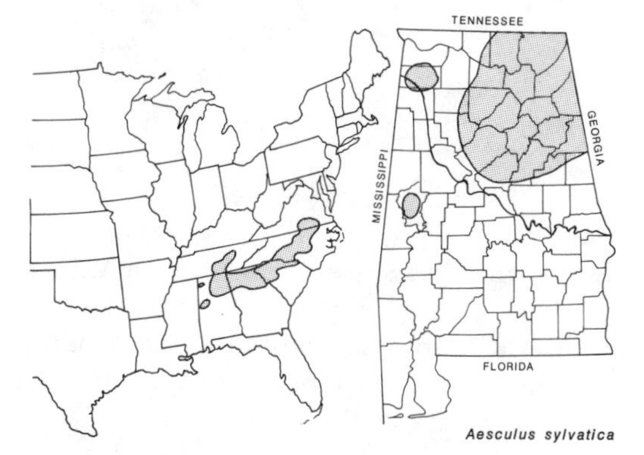

Range in eastern United States Known range in Alabama

Habitat Occurrence: Painted buckeye is most common in rich woods and along the banks of streams.

Family RUTACEAE

PONCIRUS MOCK ORANGE

One species occurs in Alabama.

***Poncirus trifoliata* (Plate 12)** MOCK ORANGE

Species Recognition: Mock orange is a small shrub 2.5–3.5 meters tall with green stems and very large thorns. The leaves are alternate and pinnately compound with 3 leaflets. The fruit is an orange-colored berry similar to a small orange, with bitter pulp.

Toxic Properties: Mock orange fruit contains oils, an acrid component, and a saponin. Prolonged exposure to these compounds may cause dermatitis to sensitive skin. Large amounts eaten raw may cause gastroenteric irritation. The fruit pulp, although bitter when raw, makes an excellent jelly. The large thorns can inflict a painful injury.

Geographic Distribution: Native of China introduced into the Coastal Plain from Florida to Texas. Cultivated as far north as Tuscaloosa and grows as an escape throughout the southern part of Alabama.

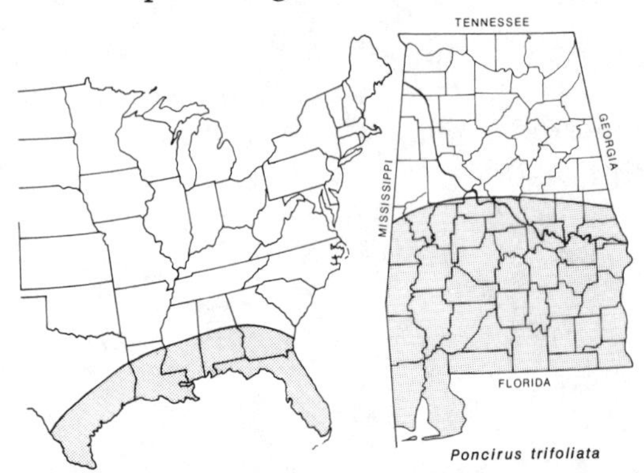

Range in eastern United States   Known range in Alabama

Habitat Occurrence: Mock orange is planted as an ornamental shrub on lawns and in gardens. It occasionally escapes into deciduous woodlands, especially near the coast.

Family MELIACEAE

MELIA CHINABERRY

One species occurs in Alabama.

***Melia azedarach* (Plate 13)** CHINABERRY

Species Recognition: Chinaberry is a tree that grows to about 10 meters in height. It often has a branched, relatively thick trunk. The bark is gray or gray-brown with fine furrows. The leaves are large, alternate, and pinnately compound. Each leaflet is serrate or lobed and is 2.5–5 cm long. The inflorescence is an axillary panicle of purplish flowers. The fruit is a smooth ovoid drupe about 1 cm in diameter; it is green at first and becomes yellow with maturity.

Toxic Properties: Parts of the plant contain a saponin and a bitter component that has not been thoroughly analyzed. The leaves also contain paraisine, an alkaloid. The fruits contain the alkaloid azaridine, a resin, and an organic acid. The bark also contains an alkaloid, margosine, plus tannins. Deaths of humans and of livestock have been reported often in Asia and Africa, especially after ingestion of the berries or of concoctions made from the leaves. Symptoms are of 2 types. One type is an irritant and produces nausea, vomiting, constipation, and stools that contain blood. In addition, there are nervous symptoms of excitement, depression, and weakened heart action. Symptoms may appear in less than an hour after ingestion of a toxic dose or after several hours. Death often occurs within 24 hours, but victims sometimes linger on for several days. People who sit in rocking chairs on the porches of old houses in the South, tell of birds that eat the berries and appear "drunk."

Geographic Distribution: Native to Asia and Africa, but locally common in the southern United States where it has often been planted. Occurs in all southern states and can be expected throughout Alabama.

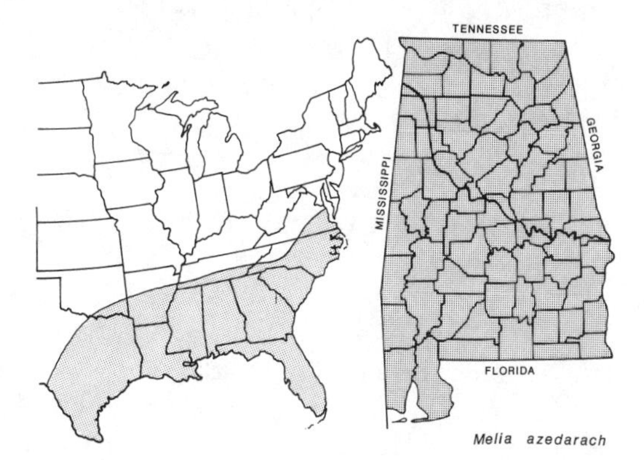

Range in eastern United States Known range in Alabama

Habitat Occurrence: Chinaberry is most common around old homes and other disturbed sites.

Family OXALIDACEAE

OXALIS **(PLATE 13)** WOOD SORREL

Oxalis is a genus of herbs with either basal or alternate palmately compound leaves, the leaves resembling those of a clover. The flowers are yellow or violet. There are 5 sepals, 5 petals, and 5 united carpels, each with a separate style. The fruit is a many-seeded capsule. Several species of *Oxalis* occur in Alabama, but they are all so small that any chance of poisoning is remote, so only the genus will be discussed and not the individual species.

Toxic Properties: All species of *Oxalis* concentrate oxalates in the leaves. Large amounts of soluble oxalates or oxalic acid may cause calcium deficiency. In sufficient quantity they will rapidly cause electrolyte imbalance, nervous symptoms, reduced blood coagulation, formation of oxalate crystals in the kidney tubes and in the urinary tract. The plants are so small, however, that it would take large quantities to detrimentally affect a person. Small amounts may even be eaten raw as a potherb. Cooking *Oxalis* leaves with a pinch of baking soda and discarding the cooking water may help to neutralize the oxalic acid.

Geographic Distribution: Genus found throughout the United States.

Habitat Occurrence: *Oxalis* occurs in open areas, lawns, and occasionally in deep woods.

Alabama species: The following species of *Oxalis* occur in Alabama: *O. corniculata* (CREEPING WOOD SORREL); *O. dillenii* (WOOD SORREL); *O. florida* (FLORIDA WOOD SORREL); *O. grandis* (WOOD SORREL); *O. rubra* (WOOD SORREL); *O. stricta* (WOOD SORREL); and *O. violacea* (WOOD SORREL).

Family BALSAMINACEAE

IMPATIENS — JEWELWEED; TOUCH-ME-NOT

Impatiens is a genus of erect annual herbs with glabrous alternate leaves. The flowers are irregular, with a yellow-to-red tubular corolla that has a curled spur at its base. The fruiting structures are green capsules that dehisce violently when touched.

Toxic Properties: *Impatiens* stems, leaves, and roots contain an unknown compound that has an acrid, burning taste and eventually promotes vomiting and diarrhea when ingested. We know of no cases of death from *Impatiens* ingestion. Small amounts have been used medicinally.

***Impatiens capensis* (Plate 14)** — SPOTTED JEWELWEED

Species Recognition: Spotted jewelweed is an erect herb with orange flowers. The spur is more than 0.5 cm long and is bent parallel to the petals.

Geographic Distribution: Newfoundland and Quebec to Alaska, and south to Florida and Texas. Occurs in the northern two-thirds of Alabama.

Habitat Occurrence: Spotted jewelweed is found in swamps, marshes, moist woods, along streams, and in seepage areas, in the open or in shade.

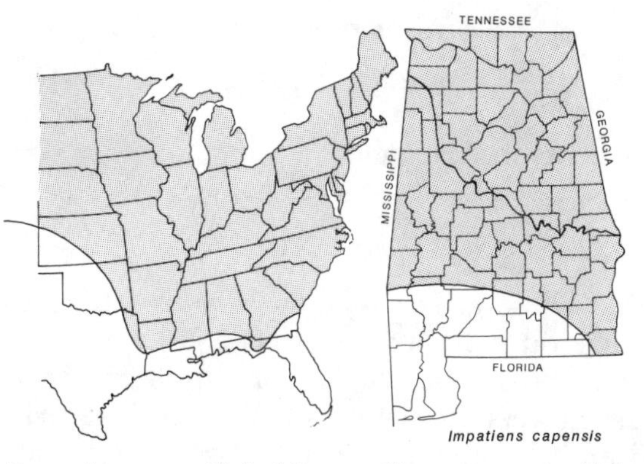

Range in eastern United States   Known range in Alabama

Impatiens pallida JEWELWEED

Species Recognition: Jewelweed is an erect herb with yellow flowers and a spur 0.5 cm or less long. The spur is bent at right angles to the petals.

Geographic Distribution: Newfoundland to Saskatchewan, south to Georgia, Alabama, Tennessee, Missouri, and Kansas. In Alabama known only from the extreme northern region.

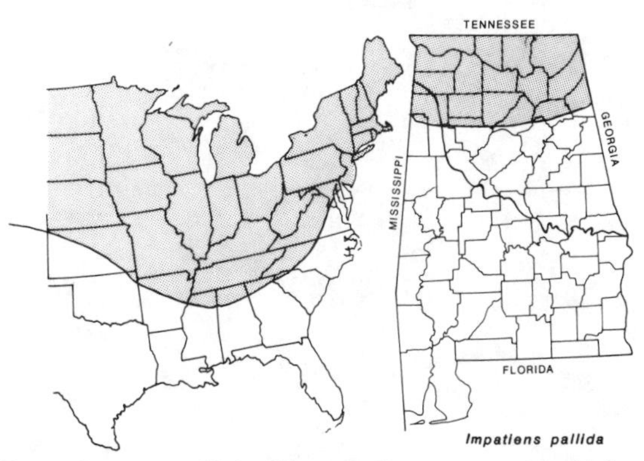

Range in eastern United States   Known range in Alabama

Habitat Occurrence: Jewelweed occurs in wet areas or spring seepages, or often in shade and chiefly in calcareous soils.

Family APIACEAE

CICUTA WATER HEMLOCK

A single species occurs in Alabama.

Cicuta maculata WATER HEMLOCK

Species Recognition: These glabrous herbs grow as tall as 2 meters, from fleshy or tuberous, fibrous roots. The leaves are pinnately compound. The leaflets are lanceolate or elliptic in shape and coarsely serrate. The flowers are white, small, and clustered into compact compound umbels. The umbel stalks are 3–12 cm long. The fruits are 2 hemispherical, ribbed, achene-like structures.

Toxic Properties: All parts of the plant are considered dangerously toxic to humans and livestock. The active compound is cicutoxin, an unsaturated alcohol known also as resinoid. The seeds and flowers have a lesser amount of the compound than does the rest of the plant. Plants of this species are a cause of many plant poisonings and deaths. A single root or a large mouthful can be a fatal dose. Symptoms appear in 15 minutes to 1 hour, and death may occur soon thereafter. Symptoms include frothing at the mouth, nausea, diarrhea, abdominal pain, dilated pupils, and periodic violent convulsions between short periods of relaxation. Death is due to respiratory or cardiac failure.

Geographic Distribution: Gaspé Peninsula of Quebec, west to Manitoba, and south to Florida and Texas. Can be expected throughout Alabama.

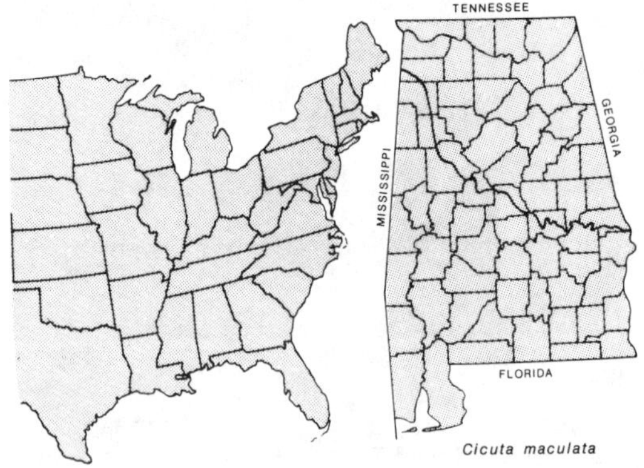

Range in eastern United States Known range in Alabama

Habitat Occurrence: Found in full sunlight in meadows, swells, low thickets, and prairies, this species is especially common along the shores and margins of small lakes and streams.

CONIUM POISON HEMLOCK

Only one species occurs in Alabama.

***Conium maculatum* (*Plate 15*)** POISON HEMLOCK

Species Recognition: Poison hemlock is a coarse, erect annual or perennial with tuberous rhizomes. The plant grows to be about 3 meters tall. The stems are stout, ridged, usually black spotted, and hollow except at the nodes. The leaves are pinnately compound, and the upper ones are much divided. The flowers are glabrous, and some are borne as a rosette at the top of the taproot. The flowers are white, small, and in umbellate clusters at the top of the plant. The fruit is a 2-seeded achene.

Toxic Properties: All parts of the plant contain at least 5 distinct, closely related alkaloids which include coniine, N-methyl coniine, conhydrine, lambda-coniceine, and pseudoconhydrine. Based upon symptoms, this is apparently the hemlock that Socrates used to poi-

Conium maculatum (Poison Hemlock)

son himself. Symptoms include bloody feces, gastrointestinal irritation, nausea, vomiting, convulsions, and respiratory congestion. Death usually follows as a result of respiratory paralysis.

Geographic Distribution: An introduced weed from Europe, found sporadically from Quebec to Iowa, south to Florida, Louisiana, and Texas. Can be expected throughout Alabama.

Habitat Occurrence: Poison hemlock occurs on waste grounds, the edges of woods, roadsides, and in fields. It has been planted in herb gardens and flower gardens for a ferny appearance.

Family LOGANIACEAE

GELSEMIUM — YELLOW JESSAMINE

Gelsemium is a genus of high-climbing or trailing vines with opposite evergreen leaves that are lanceolate to elliptic in shape. The leaves are 3–7 cm long and 1–2.5 cm wide. The flowers are tubular with 5 yellow lobes and range 2–4 cm in length. Two species occur in Alabama.

Toxic Properties: All parts of the plant, including the flower nectar, contain the alkaloids gelsemine, gelseminine, and gelsemicine. The alkaloids are cumulative. They depress and paralyze the motor-nerve endings. Symptoms in humans and animals include dizziness, dilated pupils, dry mouth, difficulty in swallowing and breathing, muscular weakness, nausea, sweating, weak pulse, and convulsions. Death occurs due to respiratory failure. The honey from the gelsemium nectar can be deadly. Repeated exposure also causes a contact dermatitis.

Gelsemium rankinii — UNSCENTED YELLOW JESSAMINE

Species Recognition: Unscented yellow jessamine has odorless flowers with sepals that are acute at the apex. The capsule has a beak more than 2 mm long, and the seeds are without wings.

Geographic Distribution: North Carolina, south to Florida, and west to Mississippi. Occurs in the extreme southern part of Alabama.

Habitat Occurrence: Unscented yellow jessamine occurs in swamps, especially along the outer Coastal Plain.

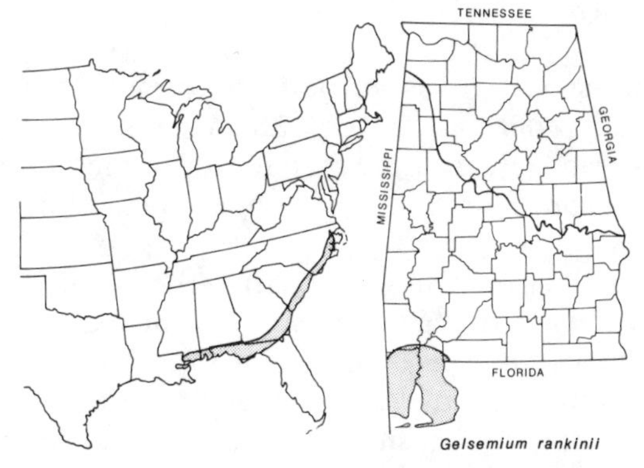

Range in eastern United States   Known range in Alabama

***Gelsemium sempervirens* (Plate 15)** YELLOW JESSAMINE

Species Recognition: Yellow jessamine has sweet-smelling yellow flowers with sepals that are obtuse at the apex. The fruits are without a beak, or the beak is shorter than 2 mm; the seeds are winged.

Geographic Distribution: Northern Virginia and Arkansas, south to Florida and Texas. Can be expected throughout Alabama.

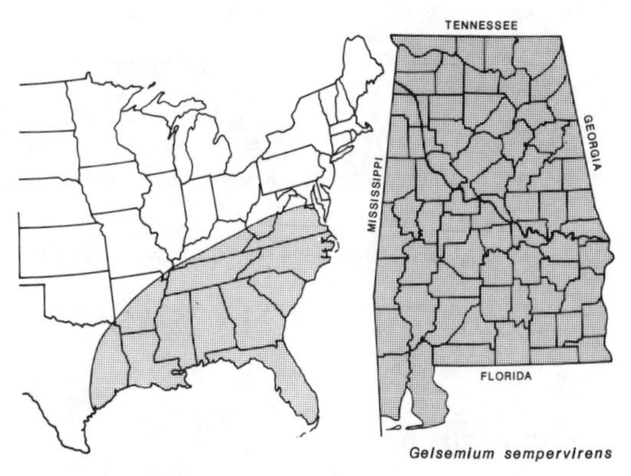

Range in eastern United States   Known range in Alabama

Habitat Occurrence: Yellow jessamine occurs in dry-to-wet woods, thickets, and sandy areas, along fencerows and roadsides. It is especially common in the Coastal Plain and Piedmont.

SPIGELIA INDIAN PINK

Only one species occurs in Alabama.

Spigelia marilandica INDIAN PINK

Species Recognition: *Spigelia* is an herb that grows to about 1 meter in height and has opposite leaves with a terminal cluster of red flowers that are secund (all arranged on one side of the flowering stalk). The leaves are about 8 cm long and about 2.5 cm wide.

Toxic Properties: All parts of *Spigelia* contain the alkaloid spigeline as well as an unknown acrid, bitter substance. The plant has been used medicinally to kill or expel intestinal worms, but overdoses frequently occur. Symptoms of overdose include increased circulation, dizziness, dilated pupils, spasms of the eye and facial muscles, convulsions, and occasionally death results.

Geographic Distribution: Maryland, Ohio, Indiana, Missouri, Oklahoma, south to Florida and Texas. Can be expected throughout Alabama.

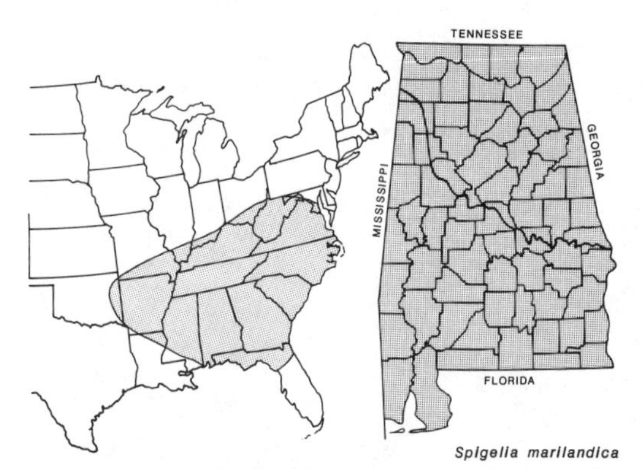

Range in eastern United States Known range in Alabama

Habitat Occurrence: Indian pink occurs in rich woodlands.

Family APOCYNACEAE

APOCYNUM DOGBANE

One species occurs in Alabama.

Apocynum cannabinum CLASPING-LEAF DOGBANE

Species Recognition: These erect perennial herbs grow from spreading rootstocks with stems somewhat woody. The opposite leaves have smooth margins and milky sap. Fruits are pencil-like pods hanging in pairs. The seeds are long and narrow, each with a tuft of long white hairs, as in the milkweed, forming a cottony mass within the pod.

Toxic Properties: All parts of the plant contain the glycoside apocannoside and cymarine. In addition, all parts contain a resin (apocynin) and other substances. It is uncertain, however, if these compounds are toxic. When sufficient quantities of the plant are taken, symptoms include increased pulse, fever, dilated pupils, sweating, cold extremities, discolored and sore mouth and nose, mild diarrhea, and in some cases death follows.

Geographic Distribution: Throughout southern Canada and the United States. Occurs throughout Alabama.

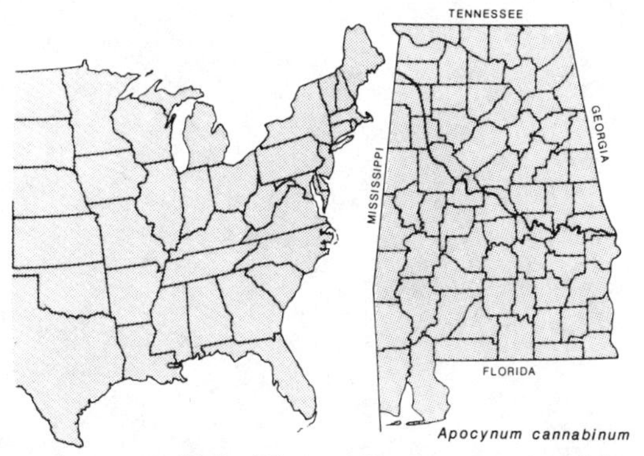

Range in eastern United States   Known range in Alabama

Habitat Occurrence: Clasping-leaf dogbane is found most commonly on open ground at the edge of thickets and along the borders of woodlands.

NERIUM OLEANDER

Only one species occurs in Alabama.

***Nerium oleander* (Plate 15)** COMMON OLEANDER

Species Recognition: *Nerium* is a genus of shrubs with opposite, evergreen, linear-to-lanceolate leaves and milky sap. The flowers are usually purplish, although they may be red or white, and are produced on cymes either in the axils of the leaves or terminally on the stem. The fruit is a follicle with many hairy seeds.

Toxic Properties: All parts of oleander contain the alkaloids nerrin and oleandrin and are poisonous to humans and livestock. One leaf is reported to be sufficient to kill an adult human, and 15–20 grams will kill mature cattle or horses. The dry leaves are as toxic as the green ones. Children may be poisoned from carrying flowers in their mouths. Deaths have been reported of persons who consumed food roasted on oleander stems. Inhaling the smoke from burning plants has caused serious poisoning. Honey made from oleander nectar is bitter and toxic. Contact with the leaves or sap may cause dermatitis to some persons. Symptoms in humans include nausea, vomiting, stomach pains, dizziness, decreased and irregular heartbeat, bloody diarrhea, dilation of the pupils, drowsiness, unconsciousness, and respiratory paralysis. Death may follow.

Geographic Distribution: Cultivated throughout the southeastern United States. Can be expected throughout Alabama.

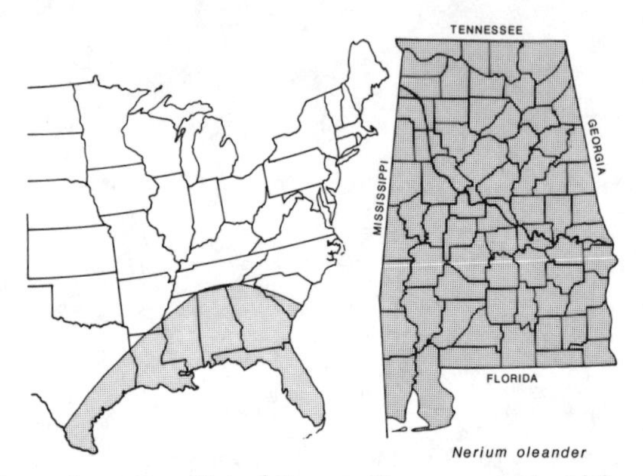

Range in eastern United States   Known range in Alabama

Habitat Occurrence: Oleander may persist around old homesites, in gardens, and is a popular ornamental shrub with highway departments.

VINCA PERIWINKLE

One species persists following cultivation; another, *Vinca minor*, is cultivated occasionally.

Vinca major BLUE PERIWINKLE

Species Recognition: Blue periwinkle is a sprawling vine with opposite petiolate, ovate, entire leaves with 1 large blue axillary flower. The flower has 5 petals and 5 sepals.

Toxic Properties: All parts of the plant contain alkaloids, in particular reserpine. Small amounts have been used medicinally. Preliminary toxicity tests indicate that the alkaloids are toxic.

Geographic Distribution: Virginia to Florida, west to Mississippi. Occurs throughout Alabama.

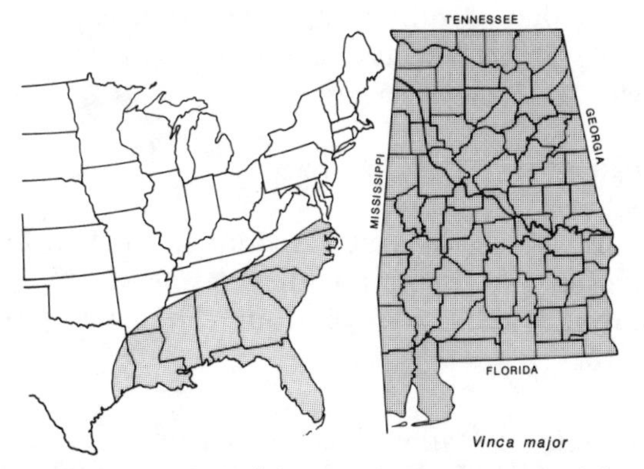

Range in eastern United States Known range in Alabama

Habitat Occurrence: Blue periwinkle is a cultivated plant. It persists around old homesites and spreads from cultivation into waste places and roadsides.

Family ASCLEPIADACEAE

ASCLEPIAS MILKWEED

Asclepias is a genus of herbs with tuberous rhizomes and erect stems with linear-to-broadly elliptic leaves that are mostly opposite, but rarely alternate or whorled. The sap is milky in all but 1 species. The flowers are in umble-like clusters in the axils of leaves or at the apex of the stem. They are highly modified, ranging in color from white to red. The fruit is an elongate follicle that has many hairy seeds. Fourteen species occur in Alabama. All species of *Asclepias* that have been tested experimentally have proven to be at least slightly toxic, so all species should be considered so unless proven otherwise.

Toxic Properties: All parts of the plant contain glycosides and resinoids. The young seeds and young pods can be eaten after being cooked, but are toxic when raw. In humans, some species have caused vomiting, stupor, and weakness. In cattle, ingestion of *Asclepias* has caused salivation, nausea, vomiting, diarrhea, uncoordination, muscular paralysis, respiratory difficulty, and cardiac disturbances. Death may result when large amounts are consumed. The milky sap is an irritant to the skin and eye. The root of *Asclepias tuberosa* has been used medicinally. Large quantities cause irritation, vomiting, and diarrhea. Great advantage is taken of milkweeds by the monarch butterfly which lays its eggs on the plant. The caterpillars grow up on a diet of milkweed toxins and when they emerge as butterflies are protected from attack by many birds because the insect itself is then poisonous to eat.

Asclepias amplexicaulis MILKWEED

Species Recognition: *Asclepias amplexicaulis* is an erect herb that reaches a height of about 1 meter with opposite clasping leaves that are about half as wide as they are long. The fruit is erect at the apex of the stem.

Geographic Distribution: Maine, Wisconsin, Iowa, and Nebraska, south to northern Florida and eastern Texas. Can be expected throughout Alabama.

Asclepias amplexicaulis (Milkweed)

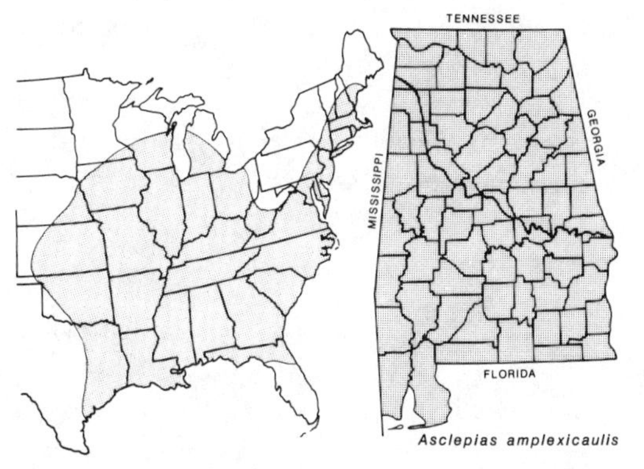

Range in eastern United States Known range in Alabama

Habitat Occurrence: *Asclepias amplexicaulis* occurs in open woods, clearings, meadows, and pastures, on prairies, old sand dunes, roadsides, and alongside railroads. It grows chiefly in sandy or gravelly soil.

Asclepias humistrata MILKWEED

Species Recognition: *Asclepias humistrata* is an erect herb about 0.5 meter tall with opposite leaves that are about half as wide as they are long. The leaves are opposite with acute apexes and do not have leaf stalks or petioles. The fruits are erect on deflexed pedicels and are about 10 cm long and 1 cm wide.

Geographic Distribution: North Carolina, south to Florida, and west to Louisiana. In Alabama occurs in Baldwin and Mobile counties.

Habitat Occurrence: *Asclepias humistrata* is found in sand dunes, dry oak woods, and pine barrens.

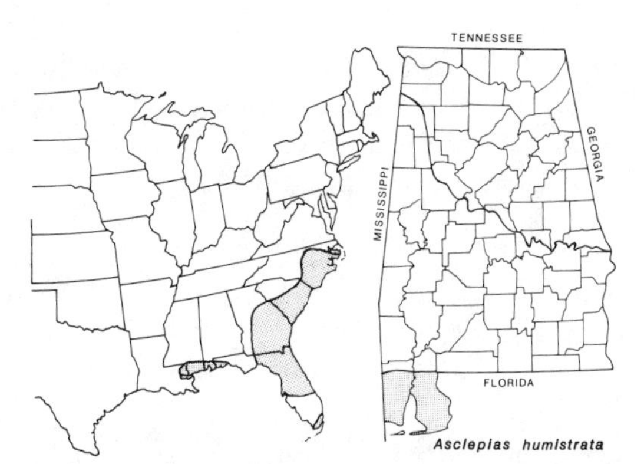

Range in eastern United States Known range in Alabama

Asclepias incarnata (Plate 16) SWAMP MILKWEED

Species Recognition: Swamp milkweed is an erect plant that grows to about 1.5 meters in height. The leaves are opposite with petioles 1–1.5 cm long and are linear lanceolate, about 3 times as long as wide, and acute at the apex. The fruits are at the terminous of the plant and are on erect pedicels.

Geographic Distribution: Northern Maine to southern Ontario and Saskatchewan, south to Florida and New Mexico. This species is rare in the southeastern coastal states. In Alabama known only from St. Clair County.

Range in eastern United States Known range in Alabama

Habitat Occurrence: Swamp milkweed occurs in swamps, wet thickets, and along shores of lakes and streams.

Asclepias lanceolata MILKWEED

Species Recognition: *Asclepias lanceolata* is an erect herb about 1 meter tall with sessile opposite leaves that are 5 times or more longer than wide. The leaves are linear to narrowly lanceolate and acute at the apex. The fruits are at the apex of the plant and are erect on deflexed petioles.

Geographic Distribution: Mostly on the outer Coastal Plain from New Jersey, south to Florida, and west to eastern Texas. In Alabama known from Mobile and Baldwin counties.

Habitat Occurrence: *Asclepias lanceolata* occurs in brackish-to-freshwater marshes, wet pine barrens, and low glades.

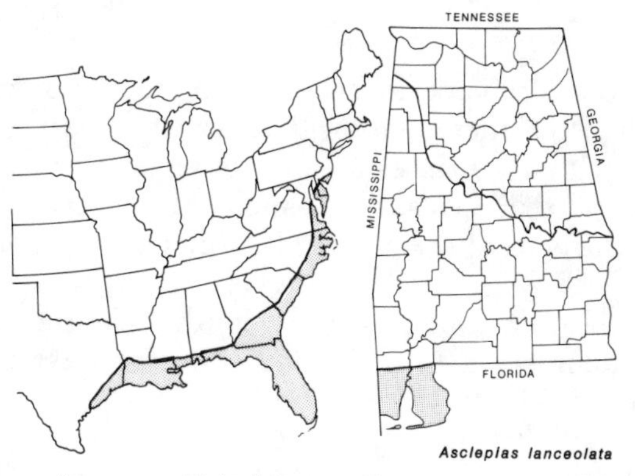

Range in eastern United States   Known range in Alabama

Asclepias longifolia MILKWEED

Species Recognition: *Asclepias longifolia* is an erect herb about 0.5 meter tall with alternate linear lanceolate leaves that are 5 times or more longer than wide with a short petiole or possibly sessile. The petiole, when present, is less than 0.5 cm long. The leaves are acute at the apex. The flowers and fruits are produced both at the apex of the plant and in the axils near the apex. The fruits are erect on deflexed pedicels.

Geographic Distribution: Delaware, south to Florida, and westward to Louisiana, mostly along the Coastal Plain. In Alabama known only from Mobile County.

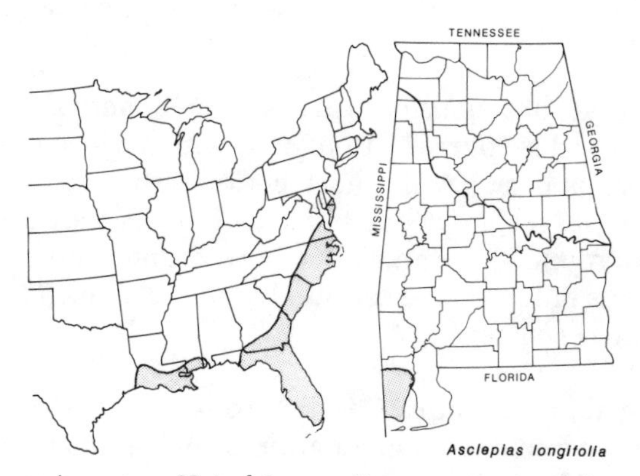

Range in eastern United States   Known range in Alabama

Habitat Occurrence: *Asclepias longifolia* occurs in flatwoods, swamps, and low pinelands.

Asclepias michauxii MICHAUX MILKWEED

Species Recognition: Michaux milkweed is an erect or sprawling herb that is less than 0.5 meter tall with alternate linear-to-filiform leaves that are at least 10 times as long as wide. The apex of the leaves is acute, and the leaves are sessile. The flowers and fruits are produced at the apex of the plant. The fruits are erect on erect pedicels.

Geographic Distribution: South Carolina, south to central Florida, and west to eastern Louisiana. In Alabama known from Escambia and Mobile counties.

Range in eastern United States Known range in Alabama

Habitat Occurrence: Michaux milkweed occurs in sandy pine barrens.

Asclepias obovata MILKWEED

Species Recognition: *Asclepias obovata* is an erect herb that grows to about 0.5 meter and is covered throughout with a dense layer of hairs. The leaves are opposite with short petioles that are less than 0.5 cm long. The leaves are about twice as long as they are wide, 3–4 cm long, and round at the apex. The flowers and fruits are produced both in the axils of leaves and terminally on the stem. The fruits are erect on deflexed pedicels.

Geographic Distribution: South Carolina to northern Florida, west to east Texas, with one known population in Arkansas. In Alabama known only from Mobile County.

Habitat Occurrence: *Asclepias obovata* occurs in sandy soil, in oak and pine woods, spreading to fields and roadsides.

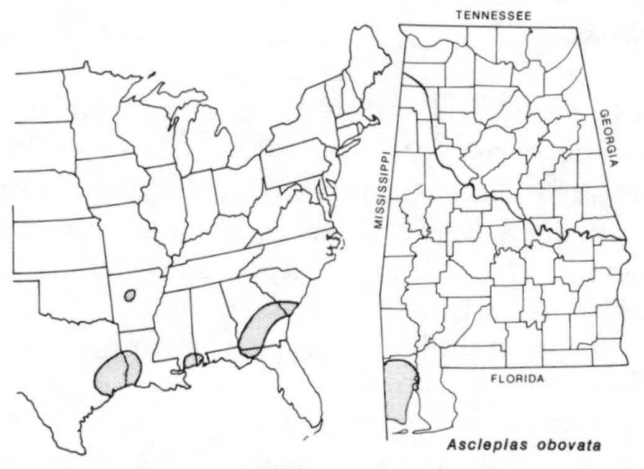

Range in eastern United States   Known range in Alabama

Asclepias perennis PERENNIAL MILKWEED

Species Recognition: *Asclepias perennis* is an erect herb that grows to a height of about 0.5 meter, with opposite petiolate leaves, the petioles being about 1 cm long. The leaves are acuminate at the apex and about 3 times as long as wide. The flowers and fruits are in both terminal and axillary inflorescence clusters. The fruits are pendulous on deflexed peduncles.

Geographic Distribution: South Carolina, south to central Florida, west to central Texas, and inland along the Mississippi River to Missouri and Illinois. In Alabama can be expected in the southern two-thirds of the state.

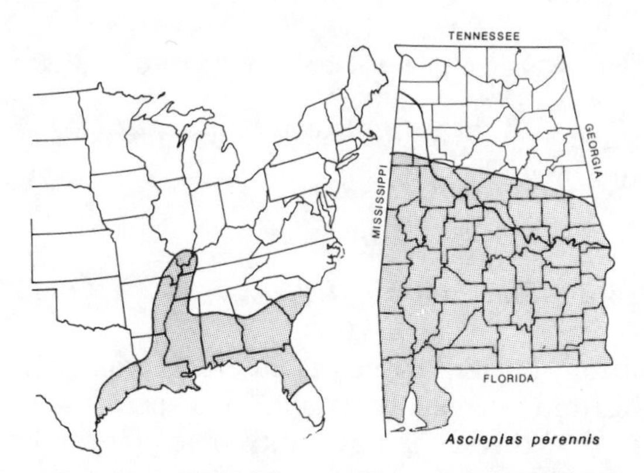

Range in eastern United States   Known range in Alabama

Habitat Occurrence: *Asclepias perennis* occurs in low swampy ground, especially where bald cypress occurs, and also in alluvial woods, sloughs, and ditches.

Species Recognition: Four-leaf milkweed is a perennial from a long fleshy rootstalk. The stem is erect and slender and less than 0.5 meter tall. The leaves are opposite, but occasionally 2 opposite leaves are condensed to form a whorl of 4 leaves, therefore the name, *quadrifolia*. The leaves are long and acute at the apex, lanceolate, 2–3 times as long as wide, and petiolate with the petiole 1–1.5 cm long. The flowers and fruits are terminal clusters. The fruits are erect on erect pedicels.

Geographic Distribution: Southern New England, west to central Illinois, south to the Atlantic Coast as far as New Jersey, and then inland on the west slope of the Smoky Mountains to northern Georgia, northern Alabama, and eastern Oklahoma. Known from the northeastern corner of Alabama.

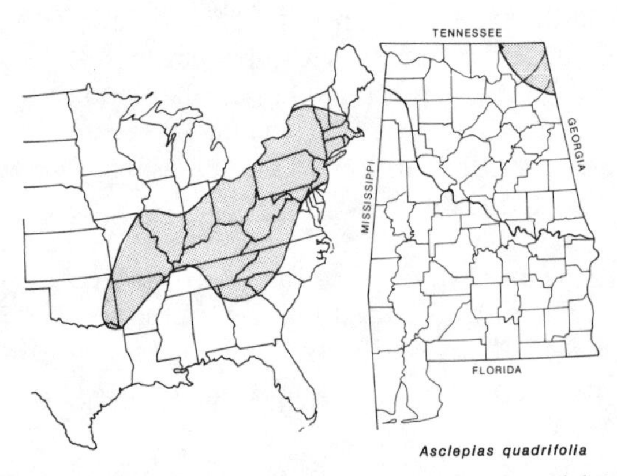

Range in eastern United States Known range in Alabama

Habitat Occurrence: Four-leaf milkweed occurs in woods and thickets, usually in dry or rocky soil.

Asclepias tuberosa (*Plate 16*) BUTTERFLY WEED

Species Recognition: Butterfly weed is an erect, stout herb about 1 meter in height from a woody rootstalk. This species is unusual for milkweeds in that the sap is not milky. The plant is hairy throughout. The alternate leaves are on short petioles or nearly sessile. The petiole is less than 0.5 cm long, and the leaves are about 3 times as long as broad and acute at the apex. The flowers and fruits are produced in terminal and axillary clusters. The fruits are erect on deflexed pedicels.

Geographic Distribution: Southern New England, west to northern Minnesota, south to southern Florida, and west to New Mexico. Occurs throughout Alabama.

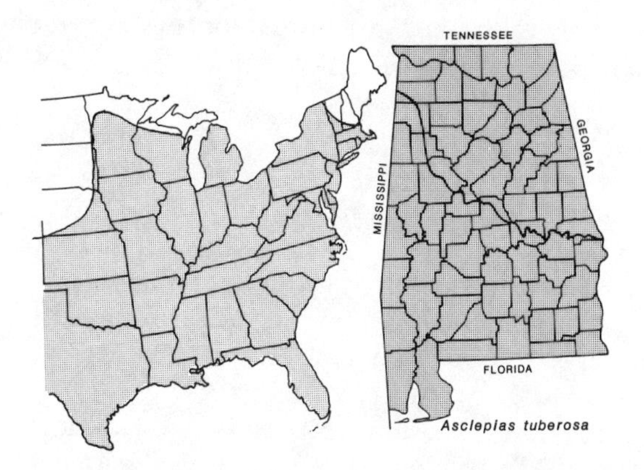

Range in eastern United States   Known range in Alabama

Habitat Occurrence: Butterfly weed is found in dry fields, thickets, open woods, on hillsides, and sand dunes, especially on rocky soil, or in pine barrens and flatwoods.

Asclepias variegata VARIEGATED MILKWEED

Species Recognition: Variegated milkweed is an erect herb 1–1.5 meters tall with slender stems rising from a fusiform rootstalk. The leaves are opposite petiolate; they are 1–2 times as long as wide and are round acute to round and only rarely acute. Also, the margins of the leaves are slightly serrate. The fruits and flowers are produced in terminal and axillary inflorescences. The fruits are erect on deflexed pedicels.

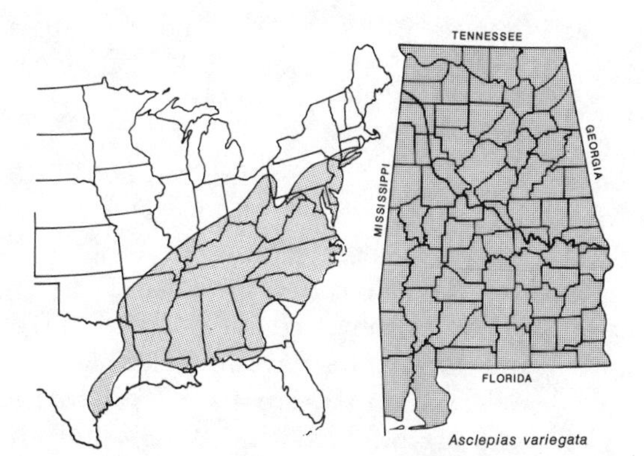

Range in eastern United States   Known range in Alabama

Geographic Distribution: Southern Massachusetts to northern Florida, west to southern Indiana and central Texas. Can be expected throughout Alabama.

Habitat Occurrence: Variegated milkweed occurs in thickets and open woods usually in sandy or rocky soil.

Asclepias verticillata WHORLED MILKWEED

Species Recognition: The whorled milkweed is an erect slender herb that grows to 0.75–1 meter tall from a very small rootstalk. Its leaves are 20 or more times as long as wide. The flowers are whorled, acute, sessile and are produced in axillary or terminal inflorescences. The fruits are erect on erect pedicels.

Geographic Distribution: Massachusetts to North Dakota, south to southern Florida, and west to east-central New Mexico. Occurs throughout Alabama.

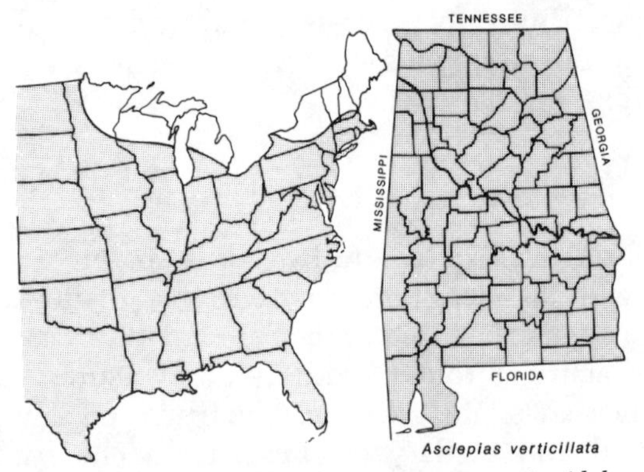

Range in eastern United States Known range in Alabama

Habitat Occurrence: Whorled milkweed is found in prairies, thickets, and open woods, usually in rather dry soil, and also on sand dunes, spreading to roadsides and fence corners.

Asclepias viridiflora GREEN-FLOWERED MILKWEED

Species Recognition: Green-flowered milkweed is an erect herb with a rather stout stem that is covered with a velvety pubescence. The stem grows to 1 meter from a coarse bulbous rhizome. The leaves are opposite or irregularly alternate and 0.5–2 times as wide as long. The petioles are less than 0.5 cm long. The apex of the leaves is acute. The flowers are produced in axillary and terminal clusters. The fruits are erect on deflexed pedicels.

Geographic Distribution: Southern Ontario and Manitoba, Connecticut to Georgia, and westward to Montana, Arizona, and Coahuila. In Alabama, can be expected in the northern two-thirds of the state, especially in the Black Belt counties.

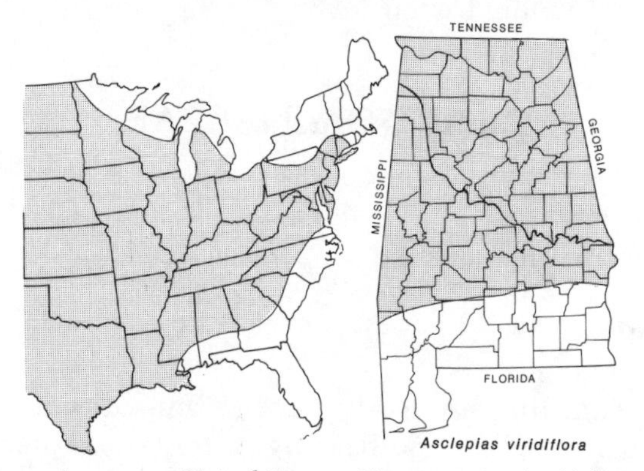

Range in eastern United States   Known range in Alabama

Habitat Occurrence: Green-flowered milkweed occurs in glades, prairies, plains, and rocky or sandy hillsides, spreading to old fields and roadsides.

Asclepias viridis GREEN MILKWEED

Species Recognition: The green milkweed is a coarse erect perennial from a stout, fusiform rootstalk. The stems and leaves are completely glabrous, and the leaves are opposite, short, and petiolate, the petioles being about 0.5 cm long. The leaf blades are lanceolate, broadest at the base, and then tapering to a long, rounded, subacute apex. The flowers and fruits are at the apex of the stem; the fruits are erect from a deflexed pedicel.

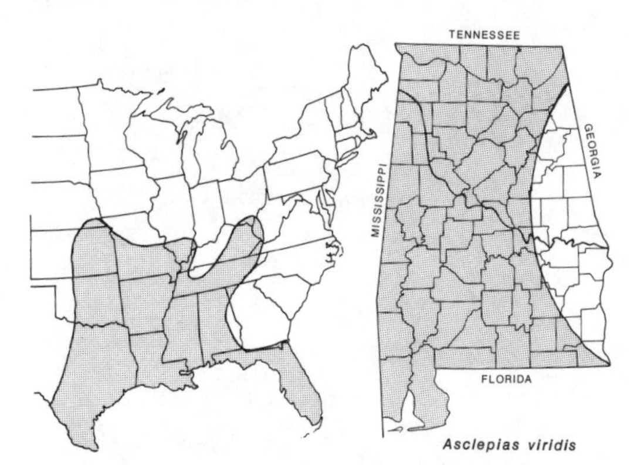

Range in eastern United States   Known range in Alabama

Geographic Distribution: Kentucky, south to southern Florida, and west to Nebraska and central Texas. Can be expected throughout Alabama.

Habitat Occurrence: The green milkweed occurs in glades, prairies, dry hillsides, and dry pine barrens.

Family SOLANACEAE

DATURA JIMSONWEED; THORN APPLE

One species occurs in Alabama.

Datura stramonium JIMSONWEED

Species Recognition: Jimsonweed is a coarse annual that grows to around 1.5 meters in height and has large alternate, coarsely lobed leaves and large tubular showy flowers. The flowers are a cream color. The petal tube might be as much as 15 cm long. The fruit is a spiny capsule with many black seeds.

Toxic Properties: All parts of the plant, including the pollen and especially the seeds, contain varying amounts of atropine, scopolamine, and other alkaloids. Ingestion of small amounts will produce undesirable symptoms, and larger amounts can cause death if treatment is not swift and successful. It has been calculated on the basis of the toxicity of pure atropine, that 4–5 grams of crude leaf or seed of *Datura* approximate a fatal dose for a child. Symptoms include dilated pupils, thirst, fever, dry and flushed skin, rapid and weak pulse, headache, lack of coordination, and confusion. Sometimes delirium, hallucinations, convulsions, stupor, coma with low temperature, labored respiration due to oxygen deficiency, and, rarely, vomiting

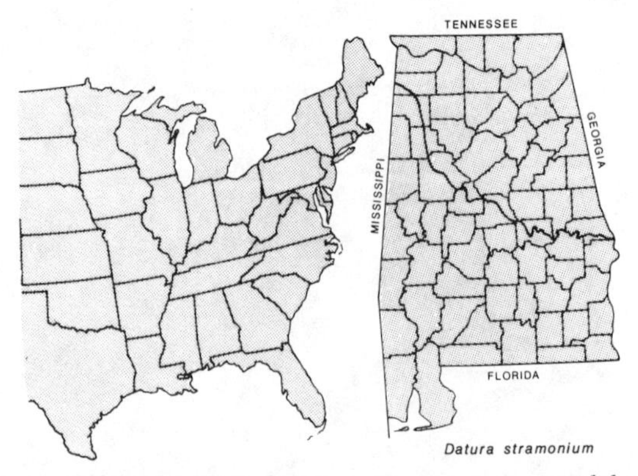

Range in eastern United States Known range in Alabama

occur. Deaths are rare and recovery is usually within several days, if the stomach contents are removed.

Geographic Distribution: Widely distributed from Florida to Texas, and north into Canada and the far western states. Can be expected throughout Alabama.

Habitat Occurrence: Jimsonweed is most common in waste areas, especially those with rich soils of barnyards and heavily used portions of pastures.

ura stramonium (Jimsonweed)

One species occurs in Alabama.

Lycopersicon esculentum GARDEN TOMATC

Species Recognition: The tomato is a sprawling viscid-scented annual her with stems that freely branch. The leaves are up to 50 cm long an 6 cm wide. The flowers are white with evident yellow anthers stick ing up above the petals. The fruit is a red berry 2–10 cm in diamete

Toxic Properties: The ripe fruits are, of course, familiar to all and eaten b most. However, the tomato has solanine alkaloids in the leaves an all other parts of the plant. Cattle, horses, hogs, as well as childre have been poisoned and have died from eating immature fruits anc or leaves or suckers from the plant. Symptoms include salivatio weakness, abdominal pain, and gastroenteritis. Also, contact derm titis frequently develops when plants are handled frequently. Frie green tomatoes and green tomato pickle can be safely eaten becaus they are cooked.

Geographic Distribution: Cultivated throughout the United States, occu ring rarely as an escape. Cultivated in every county of Alabama.

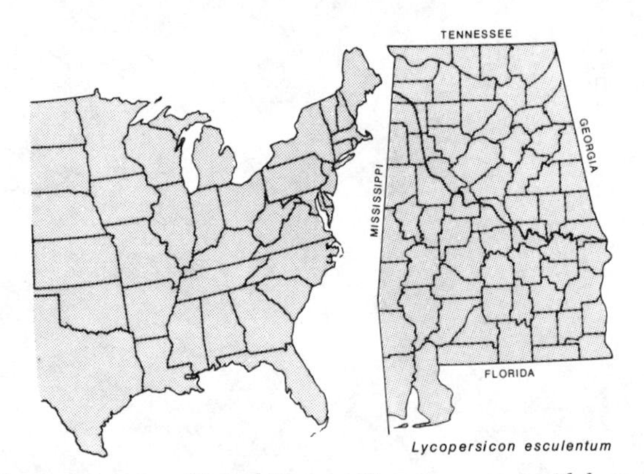

Range in eastern United States Known range in Alabama

Habitat Occurrence: A rare escape, but found most commonly in gardens

SOLANUM NIGHTSHAD

Solanum is a genus of annual or perennial herbs or small shrubs with a ternate simple or compound leaves, axillary inflorescences, regularly sym metrical flowers composed of 5 free sepals and 5 petals that are united a the base and radiate out to form a fairly flat surface. The fruit is a berry that may or may not be covered with prickles.

Toxic Properties: All parts of the plant contain the alkaloid solanine. Ingestion of this compound can cause gastroenteritis, with blood and mucus in the stool, and may eventually result in death. The fully ripe berries of some species are reportedly edible and are made into a jam. This practice should be carried out with great caution, if at all, since the unripe berries are highly toxic. The edible potato, *Solanum tuberosum*, is highly toxic in all conditions other than the tuber which is mostly nontoxic. However, even the green spots on a potato are toxic and have caused death to humans and livestock. The potato tuber should therefore be cooked before it is eaten.

Solanum americanum BLACK NIGHTSHADE

Species Recognition: Black nightshade is an erect herb with alternate, simple, coarsely toothed leaves, glabrous foliage, and black fruits.

Geographic Distribution: Maine to North Dakota, south to Florida, Louisiana, and Texas. Can be expected throughout Alabama.

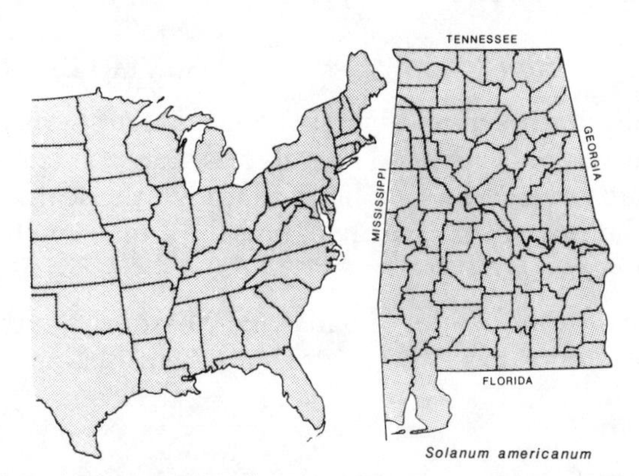

Range in eastern United States Known range in Alabama

Habitat Occurrence: Black nightshade occurs in rocky or dry, open woods, thickets, shores, and openings, often spreading to cultivated or waste ground.

Solanum carolinense* *(Plate 16) CAROLINA HORSE NETTLE

Species Recognition: Carolina horse nettle has stems and leaves covered with spines or prickles. The berry is not covered with a spiny calyx and is less than 2 cm in diameter. The leaves are covered with branched hairs that give a gray appearance. The fruit is yellow when mature.

Geographic Distribution: New England to southern Ontario, Illinois, Iowa, and Nebraska, south to Florida and Texas. Can be expected throughout Alabama.

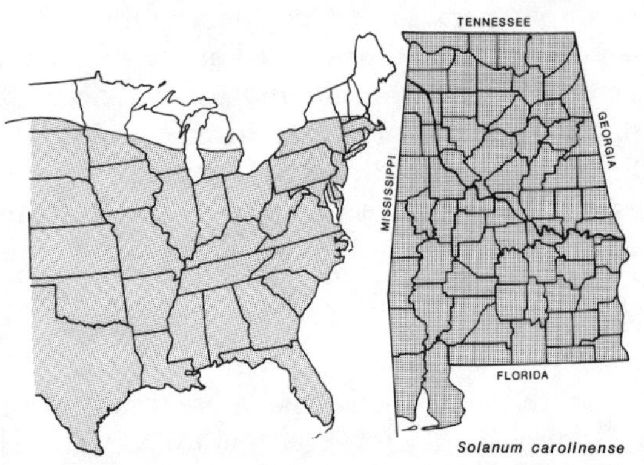

Range in eastern United States   Known range in Alabama

Habitat Occurrence: Carolina horse nettle occurs along roadsides and in old fields, farm lots, and waste places.

Solanum elaeagnifolium SILVERLEAF NIGHTSHADE

Species Recognition: Silverleaf nightshade is an erect herb with or without spiny leaves and stems. The calyx is not covered by spines. The leaves are covered by fine hairs that give the leaf a silvery appearance. The flowers are purplish, and the fruit is yellow or brownish when mature.

Geographic Distribution: Ohio, Indiana, and Missouri, south to Florida and Mexico. Can be expected throughout Alabama.

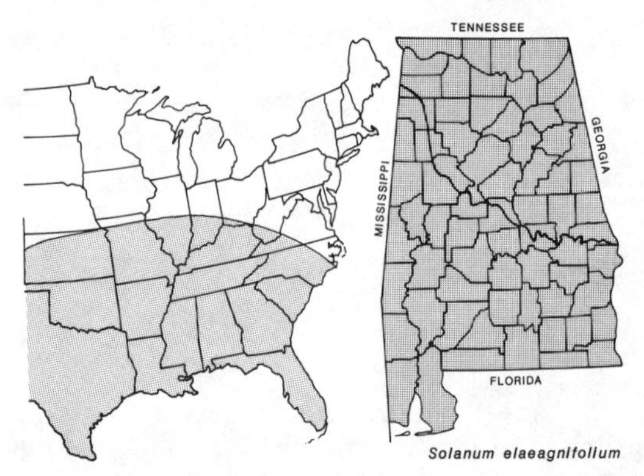

Range in eastern United States   Known range in Alabama

Habitat Occurrence: Silverleaf nightshade occurs in dry, open woods, prairies, waste places, and disturbed soil.

Solanum gracile DEADLY NIGHTSHADE; COASTAL-DUNE NIGHTSHADE

Species Recognition: Deadly nightshade has stems and leaves not covered by spines or prickles and leaves that do not appear to be compound, but are entire and simple and have entire to rarely irregularly toothed margins. They are covered by pubescence. The hairs are simple and the fruit is black.

Geographic Distribution: North Carolina and Florida, west to Alabama. In Alabama known only from the Outer Coastal Plain.

Habitat Occurrence: Deadly nightshade occurs on coastal dunes and margins of maritime forests and brackish marshes.

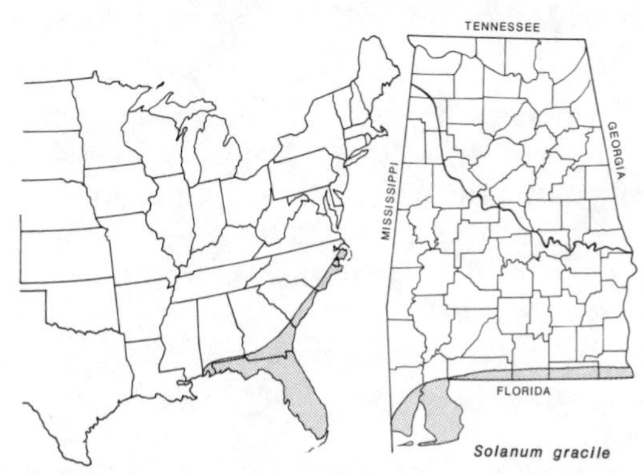

Range in eastern United States Known range in Alabama

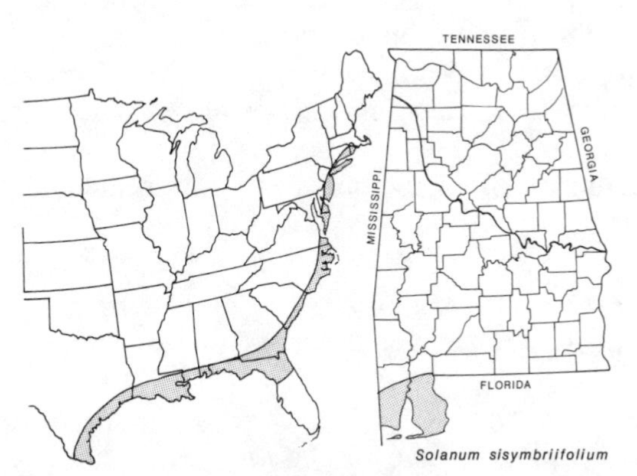

Range in eastern United States Known range in Alabama

Species Recognition: Spiny nightshade is covered by long orange spines on both the leaves and the stem. The leaves are pinnatifid and are occasionally covered with spines and hairs. The fruit is red when mature.

Geographic Distribution: Massachusetts, south to Florida and Texas. In Alabama known only from Mobile County.

Habitat Occurrence: Spiny nightshade occurs on roadsides and waste places.

Family CUSCUTACEAE

CUSCUTA DODDER

Cuscuta is a genus of sprawling vines, an obligate parasite on other plants. The leafless stems are golden yellow without chlorophyll. The flowers are clustered in a small group near the top of the host plant. Haustoria penetrate the host plant and extract the nutrients from it. Six species occur in Alabama. The genus is very difficult to deal with taxonomically, and separating the 6 species requires close examination of the sepals. Since it would be a lengthy process to give enough characteristics to separate the species, these plants will be considered at the generic level only.

Toxic Properties: All parts of the plant contain unknown toxins. Extracts of the plants have been used by Asian Indians to induce abortion. Side effects cause depression, nausea, and vomiting. It is suspected that *Cuscuta* growing on clover and then ingested by horses has caused digestive upset. However, there are no reports of human or livestock deaths in the United States from ingestion of *Cuscuta*.

Geographic Distribution: Throughout the continental United States. In Alabama can be expected in every county.

Habitat Occurrence: *Cuscuta* is most common in marshy habitats, but it does occur near roadsides, in weedy places, and in pastures.

Alabama species: The following species of *Cuscuta* occur in Alabama: *Cuscuta campestris* (love vine), ***Plate 17***; *C. compacta*; *C. gronovii*; *C. harperi*; *C. indecora*; and *C. pentagona*.

Family VERBENACEAE

LANTANA LANTANA

Lantana is a perennial shrub with opposite, simple, petiolate, dentate leaves. The stems are covered with small prickles. The flowers are slightly irregular and tubular. Several are clustered together at the apex of the plant. The fruit is a 2–4-seeded nutlet. Two species occur in Alabama, although only *Lantana horrida* is known to be toxic. They are quite similar and difficult to separate so an account will be given of each.

Toxic Properties: All parts of the plant especially the leaves and green fruits contain a glycoside called lantanin. Large amounts of stems, leaves, and berries consumed by cattle and sheep cause acute poisoning. In 12–24 hours there is weakness, gastroenteritis with bloody diarrhea, loss of appetite, eye irritation, blindness, sores in the mouth, and partial paralyses. When exposed to the sun, the tender areas become swollen, yellow, hard, cracked, and peeled to expose raw bleeding areas. Children have died after eating the green berries. In 2–5 hours symptoms appear that include vomiting, lethargy, dilated pupils, weakness, slow respiration, diarrhea, circulatory disorders, and collapse.

***Lantana camara* (*Plate 17*)** YELLOW SAGE

Species Recognition: Yellow sage is an erect shrub with pricklish stems. The calyx lobes are acute and as long as or longer than the calyx tube.

Geographic Distribution: Escaped or persistent from cultivation in Florida to South Carolina, west to Texas. This species is cultivated farther north. In Alabama known only from Mobile and Baldwin counties.

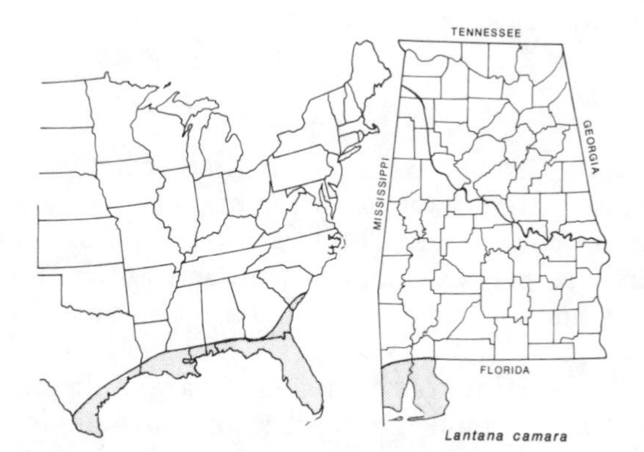

Range in eastern United States Known range in Alabama

Habitat Occurrence: Yellow sage is persistent after cultivation on roadsides and in waste places.

Lantana horrida LANTANA

Species Recognition: *Lantana horrida* is similar to *L. camara* except that the calyx lobes are obtuse and shorter than the calyx tubes.

Geographic Distribution: Escaped or persistent from cultivation in South Carolina, Florida and west to Texas. In Alabama known only from Mobile and Baldwin counties.

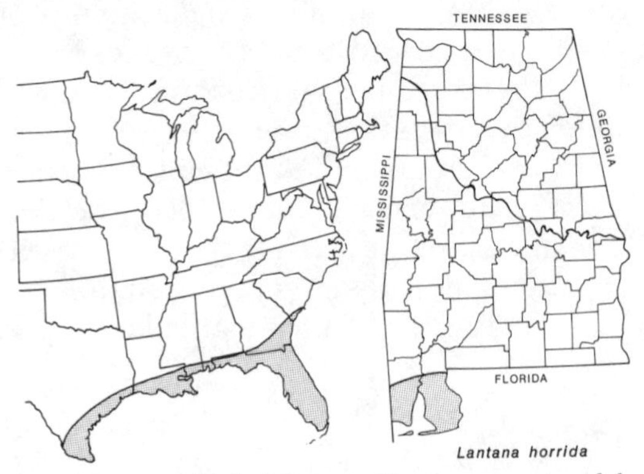

Range in eastern United States   Known range in Alabama

Habitat Occurrence: *Lantana horrida* persists after cultivation on roadsides and in waste places.

Family LAMIACEAE

GLECHOMA GROUND IVY

One species occurs in Alabama.

Glechoma hederacea GROUND IVY

Species Recognition: Ground ivy is a prostrate perennial herb forming ground cover, often to the exclusion of other vegetation. The stems are prostrate and square. The leaves are opposite and orbicular in shape, crenate, petiolate, and 1–2 cm across. The flowers are blue, two-lipped, and formed in the leaf axis. The fruit is a nut.

Toxic Properties: All parts on the plant contain a volatile oil that is physiologically active in moderate-to-large amounts and is assumed to be the source of toxicity after ingestion. Symptoms include salivation, sweating, difficulty in breathing, panting, dilation of the pupils, and sometimes signs of pulmonary edema. In North America there have been reports of horses that died from eating ground ivy.

Geographic Distribution: Introduced from Europe and now occurring from Newfoundland to Ontario, south to South Carolina, Alabama, Virginia, Tennessee, and Missouri. Can be expected in the northern half of Alabama.

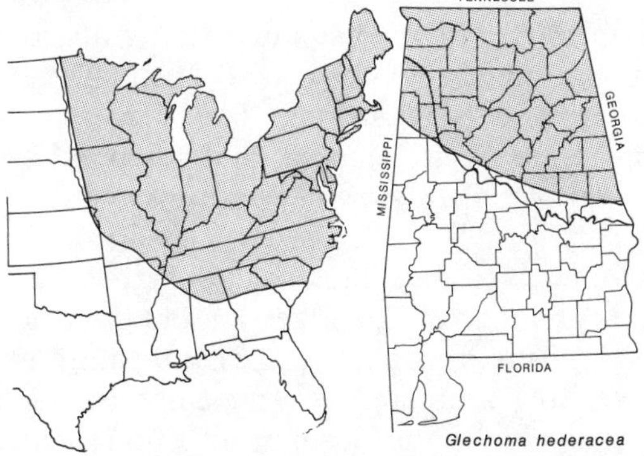

Range in eastern United States   Known range in Alabama

Habitat Occurrence: Ground ivy occurs on roadsides, in yards, damp shady places, and in open areas.

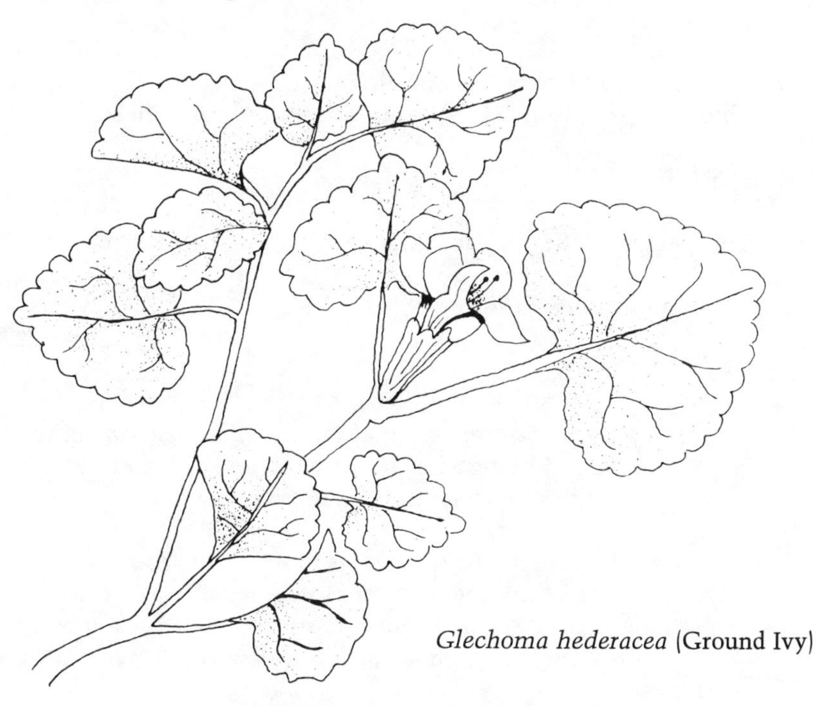

Glechoma hederacea (Ground Ivy)

Family OLEACEAE

LIGUSTRUM PRIVET

One species, *Ligustrum vulgare*, has been reported to be toxic in the United States. Two other species known as escapes in Alabama will not be discussed.

Ligustrum vulgare EUROPEAN PRIVET

Species Recognition: Privet is an erect shrub that grows to a height of 3 meters, with opposite, deciduous leaves. The blades are elliptic to elliptic lanceolate and 1–5 cm long. They have slender petioles. The flowers are small, white, perfect, and clustered into terminal panicles. The fruit is a dark subglobose drupe.

Toxic Properties: The glycosides ligustrin and syringin are known to occur in the fruits and leaves. The ingestion of either causes gastroenteric irritation resulting in vomiting, diarrhea, pain, loss of coordination, weak pulse, low temperature, and convulsions. Fatalities are reported among children in Europe and among animals in the United States.

Geographic Distribution: Maine to Ontario, south to North Carolina and Texas. Cultivated throughout Alabama and known to escape in many localities.

Habitat Occurrence: Privet occurs along roadsides, in old fields, and in thickets, persisting and spreading from cultivation.

CHIONANTHUS FRINGE TREE

A single species occurs in Alabama.

Chionanthus virginicus* *(Plate 17) CHIONANTHUS; GRANDSIR-GRAYBEARDS

Species Recognition: A small tree that grows to a height of 8 meters, with opposite leaves that are entire along the margins and acute at both apex and base. The flowers are in panicles with long white petals. The fruit is a dark purple drupe, 1–1.5 cm long.

Toxic Properties: An unknown chemical is found in the roots. Extracts from the roots have been used medicinally, especially as a tincture. Overdoses have caused severe headache, sore eyes, nausea, intestinal gas, severe vomiting, black stools, slow pulse, cold perspiration, and weakness.

Geographic Distribution: Central New Jersey and southeastern Pennsylvania to eastern Kentucky and southern Missouri, south to Florida and Texas. Can be found throughout Alabama.

Habitat Occurrence: Fringe trees are found in damp rich woods and thickets.

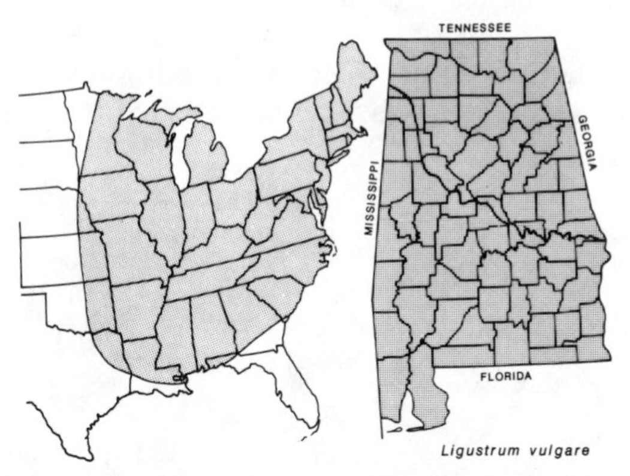

Range in eastern United States   Known range in Alabama

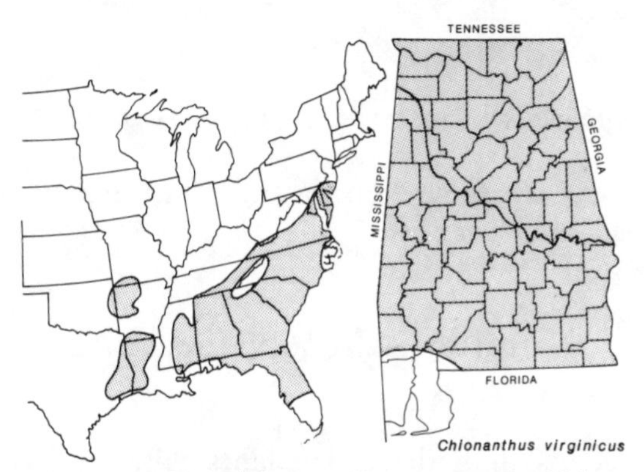

Range in eastern United States   Known range in Alabama

Family BIGNONIACEAE

ANISOSTICHUS CROSS VINE

A single species occurs in Alabama.

Anisostichus capreolata CROSS VINE

Species Recognition: Cross vine is a woody vine that climbs by tendrils. The leaves are divided into 2 leaflets, between which the terminal tendril arises. The flowers are showy, 4–5 cm long, with the corolla ranging from dull red to orange outside and yellow or red inside. The fruit is a sword-shaped capsule about 15 cm long and 2 cm wide with numerous papery, winged seeds.

Toxic Properties: The active ingredients are unknown. Extracts from all parts of the plants are used as a fish poison, and cross vine is assumed to be toxic to humans.

Geographic Distribution: Eastern Maryland to Illinois and Missouri, south to Florida and east Texas. Can be expected throughout Alabama.

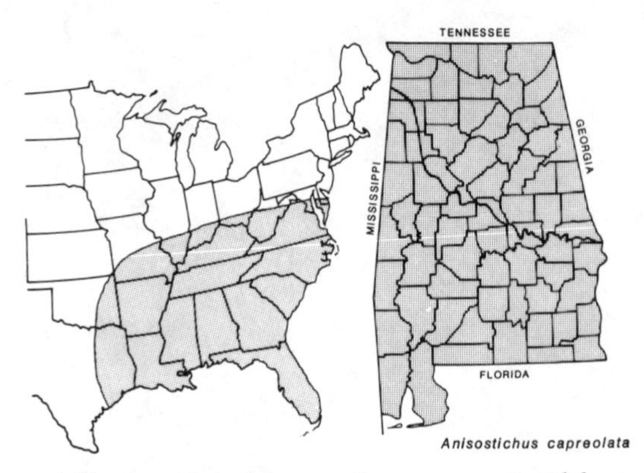

Range in eastern United States   Known range in Alabama

Habitat Occurrence: Cross vine climbs trees in thickets, mixed or deciduous forests, and swamps.

Family CAMPANULACEAE

LOBELIA LOBELIA

Lobelia is a genus of herbaceous plants with milky sap and few-to-numerous leaves dispersed more or less evenly along the stem. The flowers are mostly stalked and are arranged in a simple elongate cluster at the apex of the stem. The petals are fused into a tube that has a split along one side. At the top of the tube are 5 petal lobes, 2 directed upward and 3 downward.

Toxic Properties: The active ingredient is the alkaloid lobeline, which is present in all parts of the plant. Extracts from these plants have been used in home medicines, and deaths have resulted from overdoses. Symptoms include nausea, vomiting, headache, sweating, rapid pulse, collapse, convulsions, and coma.

***Lobelia cardinalis* (Plate 18)** CARDINAL FLOWER

Species Recognition: Cardinal flower is an erect herb that grows to a height of 2 meters. The leaves are numerous, scattered throughout the length of the stem, and widest near or just above the base. They taper to an acute point on both ends and are irregularly toothed. The flowers are numerous, fairly close together, quite showy, and have bright red petals. This is the only member of this genus with red flowers.

Geographic Distribution: Quebec to Minnesota, and south to northern Florida and eastern Texas. Can be expected throughout Alabama.

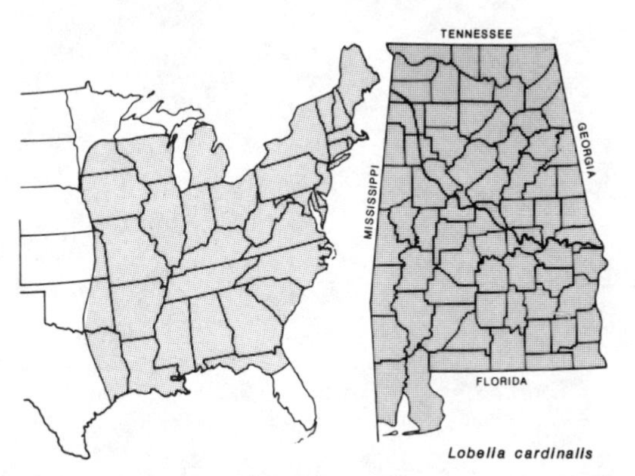

Range in eastern United States Known range in Alabama

Habitat Occurrence: Cardinal flowers occur in damp-to-wet soils of lake or river margins, swales, meadows, or swamps.

Lobelia puberula LOBELIA

Species Recognition: *Lobelia puberula* is an erect herb that grows to a height of 2 meters and has stiff hairs covering the entire stem. The leaves are numerous, scattered throughout the length of the stem, widest at the middle, acute to rounded at the apex, and are usually tapered to the stem at the base. They are irregularly and coarsely toothed. The flowers are numerous, spaced close together, blue, 1 cm or more long, and without auricles at the base of the sepals.

Geographic Distribution: West Virginia and Ohio to western Kentucky and eastern North Carolina, southwestern Georgia, and southeastern Louisiana. A disjunct population occurs in northwestern Illinois. Can be expected throughout Alabama.

Lobelia puberula (Lobelia)

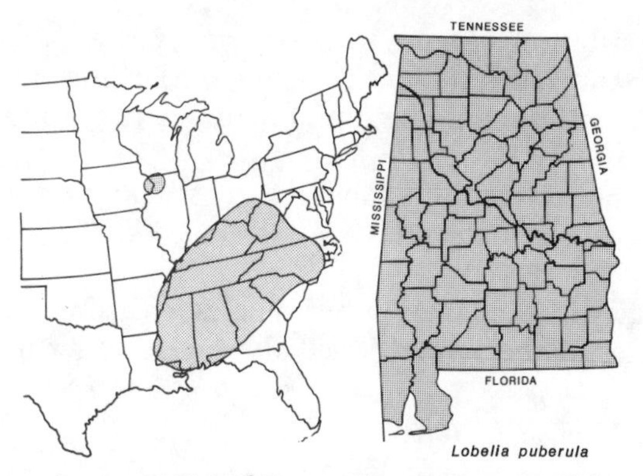

Range in eastern United States Known range in Alabama

Habitat Occurrence: *Lobelia puberula* occurs in a wide variety of habitats, ranging from dry open meadows and dry woodlands to swamps. The species also occurs commonly along the margins of lakes and streams.

Lobelia siphilitica GREAT BLUE LOBELIA

Species Recognition: Great blue lobelia is an erect herb that grows to a height of 1.5 meters. The leaves are numerous, scattered throughout the length of the stem, widest at or just above the middle. They taper to an acute angle at both top and bottom and are irregularly toothed. The flowers are numerous, showy, blue, 2–4 cm long, and have auricles at the base of the sepals. The specific name *siphilitica* comes from the use of this species by the Herbalists, physicians of the fifteenth and sixteenth centuries, who used great blue lobelia as a treatment for syphilis.

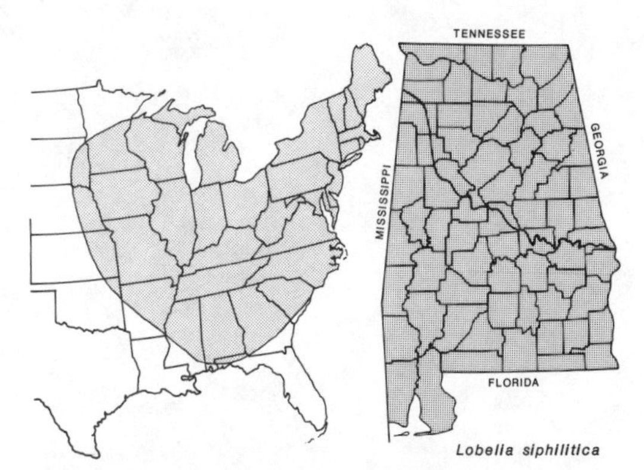

Range in eastern United States Known range in Alabama

Geographic Distribution: Maine to northern Minnesota and eastern South Dakota, and south to South Carolina, Alabama, and Arkansas. Rare in Alabama but could be found in any part of the state.

Habitat Occurrence: Great blue lobelia is most common in rich low woods with poorly drained soil.

Additional Species in Alabama: Other species with similar appearances and properties as those mentioned above also occur in this region.

Lobelia amoena LOBELIA

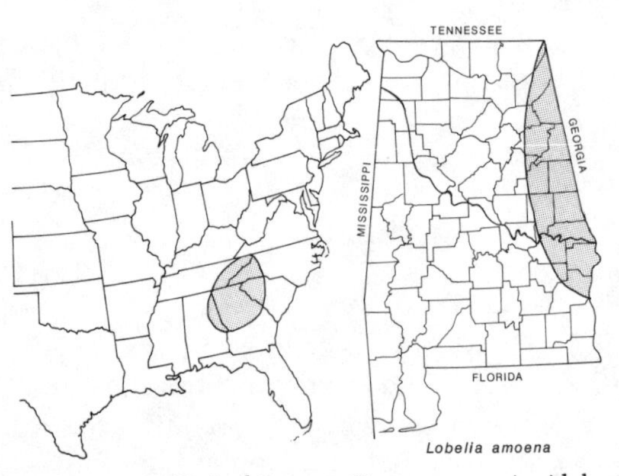

Range in eastern United States Known range in Alabama

Lobelia appendiculata LOBELIA

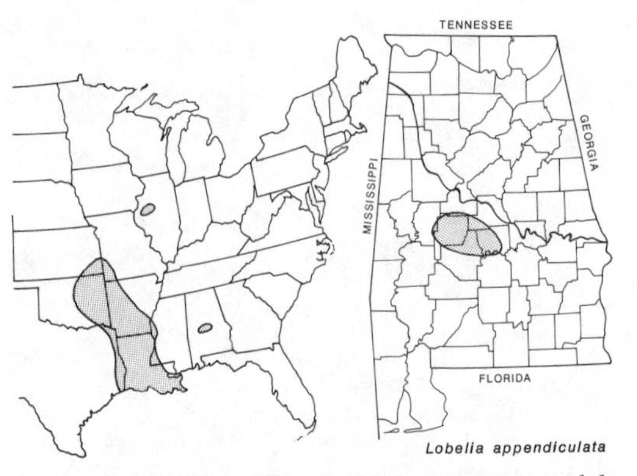

Range in eastern United States Known range in Alabama

Lobelia brevifolia LOBELIA

Range in eastern United States Known range in Alabama

Lobelia floridana LOBELIA

Range in eastern United States Known range in Alabama

Lobelia inflata INDIAN TOBACCO

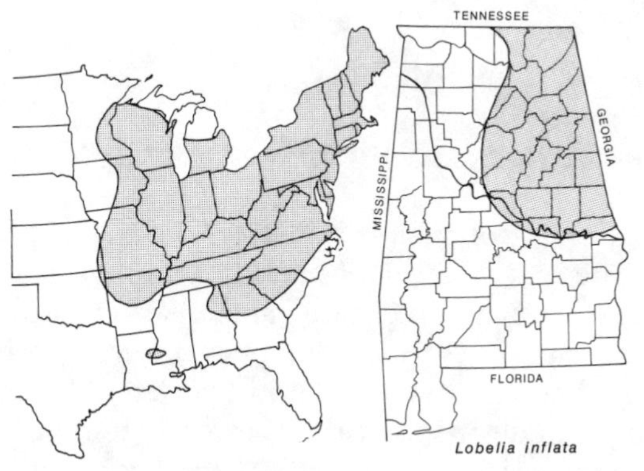

Range in eastern United States Known range in Alabama

Lobelia nuttallii LOBELIA

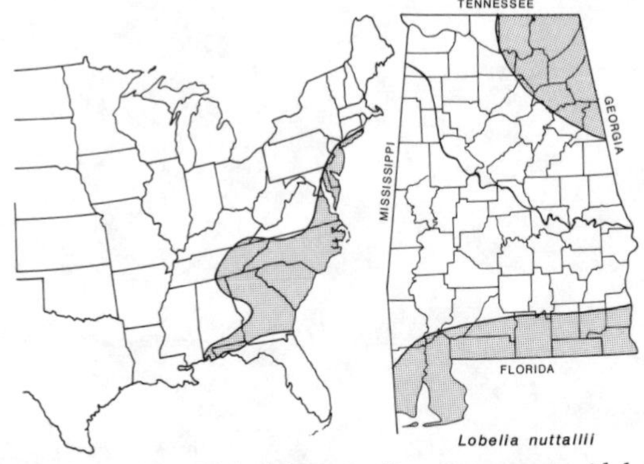

Range in eastern United States   Known range in Alabama

Lobelia spicata LOBELIA

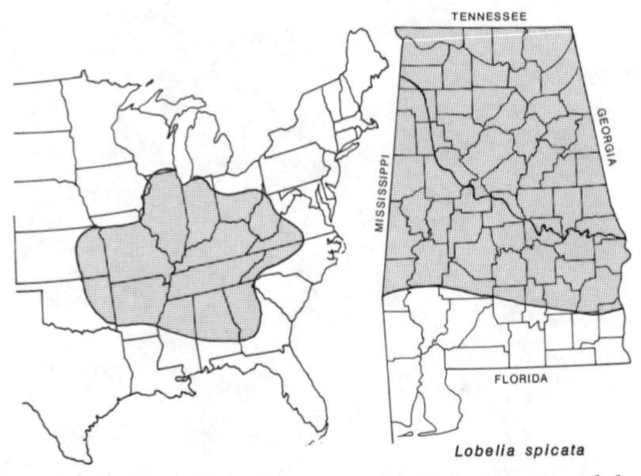

Range in eastern United States   Known range in Alabama

Family RUBIACEAE

CEPHALANTHUS BUTTONBUSH

A single species occurs in Alabama.

Cephalanthus occidentalis *(Plate 18)* COMMON BUTTONBUSH

Species Recognition: Buttonbush is a small shrub that grows as tall as 1–1.5 meters. The leaves are opposite and whorled, usually both on the same stem, are ovate to elliptic, and are 6–15 or more cm long. The flowers are small, less than 1 cm long, and are tightly compacted into a spherical head that is 2–4 cm in diameter. Several of these heads are arranged in a panicle at the apex of a stem or branch. The fruits are elongate nutlets, obypyramidal in shape, and 0.4–0.7 cm long, that at first remain in the compact head but eventually fall free.

Toxic Properties: The active ingredients are 2 glycosides, cephalin and cephalanthin, which occur in all parts of the plant. Consumption has caused vomiting, paralysis, spasms, and destruction of red blood cells.

Geographic Distribution: Maritime provinces of Canada to southern Quebec, southern Ontario, to Wisconsin, Kansas, and Texas, south to Florida and Mexico. Can be expected in all counties of Alabama.

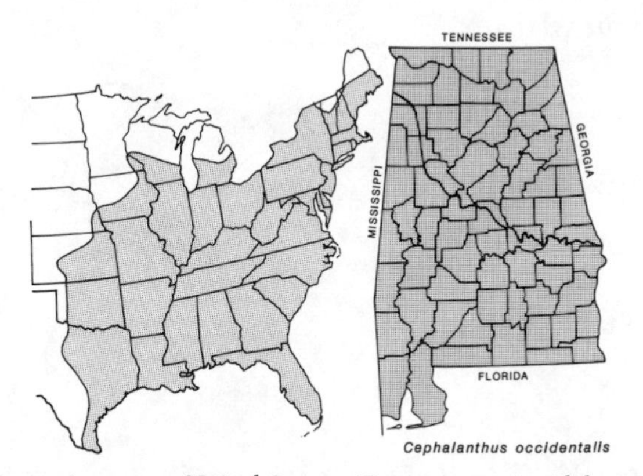

Range in eastern United States Known range in Alabama

Habitat Occurrence: Buttonbush is common in and along the water's edge of creeks, rivers, lakes, ponds, and swamps. The species usually occurs in full sun, although it will occasionally grow in shaded parts of swamps. It is rarely found far from water.

Family CAPRIFOLIACEAE

SAMBUCUS ELDERBERRY

One species occurs in Alabama.

Sambucus canadensis ELDERBERRY; COMMON ELDER

Species Recognition: Elderberry is a coarse shrub or small bush with opposite compound leaves that have serrate leaflets. The flowers are small and white and grouped into large, conspicuous, flat-topped inflorescences. The fruit is purple-black and is also grouped in these large, conspicuous masses. The stems are about 18 mm in diameter, or more, and have large openings filled with light pithy wood.

Toxic Properties: The roots, bark, stems, leaves, and, to a less extent, flowers and fruits contain an unknown alkaloid and an unknown cyanogenic glycoside. These compounds cause nausea, vomiting, and diarrhea. The roots have the greatest concentration of these compounds, and livestock have died after eating roots. The fruits are used in making pies, jellies, and wines. Once they have been cooked, the fruits are perfectly harmless. However, uncooked fruits can cause nausea.

Geographic Distribution: Elderberry occurs from Quebec to Manitoba, south to Florida, Texas, Oklahoma, and Iowa. Can be expected throughout Alabama.

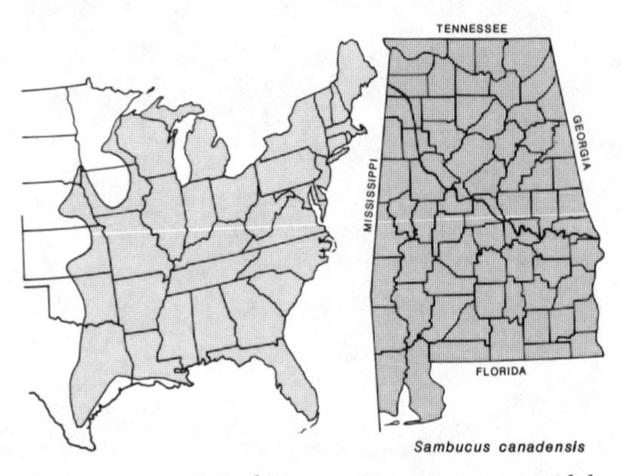

Range in eastern United States Known range in Alabama

Habitat Occurrence: Elderberry occurs in swamp forests, alluvial woods, pastures, and is especially common in wet, open habitats.

LONICERA HONEYSUCKLE

Lonicera is a genus of woody shrubs or woody climbing vines with opposite leaves and 2-lipped, yellow, pink, white, or rose flowers. The leaves are often peltate-perfoliate, that is, without petioles, and on opposite sides of the stem. The bases are fused, making the stem appear as if it goes through the leaf. The fruit is a red berry.

Toxic Properties: Honeysuckle has some unknown properties, especially in the leaves, but also occasionally in the fruit, which have resulted in deaths among cattle and humans in Europe. Symptoms include colic, diarrhea, cardiac arrhythmias, muscular twitching, convulsions, and finally respiratory failure. Four species occur in Alabama. None of these has proven to be toxic, but they should be considered so unless shown otherwise.

Lonicera flava YELLOW HONEYSUCKLE

Species Recognition: Yellow honeysuckle is a climbing vine with a terminal cluster of flowers that is subtended by 2 pairs of connate leaves. The flowers are yellow and strongly 2-lipped.

Geographic Distribution: North Carolina to Missouri, south to Georgia, Alabama, Arkansas, and Oklahoma. Can be expected in the mountains of the northeastern quarter of Alabama.

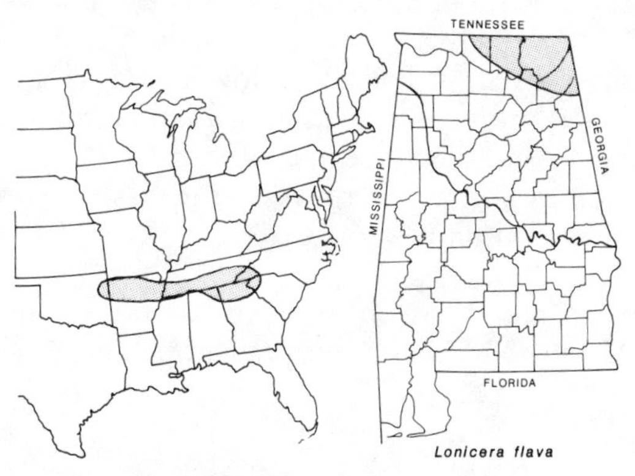

Range in eastern United States Known range in Alabama

Habitat Occurrence: Yellow honeysuckle occurs in woodlands and thickets, most commonly in the mountainous regions and along rocky bluffs.

Lonicera fragrantissima SWEET BREATH OF SPRING

Species Recognition: Sweet breath of spring is an erect shrub with fragrant flowers produced in the axils of leaves.

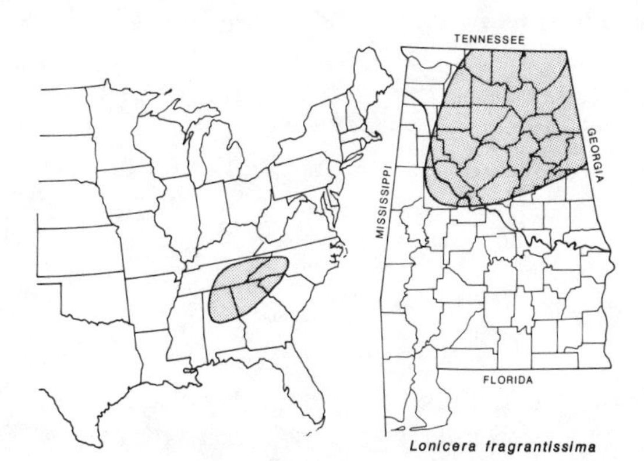

Range in eastern United States Known range in Alabama

Geographic Distribution: The Carolinas, Georgia, Alabama, and Tennessee. In Alabama can be expected from Tuscaloosa into the mountains of the northeast.

Habitat Occurrence: Sweet breath of spring occurs in woodlands and waste places. It is infrequent.

Lonicera japonica JAPANESE HONEYSUCKLE

Species Recognition: Japanese honeysuckle is a twining woody vine with axillary 2-lipped, whitish, yellowish, or reddish flowers, usually without any connate leaves. The flowers are strongly 2-lipped. The fruit is a black berry. This is the common honeysuckle with which southerners are familiar. The nectar of this species is sweet and non-toxic, and children often remove the pistils to expose a few drops that can be placed on the tongue.

Geographic Distribution: Massachusetts to Michigan, Indiana, Missouri, south to Florida and Texas. Occurs throughout Alabama.

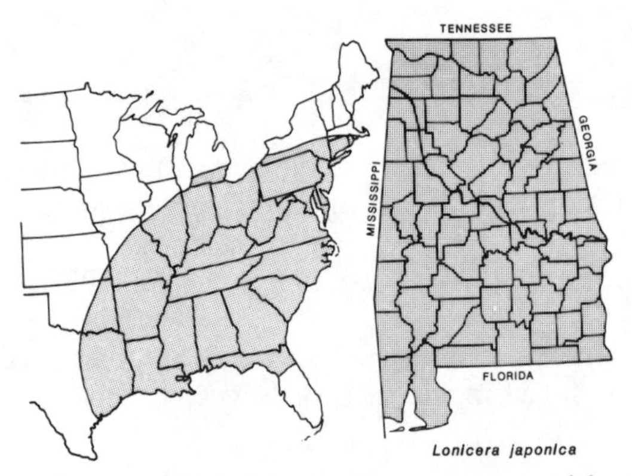

Range in eastern United States Known range in Alabama

Habitat Occurrence: Japanese honeysuckle occurs in thickets and on borders of woods and roadsides. This species is a nuisance weed, overwhelming and strangling the native flora, and is most difficult to eradicate.

***Lonicera sempervirens* (Plate 19)** CORAL HONEYSUCKLE

Species Recognition: Coral honeysuckle is a woody climbing vine with terminal flowers subtended by 2 sets of connate leaves. The flowers are red and are slightly 2-lipped. The fruit is a red berry.

Lonicera sempervirens (Coral Honeysuckle)

Geographic Distribution: Maine, New York, Ohio, and Nebraska, south to Florida and Texas. Can be expected throughout Alabama.

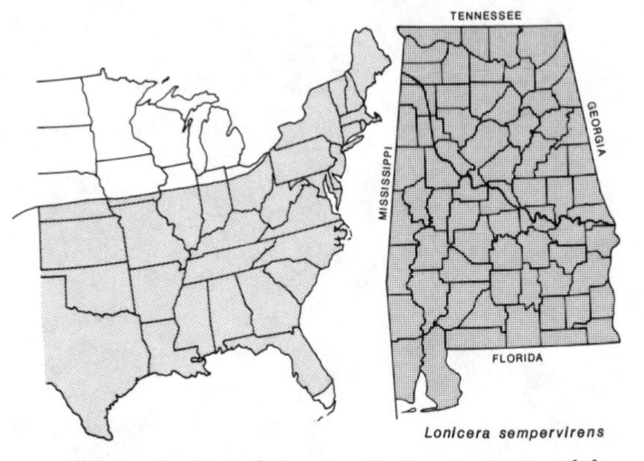

Range in eastern United States   Known range in Alabama

Habitat Occurrence: Coral honeysuckle occurs in woodlands, thickets, and along fencerows, especially in the Coastal Plain and Piedmont.

Family ASTERACEAE

ANTHEMIS CHAMOMILE

Only one species occurs in Alabama.

Anthemis cotula MAYWEED

Species Recognition: Mayweed is an annual herb that grows from a taproot to about 1 meter in height. The leaves are finely divided. The flowers are arranged in compact heads with white rays and yellow disks. The fruit is a 1-seeded achene. The vegetative material when crushed exudes an ill-smelling odor.

Toxic Properties: Mayweed contains an acrid substance in the leaves and in the flowers that is distasteful and is irritating to mucous membranes. This substance is found in both the leaves and flowers. Grazing by cattle gives the milk an undesirable flavor.

Geographic Distribution: Native to Europe but naturalized throughout the United States and Canada. Can be expected throughout Alabama.

Habitat Occurrence: Mayweed is most common along roadsides and in waste places.

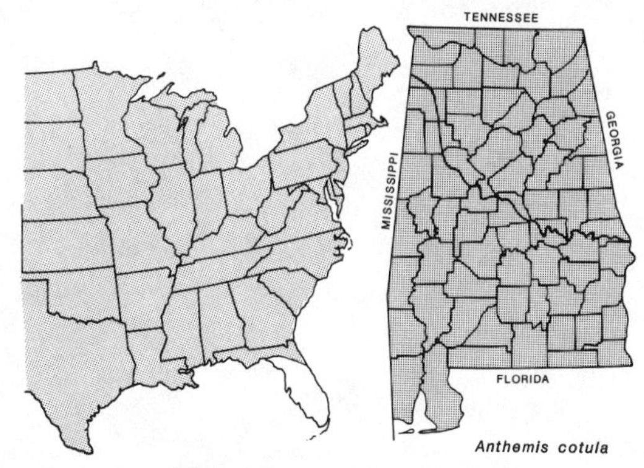

Range in eastern United States Known range in Alabama

BACCHARIS BACCHARIS

Baccharis is a genus of dioecious shrubs with alternate, fleshy, obscurely toothed leaves. The white flowers are in compact heads, each of which contains either stamens or carpels. The fruit is an achene with many bristles that cause it to be fluffy and to blow apart in the wind.

Toxic Properties: The flowers and leaves of *Baccharis* contain a glycosidal saponin that has caused death in feeding experiments with chicks and mice. Symptoms include loss of coordination, diarrhea, weakness, and heart and respiratory disturbance.

Baccharis angustifolia FALSE WILLOW

Species Recognition: False willow is a shrub 1–2 meters tall with linear leaves that are 2–4.5 cm long and less than 0.5 cm wide.

Geographic Distribution: North Carolina to Florida and Alabama. In Alabama the species is known only from Mobile and Baldwin counties.

Habitat Occurrence: The species occurs in brackish marshes in the extreme Outer Coastal Plain.

Baccharis halimifolia GROUNDSEL TREE; EASTERN FALSE WILLOW

Species Recognition: Groundsel tree is a shrub 1–4 meters tall with obovate leaves (widest above the middle) that are 1–4 cm wide and are coarsely toothed, especially near the apex.

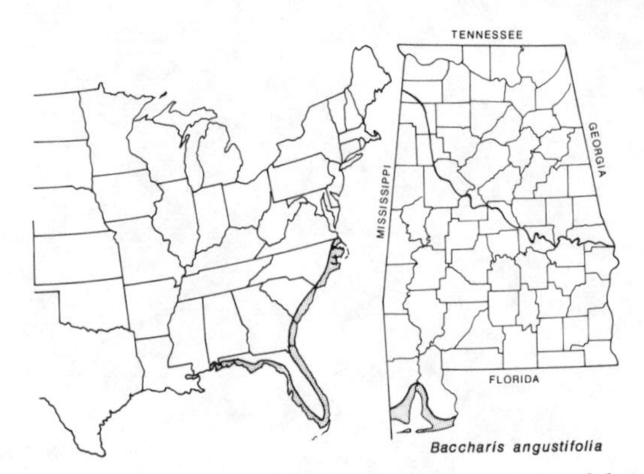

Range in eastern United States   Known range in Alabama

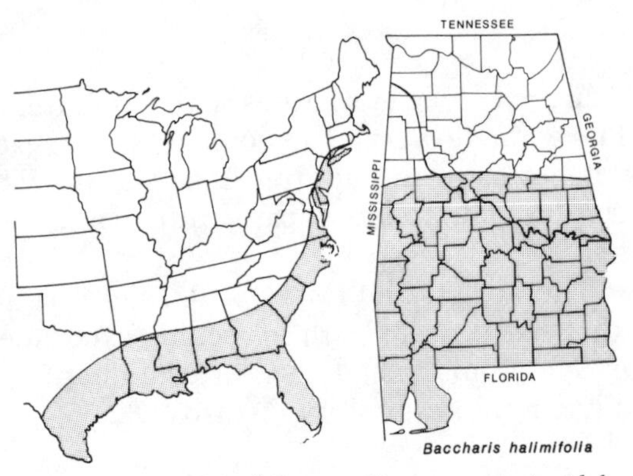

Range in eastern United States   Known range in Alabama

Geographic Distribution: Atlantic coast of Massachusetts, south to Florida, west to Texas and Mexico. Can be expected throughout the southern two-thirds of Alabama.

Habitat Occurrence: Groundsel tree is found along fencerows and in fields, brackish marshes, seepage areas, thickets, and open woods, especially near the coast.

ERECHTITES FIREWEED

A single species occurs in Alabama.

Erechtites hieracifolia FIREWEED

Species Recognition: Fireweed is an erect annual herb with a stubby taproot. It grows to a height of about 1 meter. The leaves are linear-lanceolate, coarsely toothed, and they clasp the stem. The flowers

Plate 1

Amanita, p. 26

Pteridium aquilinum, p. 40
Bracken Fern

Juniperus virginiana, p. 222
Eastern Red Cedar

Plate 2

Asimina triloba, p. 225
Common Pawpaw

Calycanthus floridus, p. 42
Eastern Sweetshrub

Aristolochia serpentaria, p. 43
Snakeroot

Plate 3

Delphinium tricorne, p. 52
Dwarf Larkspur

Actaea pachypoda, p. 45
White Baneberry

Podophyllum peltatum, p. 55
Mayapple

Plate 4

Cocculus carolinus, p. 56
Coral Beads

Sanguinaria canadensis, p. 59
Bloodroot

Dicentra cucullaria, p. 61
Dutchman's Britches

Plate 5

Cannabis sativa, p. 64
Marijuana

Juglans nigra, p. 233
Black Walnut

Phytolacca americana, p. 65
Common Pokeweed

Chenopodium ambrosioides, p. 71
Mexican Tea

Plate 6

Batis maritima, p. 78
Saltwort

Rumex acetosella, p. 80
Sheep Sorrell

Modiola caroliniana, p. 82
Carolina Bristle Mallow

Rhododendron catawbiense, p. 90
Pink Rhododendron

Plate 7

Rhododendron flammeum, p. 90
Flame Azalea

Hydrangea quercifolia, p. 93
Oak-Leaf Hydrangea

Prunus serotina, p. 96
Wild Black Cherry

Plate 8

Erythrina herbacea, p. 106
Coral Bean

Lupinus villosus, p. 110
Lady Lupine

Sesbania drummondii, p. 113
Drummond's Rattlebox

Plate 9

Trifolium incarnatum, p. 117
Crimson Clover

Dirca palustris, p. 122
Leatherwood

Phoradendron serotinum, p. 123
American Mistletoe

Plate 10

Euonymus americanus, p. 124
Strawberry Bush

Croton alabamensis, p. 236
Alabama croton

Rhamnus caroliniana, p. 137
Carolina Buckthorn

Plate 11

Parthenocissus quinquefolia, p. 138
Virginia Creeper

Aesculus octandra, p. 141
Yellow Buckeye

Acer rubrum, p. 241
Red Maple

Plate 12

Toxicodendron vernix, p. 244
Poison Sumac

Ailanthus altissima, p. 245
Tree-of-Heaven

Poncirus trifoliata, p. 143
Mock Orange

Plate 13

Melia azedarach, p. 144
Chinaberry

Melia azedarach, p. 144
Chinaberry

Oxalis, p. 145
Wood Sorrel

Plate 14

Impatiens capensis, p. 146
Spotted Jewelweed

Aralia spinosa, p. 246
Devil's-Walking-Stick

Hedera helix, p. 247
English Ivy

Plate 15

Conium maculatum, p. 149
Poison Hemlock

Gelsemium sempervirens, p. 151
Yellow Jessamine

Nerium oleander, p. 153
Common Oleander

Plate 16

Asclepias incarnata, p. 158
Swamp Milkweed

Asclepias tuberosa, p. 162
Butterfly Weed

Solanum carolinense, p. 169
Carolina Horse Nettle

Plate 17

Cuscuta campestris, p. 172
Love Vine

Lantana camara, p. 173
Yellow Sage

Chionanthus virginicus, p. 176
Grandsir-Graybeards

Plate 18

Campsis radicans, p. 248
Trumpet Creeper

Lobelia cardinalis, p. 179
Cardinal Flower

Cephalanthus occidentalis, p. 184
Common Buttonbush

Plate 19

Lonicera sempervirens, p. 188
Coral Honeysuckle

Eupatorium perfoliatum, p. 193
Boneset

Senecio glabellus, p. 198
Butterweed

Plate 20

Sorghum halepense, p. 200
Johnson Grass

Sorghum halepense, p. 200
Johnson Grass

Arisaema triphyllum, p. 202
Swamp Jack-in-the-Pulpit

Plate 21

Colocasia esculenta, p. 203
Elephant's Ear

Crinum americanum, p. 205
Southern Swamp Lily

Narcissus pseudo-narcissus, p. 208
Common Daffodil

Plate 22

Trillium erectum, p. 211
Wake-Robin

Trillium reliquum, p. 208

Iris cristata, p. 219
Crested Dwarf Iris

Lachnanthes caroliniana, p. 220
Carolina Redroot

Plate 23

Automeris io, p. 269
Io Moth Caterpillar

Automeris io, p. 269
Io Moth

Agkistrodon contortrix, p. 304
Copperhead

Agkistrodon piscivorus, p. 305
Cottonmouth Moccasin

Plate 24

Crotalus horridus, p. 309
Canebrake (Timber) Rattlesnake

Crotalus adamanteus, p. 307
Eastern Diamondback Rattlesnake

Sistrurus miliarius, p. 310
Pygmy Rattlesnake

Micrurus fulvius, p. 313
Eastern Coral Snake

are arranged in terminal clusters, primarily in heads with bracts that are of 2 lengths. The length of the outer whorl of bracts is very short; the inner whorl is longer. All bracts of each whorl are about the same length. The flowers are yellow, and the fruit is an achene.

Toxic Properties: Fireweed contains the pyrrolizidine alkaloid that is hieracifoline. All pyrrolizidine alkaloids are toxins that cause gastrointestinal upset, liver damage, often liver cancer and cirrhosis of the liver, and, many times, death. These alkaloids, which are in all parts of the plant, may also cause dermatitis.

Geographic Distribution: Florida to Texas, and north to Virginia. Occurs throughout Alabama.

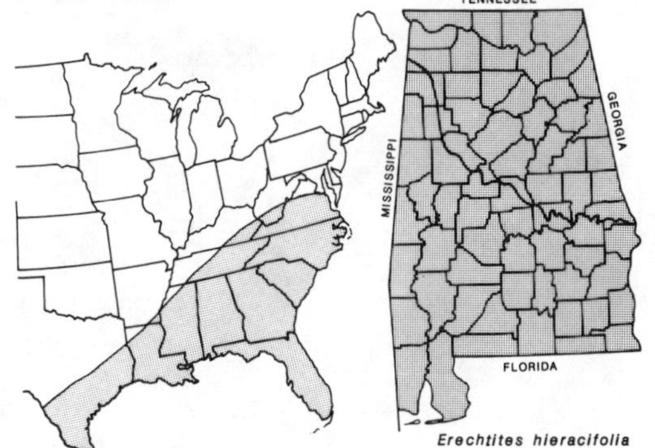

Range in eastern United States   Known range in Alabama

Habitat Occurrence: Fireweed can be found in disturbed weedy areas, usually where the soil is fairly muddy. It is most common along graded roadsides.

EUPATORIUM THOROUGHWORT

Eupatorium is a genus of erect or, rarely, sprawling herbs with alternate, opposite, or whorled simple leaves that are usually serrate to dentate along the margin. The leaves may or may not have a petiole. The flowers are clustered into compact heads at the apex or at the top of the plant. These heads are grouped into larger inflorescences. The flowers vary in color from white to purple. The fruit is an achene. Two species that have proved toxic occur in Alabama.

Eupatorium perfoliatum ***(Plate 19)*** BONESET

Species Recognition: Boneset is an erect herb about 1.5 meters tall with opposite, perfoliate, serrate-to-crenate leaves. The flowers are whitish or, rarely, purple-tinged.

Toxic Properties: Boneset has been used as a popular medicinal herb as a laxative, emetic, tonic, stimulant, and to break up colds and fevers. An additional use, as the common name implies, is to aid the setting of bones. Overdoses may cause abdominal pain, vomiting, fever, sleepiness, circulatory and cardiac disturbances, severe headaches, and body aches. It is not known what compounds are useful in the plant for medicinal purposes, but they occur in the stems and leaves. Boneset does concentrate nitrates in sufficient quantity to be toxic when large quantities of the plant are ingested.

Geographic Distribution: Quebec to southeastern Manitoba, south to Nova Scotia, Florida, and Texas. Can be expected throughout Alabama.

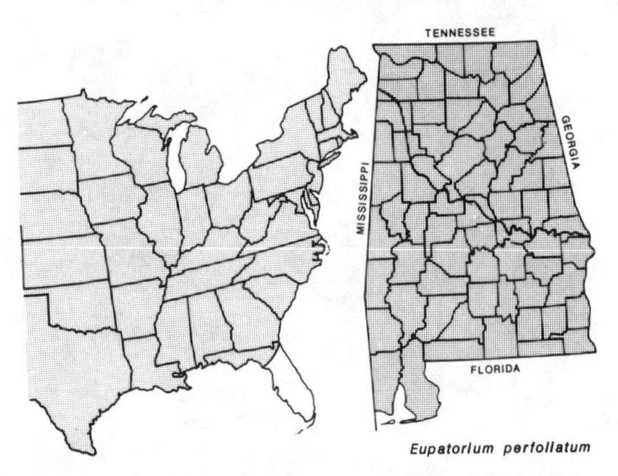

Range in eastern United States   Known range in Alabama

Habitat Occurrence: Boneset occurs in low woods, thickets, on wet shores, prairies, and, occasionally, in open fields.

Eupatorium rugosum WHITE SNAKEROOT

Species Recognition: White snakeroot is an erect herb that grows to about 1.5 meters in height. It has opposite petiolate leaves that are coarsely and sharply toothed on each margin. The flowers are white.

Toxic Properties: All parts of the plant contain a complex alcohol called tremetol. In addition, there are many glycosides. Since tremetol is rarely excreted from the body, toxicity or poisoning of livestock can result from daily ingestion of small amounts of the plant. The time between appearance of symptoms and death is highly variable after a lethal dose is ingested. Two conditions exist. One is called trembles, which is poisoning from ingestion of the plant. The second is milk sickness, which is a poisoning from ingestion of milk or milk products from cows that have eaten white snakeroot. Milk sickness has caused the death of a large number of humans and nursing livestock in the Southeast and Midwest. One famous victim was Nancy Hanks, the mother of Abraham Lincoln. Symptoms in humans are weakness, nausea, severe vomiting, abdominal pain, constipation, trembling, and coma. Recoveries are rare, slow, and incomplete. Modern methods of milk preparation make such poisoning highly unlikely.

Geographic Distribution: Gaspé Peninsula of Quebec to southeastern Saskatchewan, south to Georgia, Florida, Louisiana, and northeastern Texas. In Alabama uncommon, but can be expected throughout the northern two-thirds of the state.

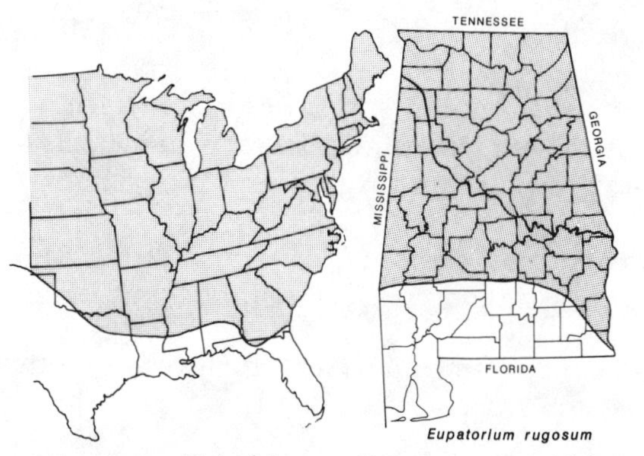

Range in eastern United States Known range in Alabama

Habitat Occurrence: White snakeroot occurs in rich woods, thickets, clearings, and chiefly in basic soils.

HELENIUM SNEEZEWEED

Helenium is a genus of perennial or annual weedy herbs with alternate, linear leaves and yellow-orange radiate heads. They have involucres of bracts in a single series, a pappus of papery scales, and fruits that are achenes. Six species occur in Alabama. Two of these, *H. amarum* and *H. autumnale,* are known to be toxic, as are additional species from the western United States. All species in Alabama should be considered toxic until proven otherwise.

Toxic Properties: All parts of the plant, especially the leaves, flowers, and stems, contain a phenol and a toxic glycoside, dugaldin. In addition, the plants contain a substance that imparts a bitter taste. A strong bitter taste is present in raw milk from cows that have eaten sneezeweeds known as bitterweed.

Helenium amarum BITTERWEED; FIVE-LEAF SNEEZEWEED

Species Recognition: Bitterweed is an erect herb with linear leaves that are less than 1 cm apart and less than 0.5 cm wide. The leaves are not decurrent on the stems.

Geographic Distribution: Virginia to Kentucky, southern Illinois, Missouri, and Kansas, and south to Florida, Texas, and Mexico. Can be expected throughout Alabama.

Habitat Occurrence: Bitterweed occurs in open areas, especially along roads and in overgrazed pastures.

Helenium autumnale COMMON SNEEZEWEED

Species Recognition: Common sneezeweed is an erect weed with leaves more than 0.5 cm wide and more than 1 cm apart on the stem. The leaves are decurrent on the stem, forming a narrow green wing.

Geographic Distribution: Western New England to Minnesota, and south to Florida and Arkansas. Occurs throughout Alabama.

Habitat Occurrence: Sneezeweed occurs in alluvial pastures, rich thickets, meadows, and along shores.

RUDBECKIA BLACK-EYED SUSAN

One species in Alabama is known to be toxic.

Rudbeckia hirta BLACK-EYED SUSAN

Species Recognition: Black-eyed susan is an erect herb with coarsely hairy leaves more than 1 cm wide. The inflorescence is a cluster of 5–7 yellow ray flowers surrounding a conical brown disk of numerous disk flowers.

Toxic Properties: Black-eyed susan is responsible for poisoning cattle and hogs. Symptoms are gastroenteritis, coma, and occasional aimless wandering. With some individuals, the species is known to produce dermatitis when touched. The toxic compound causing gastroenteritis is unknown, but oleoresin is the cause of the contact dermatitis.

Geographic Distribution: Widespread from western Massachusetts to Illinois, south to Georgia and Louisiana. Can be expected throughout Alabama, but especially common in the Black Belt counties.

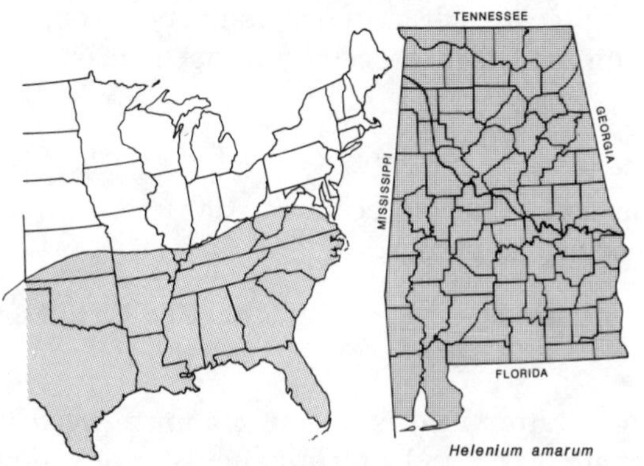

Range in eastern United States Known range in Alabama

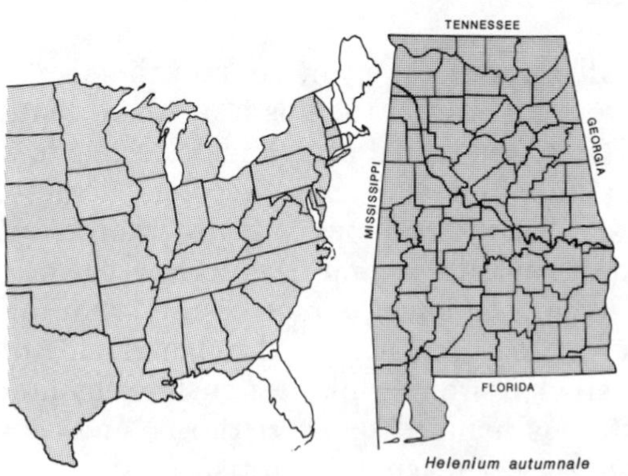

Range in eastern United States Known range in Alabama

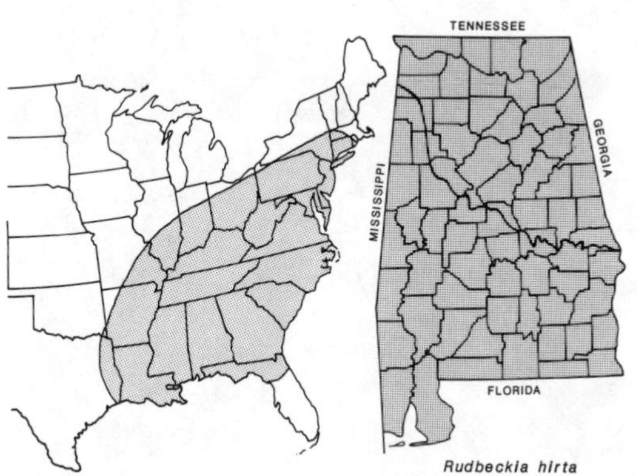

Range in eastern United States Known range in Alabama

Habitat Occurrence: Black-eyed susan occurs in open woods, thickets, open fields, and on roadsides, especially in open areas.

SENECIO GROUNDSEL

Apparently 6 species of *Senecio* occur in Alabama; however, only 1 has been reported to have toxic properties.

***Senecio glabellus* (Plate 19)** BUTTERWEED; GRASS-LEAF GROUNDSEL

Species Recognition: Butterweed is an erect annual weed with pinnately lobed alternate leaves and a terminal cluster of yellow flowers. The flowers are grouped into heads and the heads into a large panicle-like inflorescence. The heads have 2 series of subtending bracts, the inner series being about 3 times as long as the outer series. The fruit is an achene.

Toxic Properties: All parts of the plant contain the alkaloid senecionine. In acute cases of ingestion there is immediate central nervous system disturbance, abdominal pain, and death in a few hours. Smaller amounts of ingestion will cause accelerated pulse, perspiration, nervous excitement, weakness, gastroenteritis with abdominal pain, and death within several days to a week. Chronic poisoning is the most common effect, which results from ingestion of small amounts for a long time such as through grazing or hay consumption by livestock or from drinking a bush tea by humans. The most difficult effect is hemorrhaging, cirrhosis, and cancer of the liver. The sap can cause a severe dermatitis.

Geographic Distribution: North Carolina to southern Ohio, central Indiana and southern Illinois, Missouri, and Oklahoma, south to Florida and Texas. Can be expected throughout Alabama.

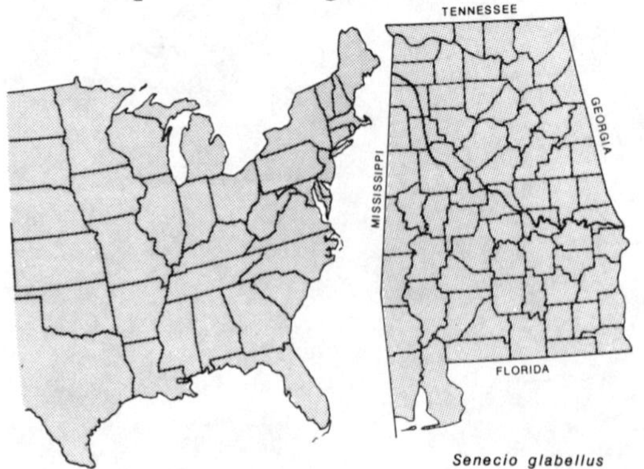

Range in eastern United States   Known range in Alabama

Habitat Occurrence: Butterweed occurs in moist and damp places such as wet woods, swamps, and ditches, occasionally standing in water about 15 cm deep. It is sometimes found in wet pastures and along the edges of rivers and lakes.

Family JUNCAGINACEAE

TRIGLOCHIN ARROWGRASS

Triglochin is a genus of herbaceous, perennial grasslike plants that form clumps of leaves from short basal stems. The leaves are long, linear, thick, and somewhat fleshy and sheathing at the base. The inflorescence is a scape with a terminal raceme. The numerous flowers are small with 6 perianth parts. The fruit is composed of 3–6 capsules, each the product of a single carpel that splits away at maturity.

Triglochin striata (Ridged Arrowgrass)

Toxic Properties: Arrowgrass contains hydrogen cyanide in all parts of th plant. Studies indicate that the hydrogen cyanide is not present i the form of a cyanogenic glycoside. The leaves especially hav caused cyanide poisoning in livestock, especially cattle and sheep Nausea and vomiting occur as initial symptoms. Small amounts ca cause weak and irregular respiration, gasping, excitement then de pression, weakness, staggering, pupil dilation with glassy, bulgin eyes, spasms, convulsions, and coma. Death, caused by respirator failure, has been known to occur.

Triglochin striata RIDGED ARROWGRAS

This is the only species in Alabama.

Geographic Distribution: New England along the Atlantic Coast to the Gu Coast, and south into tropical America. In Alabama known onl from Baldwin and Mobile counties.

Range in eastern United States Known range in Alabama

Habitat Occurrence: Ridged arrowgrass grows in salt marshes.

Family POACEAE

SORGHUM JOHNSON GRAS

One species occurs in Alabama.

Sorghum halepense (Plate 20) JOHNSON GRAS

Species Recognition: Johnson grass is an erect perennial grass with a termi nal inflorescence having leaves about 0.5–1 meter long and 2–3 c wide; often the leaves are spotted with red dots. The inflorescenc is quite often a purplish color.

Toxic Properties: All parts of the plant contain a cyanogenic glycoside called dhurrin, which degrades to form hydrocyanic acid. Although johnson grass is a common weed in fields cultivated for hay, it ordinarily is not harvested in quantities sufficient to harm livestock.

Geographic Distribution: Introduced from Europe and naturalized from New England to Ohio, Indiana, Illinois, and Iowa, south to Florida and Texas. Can be expected throughout Alabama.

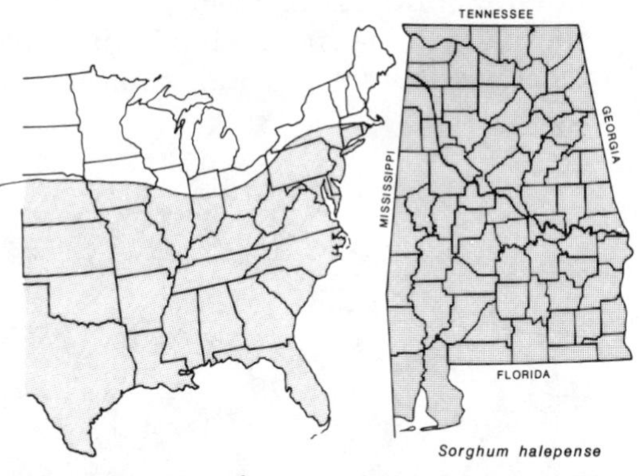

Range in eastern United States Known range in Alabama

Habitat Occurrence: Johnson grass persists in old fields, waste places, and along roadsides.

Family ARACEAE

ARISAEMA ARISAEMA

Arisaema is a genus of perennial herbs with a near spherical corm or rhizome. The stem may be up to 1 meter tall and has 1 palmately compound leaf. The minute flowers are embedded in a fleshy axis surrounded and overgrown by a green leaf called a spathe. Each flower produces a red or green fleshy fruit, and these fruits are all clustered on the fleshy axis in which the flowers were embedded. These fruits together make up a multiple fruit.

Toxic Properties: The active ingredient is a calcium oxalate crystal found in all parts of the plant, but especially in the corm. Calcium oxalate crystals are needle-like in shape and, when taken into the mouth, become embedded in the mucous membranes, provoking intense irritation and a burning sensation. Ingestion of more than one mouthful rarely occurs because of the intense pain. The juice of the plant may cause eye irritation and dermatitis to sensitive skin. No human or livestock deaths from ingestion of *Arisaema* have been reported.

Arisaema dracontium GREEN DRAGON

Species Recognition: Green dragon grows to a height of about 1 meter. The leaves have 5–15 leaflets. The spadix (the fleshy axis in which the flowers are embedded) extends greatly past the spathe. The fruiting head is conical in shape. The individual fruits are yellow-green and number 25–50 on a fruiting head.

Geographic Distribution: Southwestern Quebec to southern Ontario and Wisconsin, south to Florida and Texas. Can be expected throughout Alabama, except possibly in the southernmost counties.

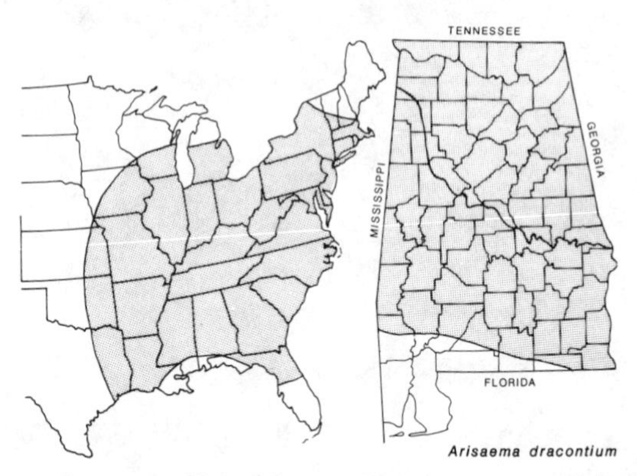

Range in eastern United States Known range in Alabama

Habitat Occurrence: Green dragon is most common in the alluvial soils of rich deciduous woods but may occur in swells and in thickets of deciduous woods.

***Arisaema triphyllum* (Plate 20)** SWAMP JACK-IN-THE-PULPIT

Species Recognition: Swamp jack-in-the-pulpit plants grow up to 0.5 meter with 1 or, rarely, 2 leaves, each having 3–5 leaflets. The spathe overtops the spadix. The fruiting head is spherical in shape; the individual fruits are reddish in color.

Geographic Distribution: Southeastern Massachusetts to Kentucky, and south to Florida and Texas. Can be expected throughout Alabama.

Habitat Occurrence: Swamp jack-in-the-pulpit occurs in rich deciduous woods.

COLOCASIA ELEPHANT'S EAR

Only one species occurs in Alabama.

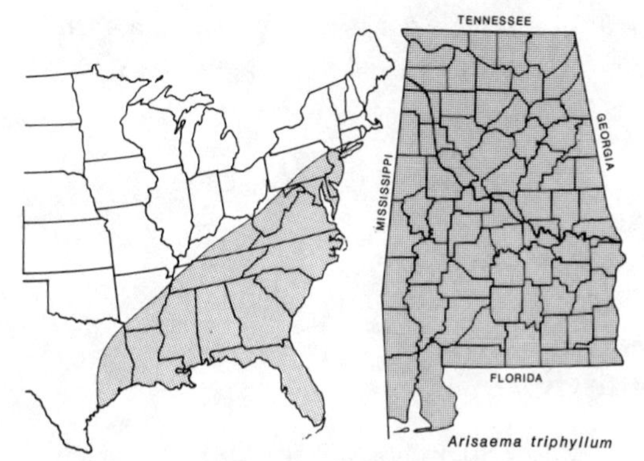

Range in eastern United States Known range in Alabama

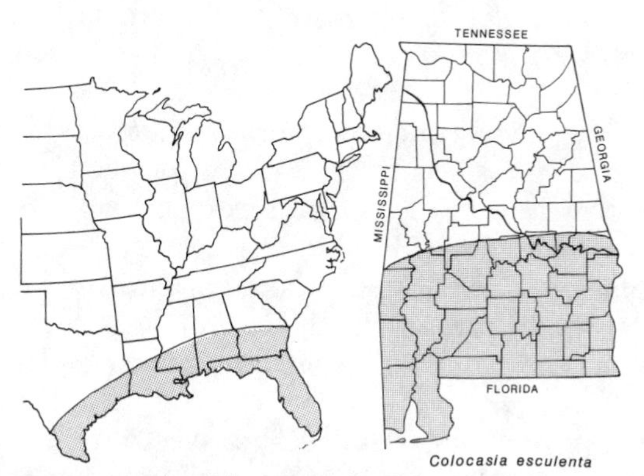

Range in eastern United States Known range in Alabama

***Colocasia esculenta* (Plate 21)** ELEPHANT'S EAR

Species Recognition: Elephant's ear is a large plant with sagittate leaves on long peduncles from a rhizome. A large yellow spathe surrounds a large greenish spadix, the structure on which the flowers are produced.

Toxic Properties: All parts of an elephant's ear plant contain calcium oxalate crystals, gastroenteric irritants that cause a burning sensation and inflammation, especially to the mouth and throat. The juice may cause dermatitis to sensitive skin. The rhizome, when specially prepared, is eaten in the tropics.

Geographic Distribution: Native to southeastern Asia but cultivated in the United States as an ornamental. Has escaped cultivation from Florida to Georgia, and west to Texas. Can be expected in central and southern Alabama.

Habitat Occurrence: Elephant's ear is an aquatic plant that grows in shallow waters of rivers and streams, or occasionally in wet ditches.

Family LILIACEAE

AMIANTHIUM FLY POISON

Only 1 species occurs in Alabama.

Amianthium muscaetoxicum FLY POISON

Species Recognition: Fly poison is a perennial herb growing from a bulb 5–8 cm below the surface of the soil. The leaves, clumped at the ground, are linear, 1.2–3 cm wide, and up to 50 cm long. The flowering stem has only reduced leaves on it and terminates by a compact raceme of white flowers. The fruit is 3-parted and recurved at the apex.

Toxic Properties: Fly poison contains 1 or more alkaloids that have not been characterized but are similar in many respects to those found in *Zigadenus*. Ingestion of half of 1 percent or more of an animal's weight can cause death. Symptoms include salivation, nausea, rapid or irregular respiration, and weakness. Death, if it does follow, is by respiratory failure. In North Carolina this species is considered to be among the 10 most dangerous poisonous plants of that state.

Geographic Distribution: Long Island west to West Virginia, south along the Coastal Plain and Piedmont to Florida, and west to Oklahoma and Missouri. Can be expected throughout Alabama, but is most common south of Tuscaloosa County in the Coastal Plain.

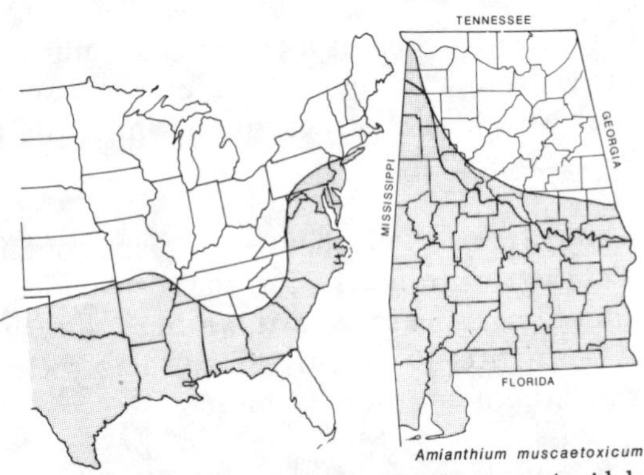

Range in eastern United States Known range in Alabama

Habitat Occurrence: Fly poison grows in sandy open woods, fields, bogs, and occasionally in swampy woods.

CRINUM SWAMP SPIDER LILY

One species occurs in Alabama.

***Crinum americanum* (Plate 21)** SOUTHERN SWAMP LILY

Species Recognition: Southern swamp lily is an herb with leathery, linear leaves arising from a subterranean bulb. The tips of the leaves are boat-shaped; the 3–4 flowers are grouped on top of a long peduncle arising from the bulb. They have 6 perianth parts that are long, linear, and white. There is no crown on the flowers. The fruit is a fleshy capsule with 3 parts and many seeds.

Toxic Properties: The bulb contains the alkaloids lycorine, crinidine, and crinamine. Raw bulbs have caused vomiting and diarrhea. Toxicities of this and other lycorine-containing plants result in gastroenteritis, vomiting, shivering, and, sometimes, diarrhea. Death occasionally occurs.

Geographic Distribution: Southern Georgia and Florida, west to Texas. Can be expected in the Lower Coastal Plain of Alabama.

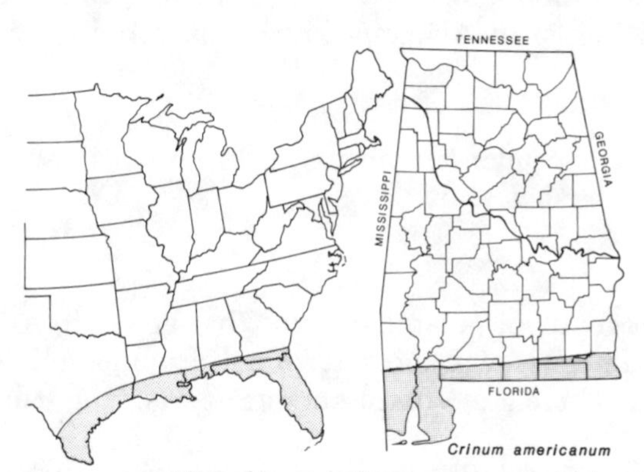

Range in eastern United States Known range in Alabama

Habitat Occurrence: Southern swamp lily is found in swamps along the major rivers of the Lower Coastal Plain.

HYMENOCALLIS SPIDER LILY

These bulbous perennials have erect, linear, leathery leaves coming from the bulb. The inflorescence is 1 scapose stalk with 5 or fewer flowers at the apex. The stamens are united into a crown. The fruit is a green fleshy capsule with 1–3 seeds.

Toxic Properties: The toxic compounds in *Hymenocallis* are various alkaloids including lycorine. Lycorine is toxic to humans, causing gastroenteritis, vomiting, shivering, and sometimes diarrhea. These compounds occur in all parts of the plant, especially the leaves. There are no records of human deaths.

Hymenocallis coronaria CAHABA LILY

Species Recognition: The cahaba lily can be identified by its 3 or fewer flowers in the inflorescence.

Geographic Distribution: Georgia and Alabama. Can be expected in the Black Warrior and Cahaba river basins of Alabama.

Habitat Occurrence: The cahaba lily is found only in the shoals and alluvial habitats of rivers.

Hymenocallis occidentalis SPIDER LILY

Species Recognition: The spider lily can be recognized by its 3–5 flowers in the inflorescence.

Geographic Distribution: North Carolina, Missouri, south to Georgia, Florida, and Texas. In Alabama known only from the Outer Coastal Plain.

Habitat Occurrence: Spider lily occurs in low woods, swamp forests, and brackish marshes.

NARCISSUS NARCISSUS

Narcissus is a genus of perennial herbs that grows from a subterranean bulb with erect linear leaves and 1 terminal druping yellow flower with a crown or corona. There are 6 fused perianth parts. The fruit is a capsule.

Toxic Properties: The bulbs of all species contain the alkaloids galanthamine, haemanthamine, homolycorine, and narcissine. Cattle fed the bulbs have died, especially in Europe. Symptoms include severe gastroenteritis, vomiting, trembling, and convulsion.

Narcissus incomparabilis NARCISSUS

Species Recognition: The narcissus has a solitary yellow flower with a corona that is about half as long as the perianth lobes.

Geographic Distribution: Cultivated throughout the eastern United States and Alabama.

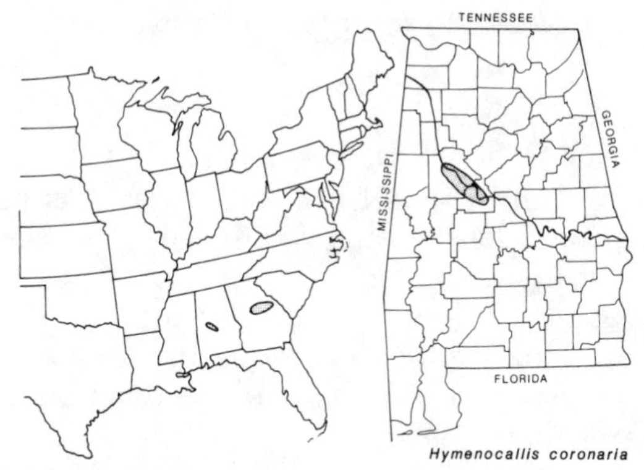

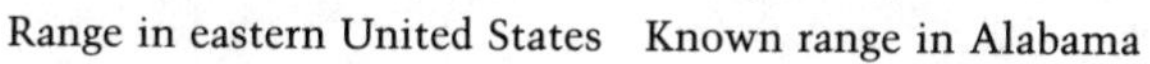
Range in eastern United States Known range in Alabama

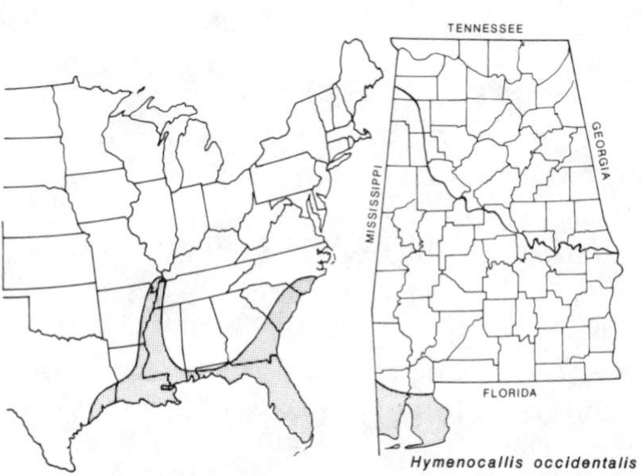

Range in eastern United States Known range in Alabama

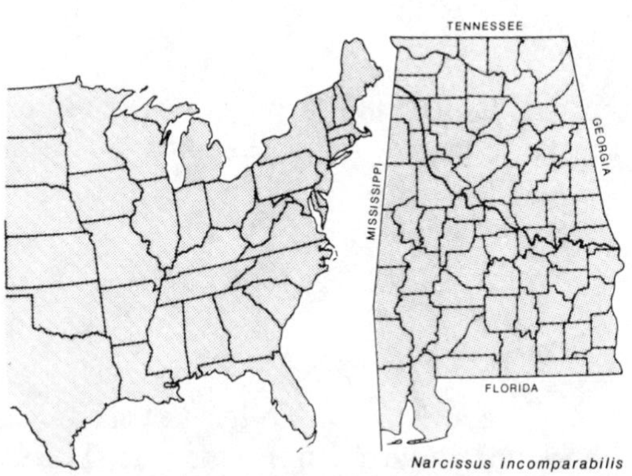

Range in eastern United States Known range in Alabama

Habitat Occurrence: *Narcissus incomparabilis* may persist after cultivation.

Narcissus jonquilla JONQUIL

Species Recognition: The jonquil has a corona less than half as long as the yellow perianth lobes, usually with 2 or more flowers. The leaves are cylindrical rather than flat.

Geographic Distribution: Cultivated throughout the eastern United States and can be expected to persist after cultivation almost anywhere. Found throughout Alabama.

Habitat Occurrence: Jonquils persist after cultivation and may be found in disturbed areas, roadsides, fields, and old homesites.

Narcissus poeticus POET'S NARCISSUS

Species Recognition: Poet's narcissus has a corona less than half as long as the perianth lobes with usually 2 or more flowers. The flowers are white and the leaves flat.

Geographic Distribution: Cultivated throughout the eastern United States. Can be expected almost anywhere in Alabama, after cultivation.

Habitat Occurrence: Poet's narcissus persists after cultivation, on roadsides, in fields, and in waste places.

Narcissus pseudo-narcissus (Plate 21) COMMON DAFFODIL

Species Recognition: The common daffodil has a corona as long as or longer than the perianth lobes. The leaves are flat and the perianth is yellow.

Geographic Distribution: Cultivated throughout the eastern United States. Can be expected to persist in any locality in Alabama.

Habitat Occurrence: Same as jonquil.

TRILLIUM ***(PLATE 22)*** TRILLIUM

Trillium is a genus of erect herbs with 3 leaves that subtend 1 flower. The stem arises from a horizontal underground rhizome. The erect or pendulant flower has 3 sepals, 3 petals, 6 stamens, and 3 carpels. The fruit is a capsule.

Toxic Properties: The rhizome and roots contain some unknown compound, possibly the alkaloid trilline, that causes them to be very acrid and when ingested causes vomiting and severe irritation. Small amounts of the root have been used medicinally.

Alabama species: Sixteen species of *Trillium* are found in Alabama.

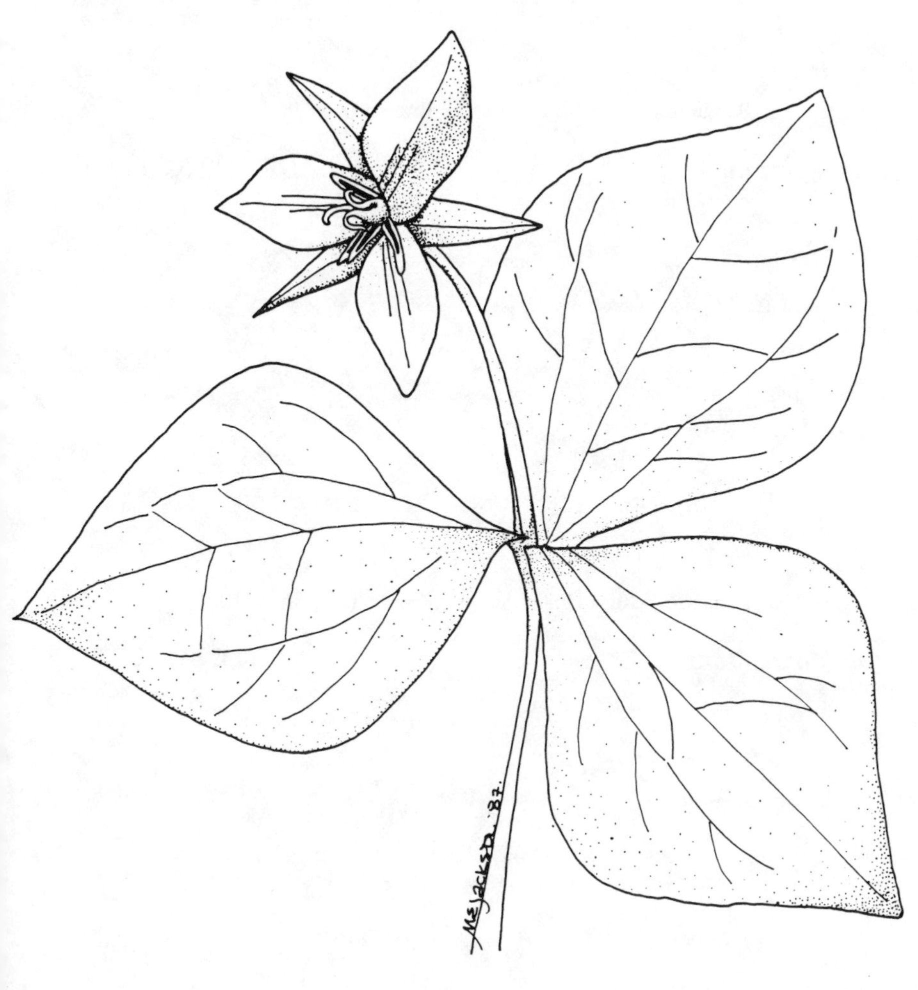

Trillium

Trillium catesbaei CATESBEY'S TRILLIUM

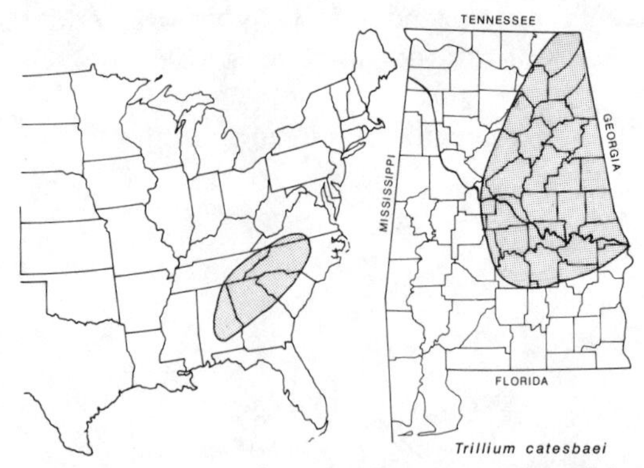

Range in eastern United States   Known range in Alabama

Trillium cernuum NODDING TRILLIUM

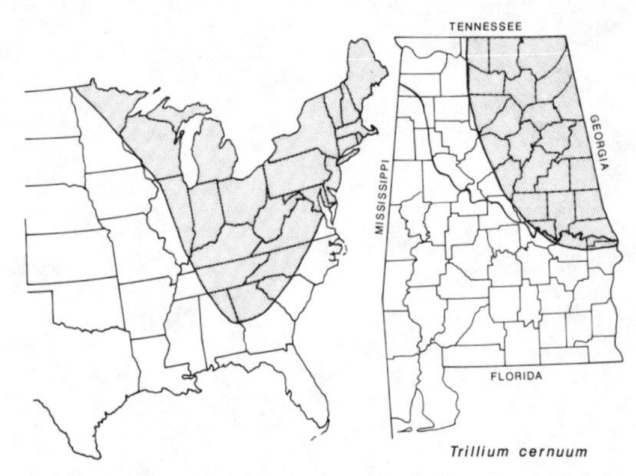

Range in eastern United States   Known range in Alabama

Trillium cuneatum LITTLE SWEET BETSY

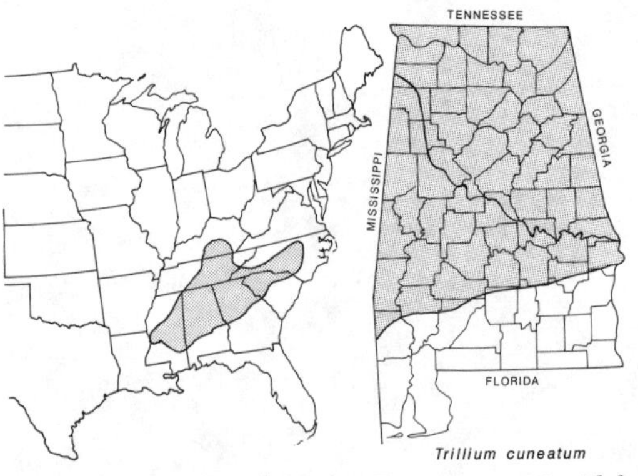

Range in eastern United States   Known range in Alabama

Trillium decipiens TRILLIUM

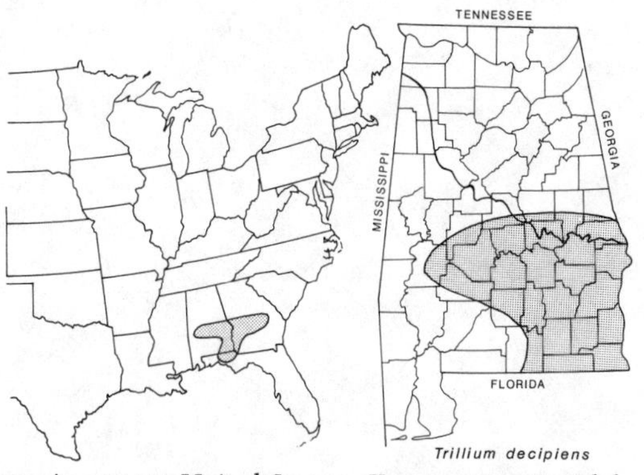

Range in eastern United States   Known range in Alabama

Trillium decumbens DECUMBENT TRILLIUM

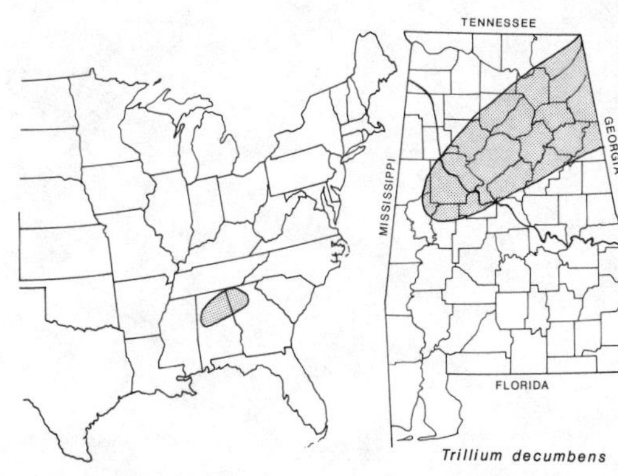

Range in eastern United States   Known range in Alabama

Trillium erectum (Plate 22) WAKE-ROBIN; RED TRILLIUM

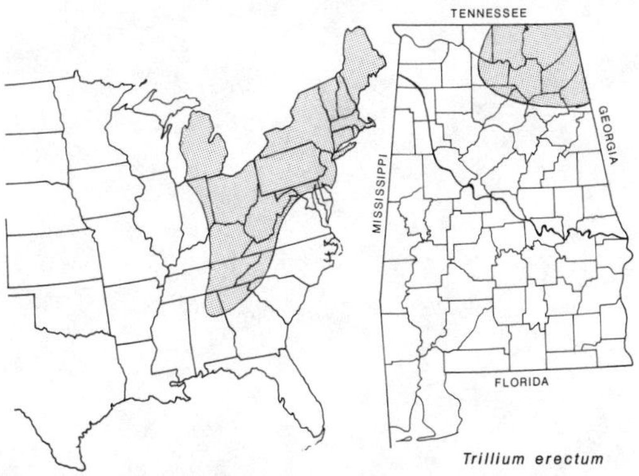

Range in eastern United States   Known range in Alabama

Trillium flexipes BENT-FOOT TRILLIUM; NODDING TRILLIUM

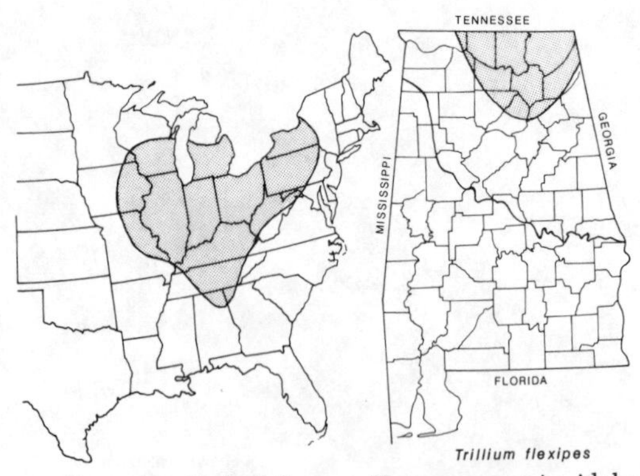

Range in eastern United States Known range in Alabama

Trillium lancifolium LANCE-LEAF TRILLIUM

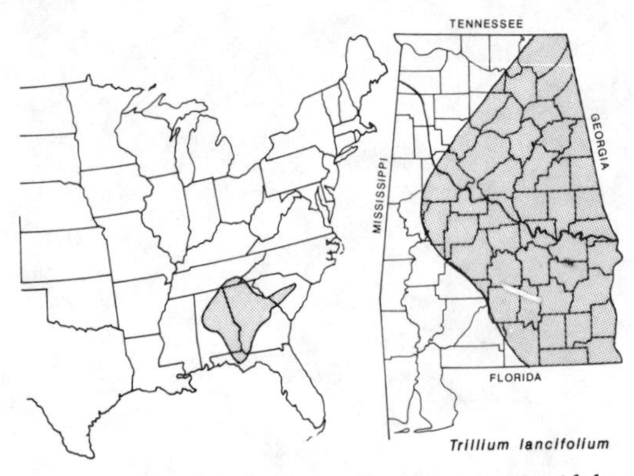

Range in eastern United States Known range in Alabama

Trillium maculatum MOTTLED TRILLIUM

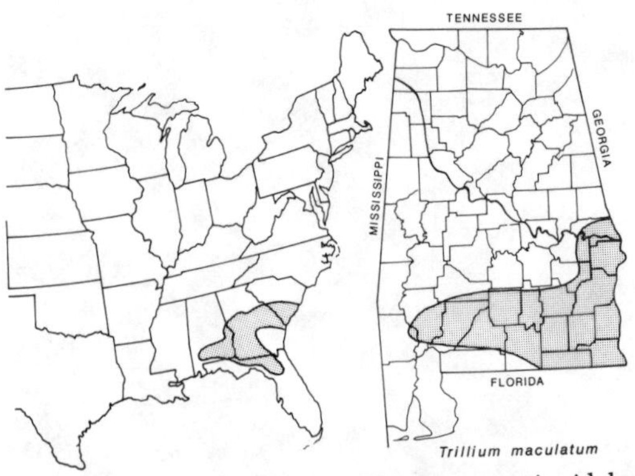

Range in eastern United States Known range in Alabama

Trillium pusillum CAROLINA TRILLIUM

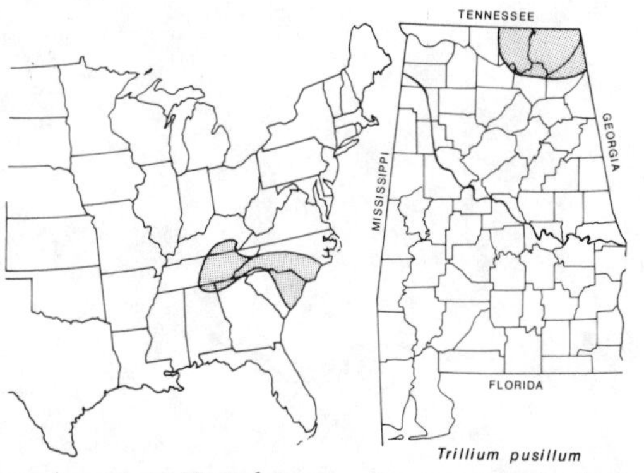

Range in eastern United States Known range in Alabama

Trillium recurvatum PRAIRIE TRILLIUM

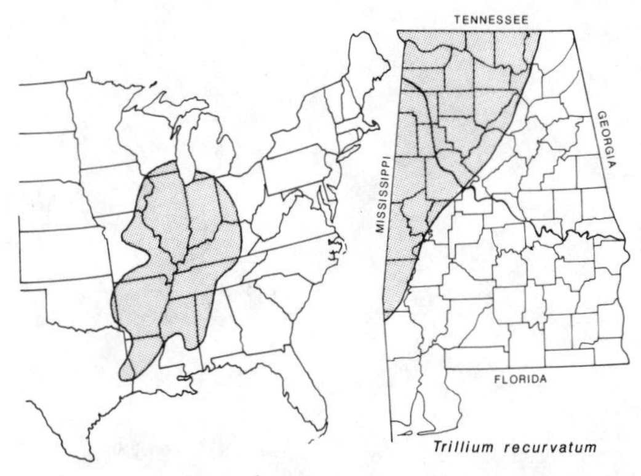

Range in eastern United States Known range in Alabama

Trillium rugelii TRILLIUM

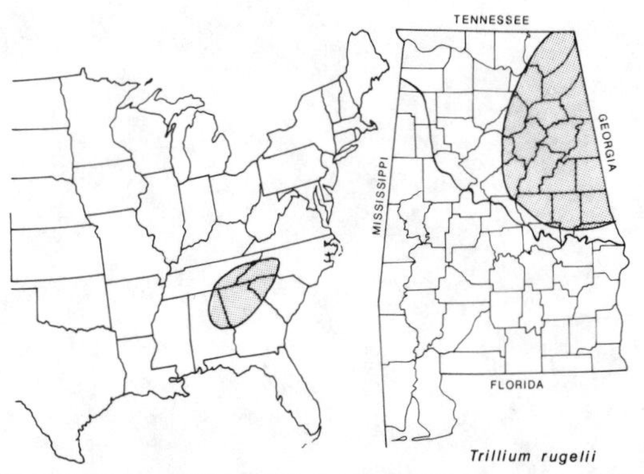

Range in eastern United States Known range in Alabama

Trillium sessile SESSILE TRILLIUM

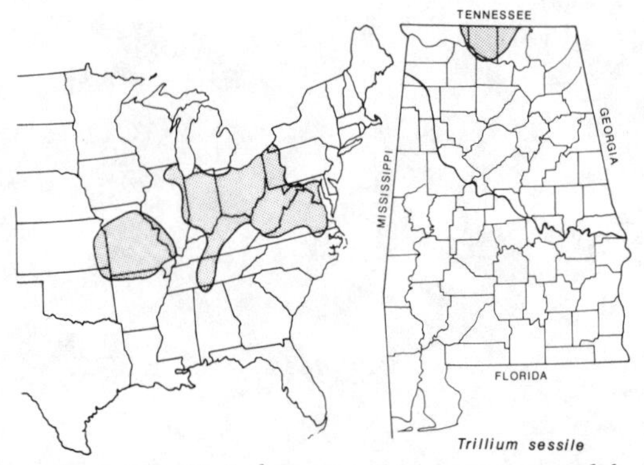

Range in eastern United States Known range in Alabama

Trillium stamineum BLUE-RIDGE TRILLIUM

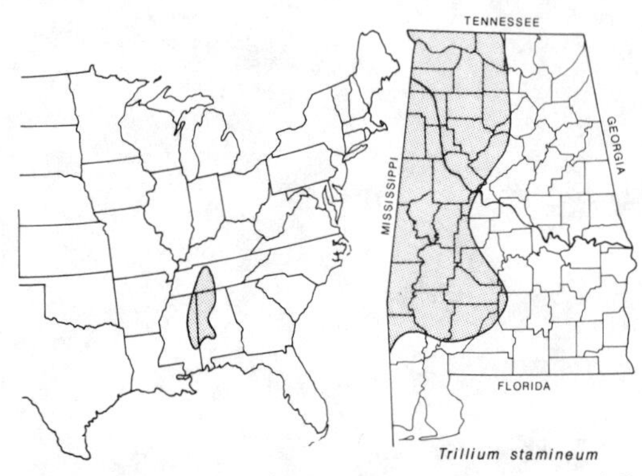

Range in eastern United States Known range in Alabama

Trillium sulcatum TRILLIUM

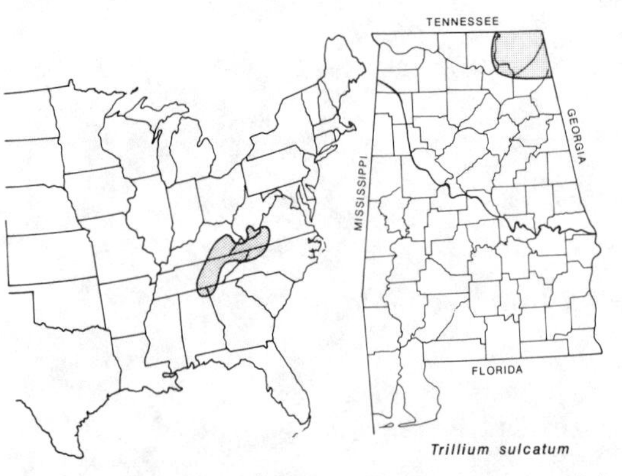

Range in eastern United States Known range in Alabama

Trillium underwoodii MOTTLED TRILLIUM

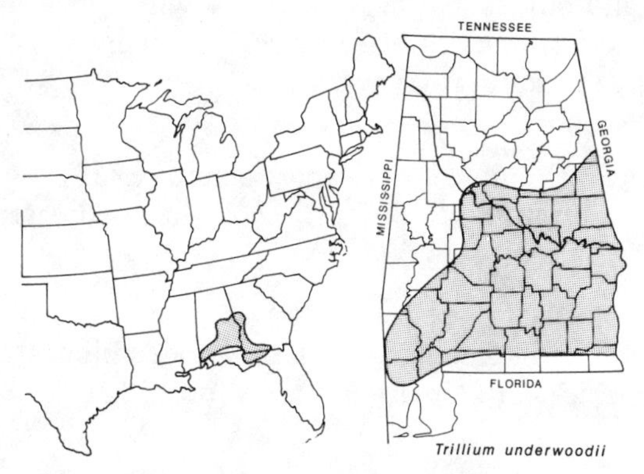

Range in eastern United States   Known range in Alabama

VERATRUM HELLEBORE

One species occurs in Alabama.

Veratrum parviflorum HELLEBORE

Species Recognition: Hellebore is an herb with large basal and stem leaves that are oval to lanceolate or elliptic in outline. They are up to 10 cm long and 5–6 cm wide. The stem grows from an erect rhizome. The inflorescence is a panicle of green-to-yellow or, rarely, purple flowers.

Toxic Properties: This species is not known to be toxic, but a related species, *Veratrum viride*, contains the alkaloid veratramine, as well as other alkaloids, and it should be expected in *V. parviflorum*. Toxicity is suspected of causing rapid and weak heartbeat, labored respira-

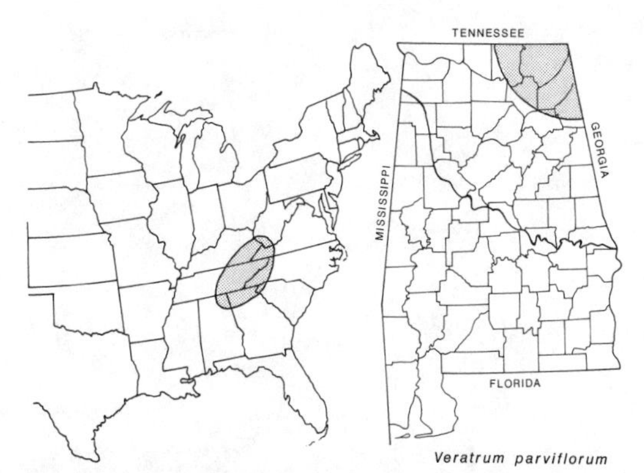

Range in eastern United States   Known range in Alabama

tion, muscular weakness, loss of appetite, and stupor to cattle, sheep, and horses. No deaths are known. It causes a burning in the mouth and throat, salivation, vomiting, diarrhea, abdominal pain, sweating, and muscular weakness.

Geographic Distribution: Georgia to Alabama and Tennessee, West Virginia, and Virginia. Known only from the northeastern corner of Alabama.

Habitat Occurrence: Hellebore occurs in wooded slopes, along streams, in the Appalachian Plateau and the Blue Ridge.

ZEPHYRANTHES — ZEPHYR LILY

One species occurs in Alabama.

Zephyranthes atamasco — ATAMASCO LILY

Species Recognition: Atamasco lily is an erect herb that grows from a subterranean bulb with erect flattened leaves and 1 terminal tubular flower without a crown. The ovary is inferior. The stalk on which the flower is produced has no leaves. All the leaves arise at the top of the bulb. The flower is white and the fruit is a capsule.

Toxic Properties: All parts of the plant, especially the bulb, contain an acrid component, possibly lycorine, and other alkaloids. Small amounts may be harmless, but large amounts affect horses, cattle, and chickens. Symptoms are soft feces with bloody mucous, staggering within 48 hours, and collapse, followed by death. Forty grams of a bulb have been fatal to a chicken and 4 kg have been fatal to a 760-kg steer.

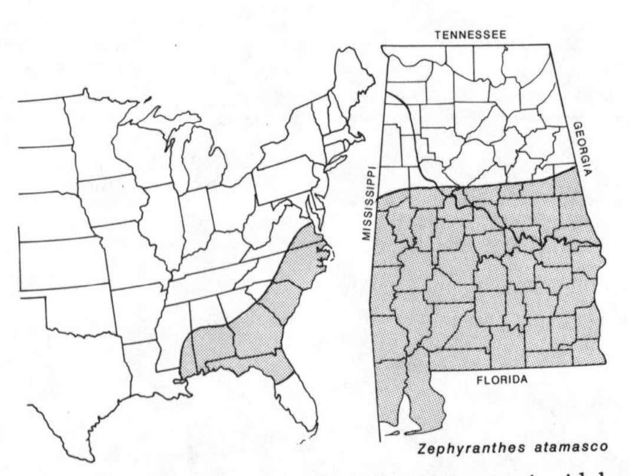

Range in eastern United States   Known range in Alabama

Geographic Distribution: Virginia to Mississippi, south to Florida. In Alabama can be expected from Tuscaloosa County south.

Habitat Occurrence: Atamasco lily occurs in rich woods and damp clearings.

ZIGADENUS DEATH CAMAS

Zigadenus is a genus of erect, rhizomatous or bulbous perennials with leafy stems. The leaves are linear, alternate, and long acute. The flowers are produced in terminal racemes, each with a small bract at its base. The 6 cream-colored petals and sepals are similar in appearance, and each has 1 or 2 glands at its base. The fruit is a 3-lobed capsule on which the style persists. Three species occur in Alabama.

Toxic Properties: Assumptions have been that all species of *Zigadenus* are toxic, but experimental investigations have not supported this. Apparently, 2 of the 3 Alabama species are toxic. Toxic properties include many alkaloids, including zygadenine, zygacine, and other compounds. These compounds are found in all parts of the plant, especially the rhizome or bulb. Symptoms include burning in the mouth and throat, salivation, vomiting, diarrhea, abdominal pain, muscular weakness, slow and weak pulse, low blood pressure, low temperature, and coma. Death has been known to occur. Recovery is likely if vomiting is early enough to empty the stomach before the alkaloids are absorbed.

Zigadenus densus BLACK SNAKEROOT; BLACK DEATH CAMAS

Species Recognition: Black snakeroot is a bulbous species with all flowers perfect and with 1 gland at the base of each sepal and petal.

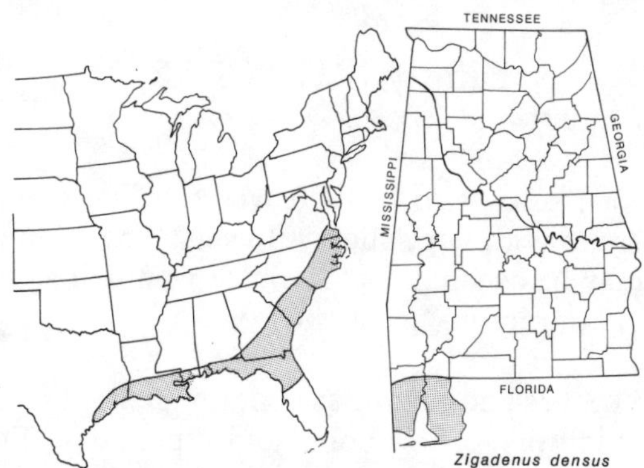

Range in eastern United States Known range in Alabama

Geographic Distribution: Southeastern Virginia to Florida and west to Texas. In Alabama occurs in the Outer Coastal Plain of Mobile and Baldwin counties.

Habitat Occurrence: Black snakeroot occurs in pine savannas, bogs, and flatwoods, especially those with a sandy substrate.

Zigadenus glaberrimus SANDBOG CAMAS

Species Recognition: Sandbog camas is a rhizomatous species with all flowers perfect and 2 glands at the base of each sepal and petal.

Geographic Distribution: Southeastern Virginia to northern Florida and west to southeastern Texas. In Alabama can be expected in the Lower Coastal Plain from Washington County south.

Habitat Occurrence: Sandbog camas occurs in bogs, shrub bogs, wet pine savannas, and flatwoods, especially those with sandy soil.

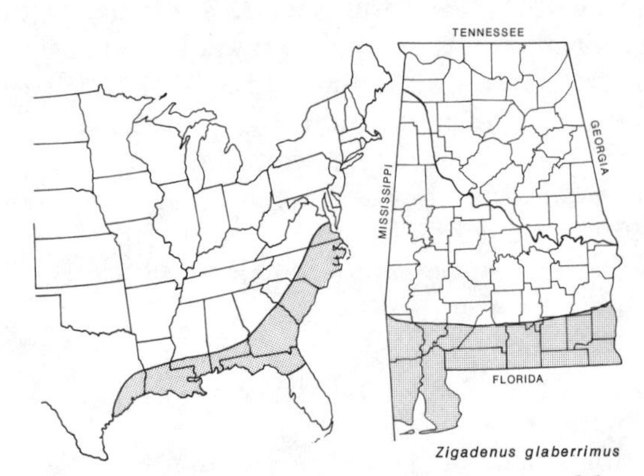

Range in eastern United States Known range in Alabama

Family IRIDACEAE

IRIS IRIS; FLAG

Iris is a genus of plants with basal leaves that are equitant. They have scapose inflorescences with large showy flowers. The diameter of the underground rhizome may be as small as 0.2 cm or as much as 0.6 cm. The fruit is a capsule bearing numerous seeds.

Toxic Properties: Leaves and rhizomes contain an acrid resinous substance, irisin, in addition to flavonoids and dipeptides. These compounds cause gastrointestinal irritation, abdominal pain, diarrhea, and vomiting. They may also affect the liver and pancreas. A contact dermatitis has developed among some florists who handle irises.

Iris brevicaulis LAMANCE IRIS

Species Recognition: The lamance iris has blue-to-purple flowers. The flowering stem is 20–40 cm, zigzagged, and shorter than the leaves.

Geographic Distribution: Ohio to eastern Kansas, south to Alabama and eastern Texas. Can be expected along major rivers of the central part of Alabama.

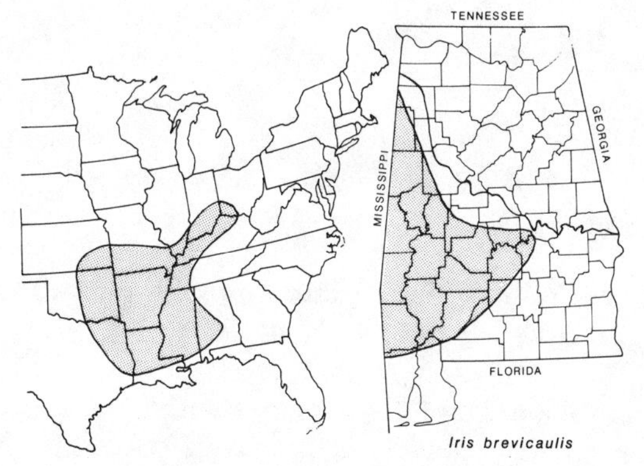

Range in eastern United States Known range in Alabama

Habitat Occurrence: The lamance iris occurs in swamps, wet woodlands, and marshy shores, especially of lakes and rivers.

***Iris cristata* (Plate 22)** CRESTED DWARF IRIS

Species Recognition: The crested dwarf iris has a flowering stem less than 10 cm tall. The rhizome is moniliform (narrow with occasional swollen portions).

Geographic Distribution: Washington, D.C., to Indiana and Missouri, south to North Carolina, Alabama, Mississippi, Arkansas, and Oklahoma. Occurs in the northern half of Alabama.

Habitat Occurrence: The crested dwarf iris occurs in deciduous woods, wooded bottoms, ravines, and along bluffs.

Iris pseudacorus YELLOW FLAG

Species Recognition: Yellow flag has yellow flowers on a stem 70–120 cm tall. The axis is straight and overtops the leaves.

Geographic Distribution: Native to Europe, Asia, and Africa, but naturalized from cultivation in the eastern United States as far north as Michigan. In Alabama known to be in Bibb County, but can be expected anywhere in the state where suitable habitat occurs.

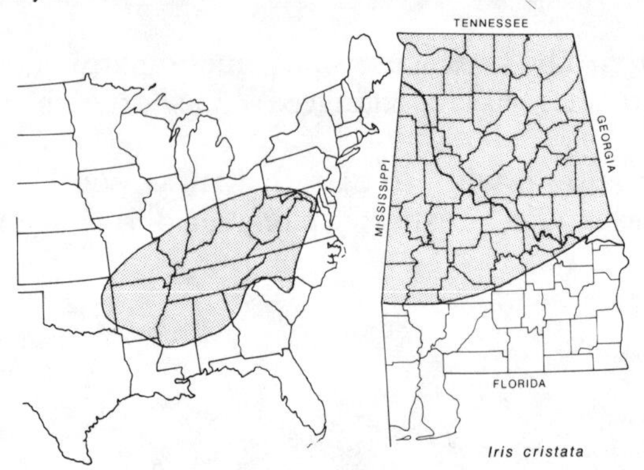

Range in eastern United States   Known range in Alabama

Habitat Occurrence: Yellow flag occurs mostly in swamp woodlands, marshes, meadows, and along small streams.

Additional Species in Alabama: Other species with appearances and properties similar to those mentioned above also occur in this region: *Iris hexagona* (angelpod blue flag); *I. verna* (dwarf iris); and *I. virginica* (southern blue flag).

Family HAEMADORACEAE

LACHNANTHES REDROOT

One species occurs in Alabama.

***Lachnanthes caroliniana* (*Plate 22*)** CAROLINA REDROOT

Species Recognition: Carolina redroot is an herb with basal leaves that are linear and equitant or nearly equitant. The flowering stem overtops small reduced leaves. The inflorescence is a cluster of many hairy flowers. The rhizome has a red latex.

Toxic Properties: The leaves, stem, flowers, and root contain some unknown toxin. Ingestion of the chemical in overdoses is reported to cause dizziness and headache. Old legends say that the root of this plant is fatal to white pigs, but not to black ones. The origin of this legend is not known, but most biologists would probably risk their reputations on denying it.

Geographic Distribution: North Carolina to Florida, west to Louisiana and central Tennessee; also in Cuba. In Alabama occurs from Houston County to Mobile County.

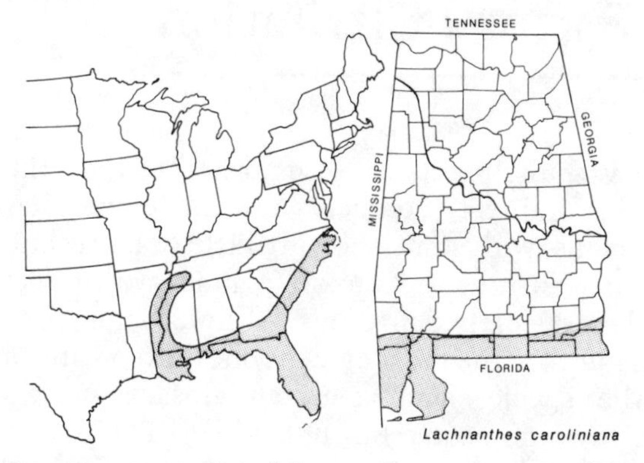

Range in eastern United States   Known range in Alabama

Habitat Occurrence: Carolina redroot occurs in wet ditches, bogs, savannas, swamps, and pine flatwoods.

Chapter 5

Vascular Plants Causing Dermatitis or Other Forms of Irritation

Some species of vascular plants cause irritation to the skin merely by direct, or sometimes indirect, contact with the plant. Sensitivity varies greatly among persons, with some showing little or no reaction even to poison ivy. But for one sensitive to its toxin, contact with the eyes or mucous membranes can be extremely dangerous. Although the species that are described in this chapter are the most evident ones responsible for contact dermatitis, many other species, such as certain grasses, will cause temporary itching or rashes in some people but not others. This selective sensitivity among individuals greatly complicates the assignment of a plant species to the poisonous category.

Family CUPRESSACEAE

JUNIPERUS JUNIPER

One species occurs in Alabama.

***Juniperus virginiana* (Plate 1)** EASTERN RED CEDAR

Species Recognition: Eastern red cedar is a small tree, up to 7 meters, with opposite, scale-like-to-awl-shaped leaves that are, at most, 1 cm long. The seeds are produced in waxy cones that resemble berries. The leaves are appressed to the stem and overlap, so that the stem appears green. Although sold as Christmas trees in some areas, red cedar is not popular with people who have a dermatitic response. The seeds of *Juniperus* are used as the principal flavoring for gin, which might be considered a poison of a different sort.

Toxic Properties: The foliage contains various oils and resins that cause dermatitis. Grazing on large amounts of red cedar by livestock has caused digestive upsets. Because of its distasteful resin, the foliage is undesirable as a forage, making poisoning unlikely.

Geographic Distribution: Eastern red cedar occurs from New England to Michigan and Missouri, south to Florida and Texas. Can be expected in any part of Alabama where proper soil occurs.

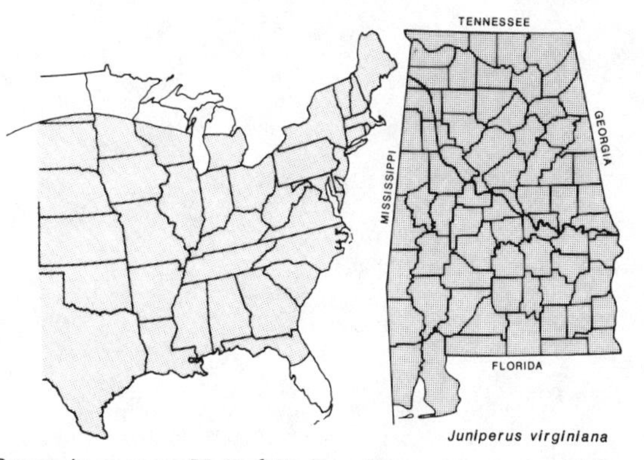

Range in eastern United States Known range in Alabama

Habitat Occurrence: Eastern red cedar occurs on dry soil in open woods, fields, on rocky slopes, and barrens, especially in soil that is calcareous.

Family ANNONACEAE

ASIMINA PAWPAW

Asimina is a genus of small shrubs that grow to a height of about 10 meters. The leaves are simple and are widest just below the apex, tapering to a narrow base. The flowers are maroon and have 3 sepals and 6 petals. They are solitary; that is, only 1 is produced at a point on the stem. The fruits are oblong, 2–8 cm in length, and have 1-to-several flattened brown seeds.

Toxic Properties: Pawpaws have fruits that are often eaten by people in the southern United States. A small percentage of the population contracts dermatitis when the skin comes in contact with the fruit. A few people also have contracted gastrointestinal problems from eating them. The toxic compounds, which apparently are found only in the fruit, have not been identified.

Asimina longifolia ASIMINA

Species Recognition: This species is a small shrub, 1–1.5 meters tall, with leaves 6 times or more longer than they are wide. The maroon flowers occasionally have white to pink streaks in them. The fruits are 4–10 cm long.

Geographic Distribution: Northern Florida, southeastern Alabama, and southern Georgia. In Alabama, known only from Dale, Geneva, and Houston counties.

Range in eastern United States Known range in Alabama

Habitat Occurrence: The species occurs in slash or longleaf pine woods, palmetto flatwoods, longleaf pine savannas, old fields, pastures, on sandy ridges, and along roadsides.

Asimina parviflora DWARF PAWPAW

Species Recognition: Dwarf pawpaw is a shrub or small tree 1–6 meters tall with leaves that are less than 4 times as long as broad. The flowers are less than 2 cm wide. The fruit is usually less than 3 cm long.

Geographic Distribution: Southeastern Virginia, west to Tennessee and eastern Texas, and south to Florida. Can be expected throughout Alabama.

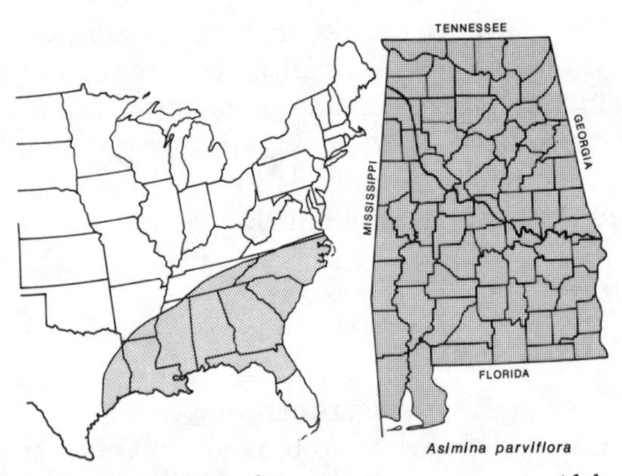

Range in eastern United States Known range in Alabama

Habitat Occurrence: Dwarf pawpaw is found in rich woods, along the borders of lime sinks, and in alluvial soil of coastal hammocks in the Coastal Plain, Piedmont, and Appalachian Mountains.

***Asimina triloba* (*Plate 2*)** COMMON PAWPAW

Species Recognition: Common pawpaw is a shrub or small tree 1.5–10 meters tall, with leaves that are less than 4 times as long as broad. The flowers are more than 2 cm wide, and the fruit is usually more than 3 cm long.

Geographic Distribution: Ontario west to Nebraska, and south to Florida and Texas. Can be expected throughout Alabama.

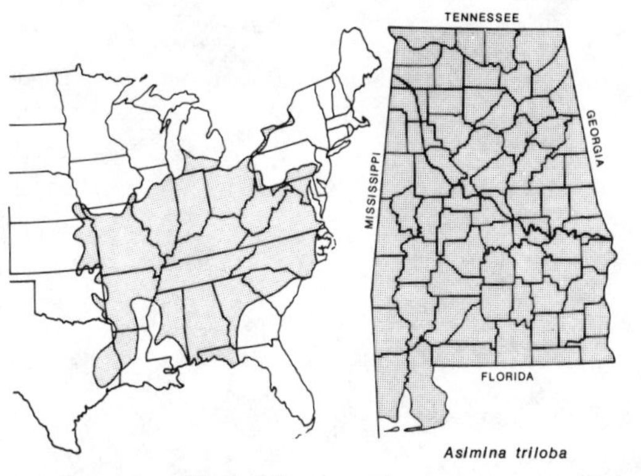

Range in eastern United States Known range in Alabama

Habitat Occurrence: Common pawpaw occurs in sandy soil of rich hardwood forests and river bottoms, usually as an understory tree growing in shade or along the edges of forests.

Family RANUNCULACEAE

CLEMATIS VIRGIN'S BOWER

Clematis is a genus of herbaceous or woody vines with opposite, pinnately compound leaves. The flowers are fairly showy with 4 enlarged sepals that look like petals. The fruits are achenes, numerous, brownish in color, and with long creamy-to-white plumose projections. Six species are known to occur in Alabama.

Toxic Properties: The leaves contain an acrid juice that causes dermatitis in some persons. Although there are no definite records of livestock or human deaths from these species in the United States or Canada, all species of *Clematis* in Europe reputedly are toxic and have caused livestock fatality upon ingestion. Early literature records indicate that alkaloids, glycosides, and saponins are present in the species. Because the sap of *Clematis* is an irritant and the species is closely related to the European forms, it is likely that several or all

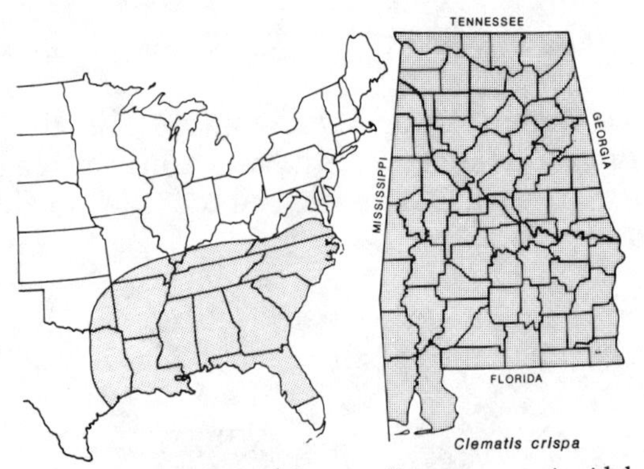

Range in eastern United States Known range in Alabama

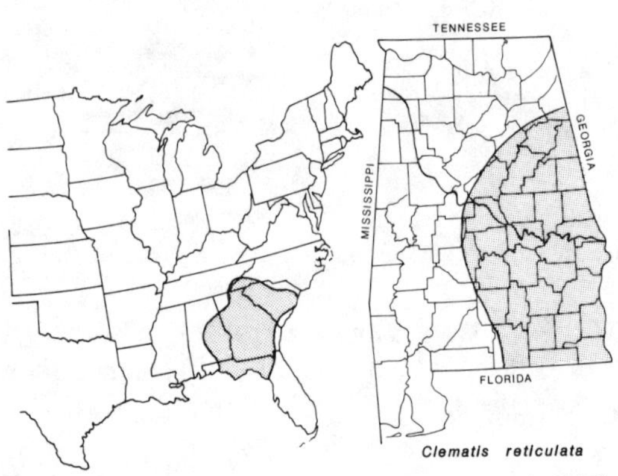

Range in eastern United States Known range in Alabama

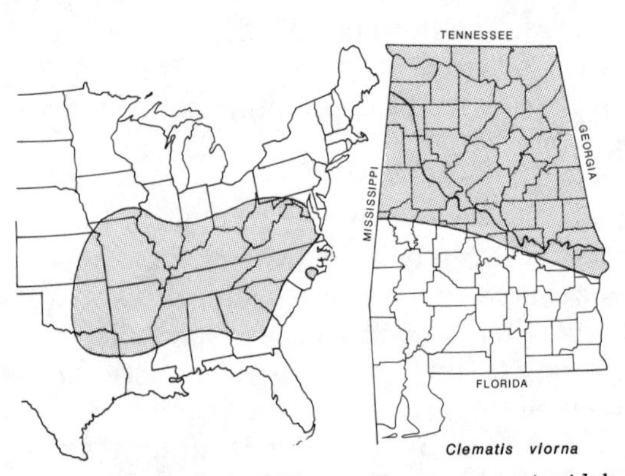

Range in eastern United States Known range in Alabama

of the species contain similar compounds and may, at least under unusual conditions, be dangerous to livestock and humans.

Clematis crispa LEATHERFLOWER; SWAMP VIRGIN'S BOWER

Species Recognition: Leatherflower is a climbing herbaceous vine with sepals that have at least some purple in them. The pedicels are without bracts.

Geographic Distribution: Southeastern Virginia to southern Illinois, and south to Texas and Florida. Can be expected throughout Alabama.

Habitat Occurrence: Leatherflower occurs in wet woods, swamps, and occasionally in marshes.

Clematis reticulata RETICULATE LEATHERFLOWER; NET-LEAF VIRGIN'S BOWER

Species Recognition: Reticulate leatherflower is an herbaceous climbing vine that has solitary flowers with purplish sepals, pedicels with 2 bracts, and leathery leaves that are coarsely reticulate. The leaves are green, not glaucous, beneath.

Geographic Distribution: South Carolina to Tennessee, south to Florida. Occurs in the eastern half of Alabama.

Habitat Occurrence: Reticulate leatherflower occurs in dry sandy woods, especially in the Coastal Plain.

Clematis viorna VASE-VINE

Species Recognition: Vase-vine is an herbaceous climbing vine with solitary flowers that have purplish sepals. The pedicels have a pair of bracts along them. The leaves are membranous and green and the stems are angled.

Geographic Distribution: Pennsylvania to Ohio, southern Missouri, and southern Iowa, south to Georgia and Mississippi. In Alabama can be expected north of the Black Belt.

Habitat Occurrence: Vase-vine occurs in rich woods and thickets.

Additional Species in Alabama: Other species with similar appearances and properties as those mentioned above also occur in this region.

Clematis glaucophylla WHITE-LEAF VIRGIN'S BOWER

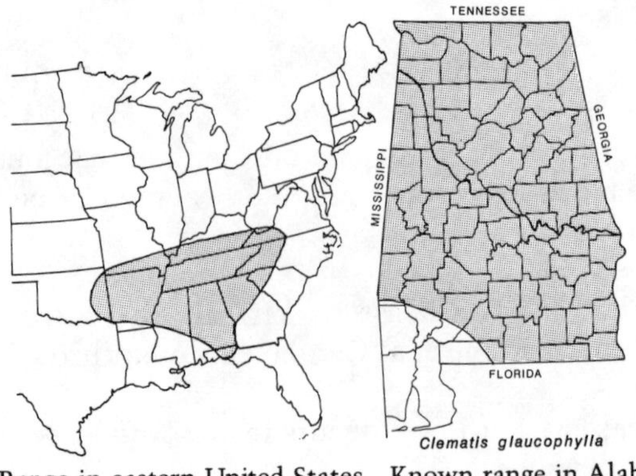

Range in eastern United States Known range in Alabama

Clematis maximowicziana CLEMATIS

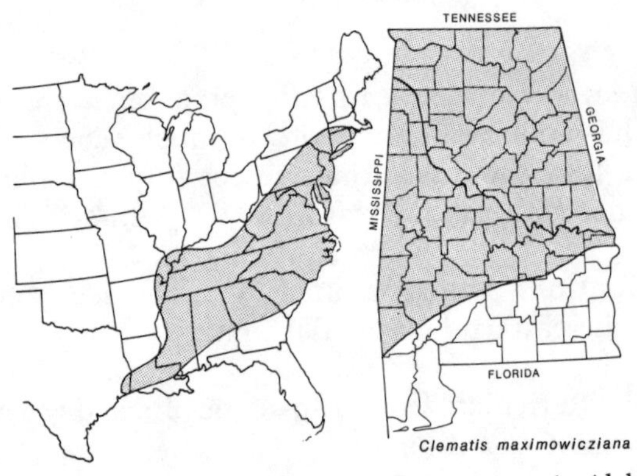

Range in eastern United States Known range in Alabama

Clematis virginiana VIRGINIA VIRGIN'S BOWER

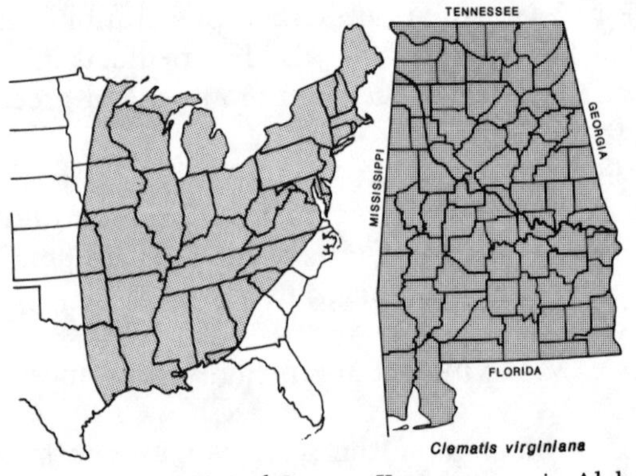

Range in eastern United States Known range in Alabama

Family URTICACEAE

LAPORTEA STINGING NETTLE

One species occurs in Alabama.

Laportea canadensis STINGING NETTLE

Species Recognition: Stinging nettle is an erect herb up to 1.5 meters tall with stinging hairs covering the stem. The leaves are alternate, ovate, and are serrate along the margin. The base is round, and the apex is long and acuminate. The flowers are imperfect with the staminate flowers in the axis of upper leaves; the carpellate flowers are in terminal panicles in the axis of the upper leaves. The fruit is an achene.

Toxic Properties: The stinging hairs contain an irritant juice. Contact with this juice can cause a painful dermatitis, red rash, and itching for a short while.

Geographic Distribution: Stinging nettle occurs from Quebec to Manitoba, south to Florida and Oklahoma. Can be expected throughout Alabama.

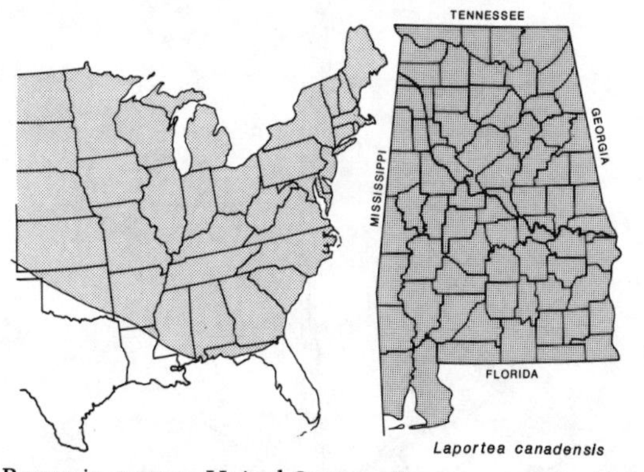

Range in eastern United States Known range in Alabama

Iabitat Occurrence: Stinging nettle occurs in low woods and along banks of streams.

JRTICA NETTLE

Urtica is a genus of monoecious or dioecious annual or perennial herbs vith stinging hairs that are evident as whitish bristles on the stems and eaves. The leaves are opposite with persistent stipules. The flowers are on xillary inflorescences.

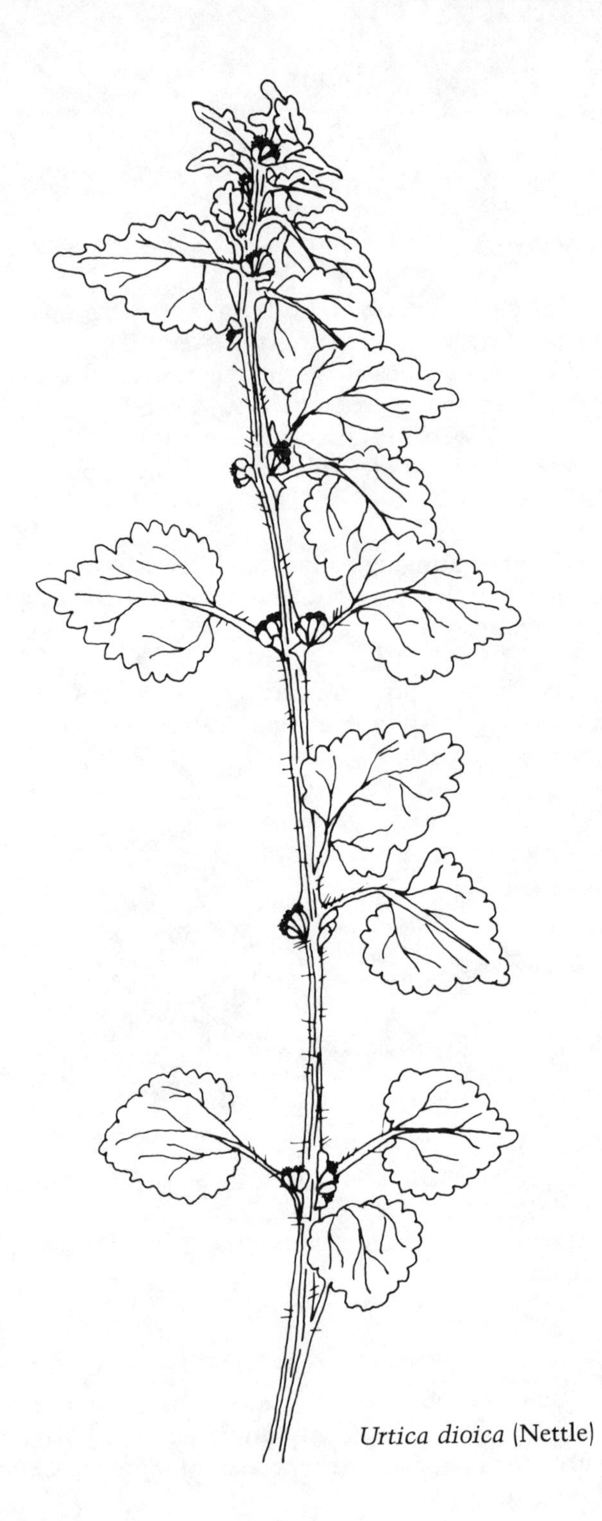

Urtica dioica (Nettle)

Toxic Properties: Injury from the stinging hairs of the nettle is primarily mechanical, although the hairs contain acetylcholine and histamine. When these compounds are injected into an organism, they cause respiratory distress, occasionally with irregular heartbeat and muscular weakness. Contact with the plant is usually not dangerous, but can cause dermatitis. Three species occur in Alabama.

Urtica chamaedryoides HEART-LEAF NETTLE

Species Recognition: Heart-leaf nettle has inflorescences that are compact, less than 1.5 cm long, and nearly spherical. The plant is sprawling or weakly ascending.

Geographic Distribution: From West Virginia and Kentucky to southeastern Missouri and Oklahoma, south to Florida and Texas. Can be expected throughout Alabama.

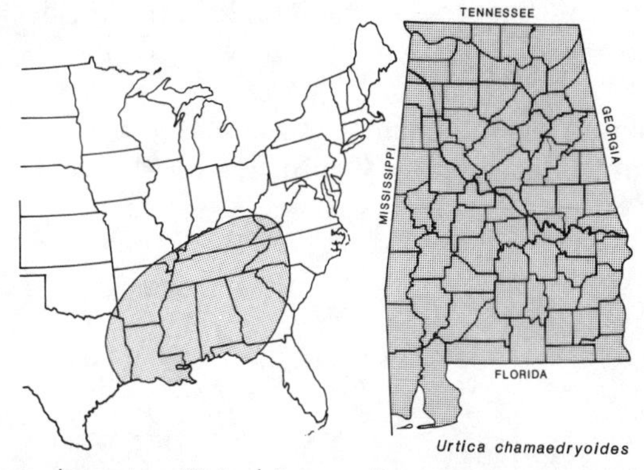

Range in eastern United States Known range in Alabama

Habitat Occurrence: Heart-leaf nettle occurs in bottomlands, rich woods, and waste places.

Urtica dioica STINGING NETTLE

Species Recognition: Stinging nettle is a stiffly erect plant with inflorescences that are more than 2 cm long.

Geographic Distribution: Newfoundland to Manitoba, south to Virginia, the Carolinas, Alabama, and Illinois. Can be expected throughout Alabama.

Habitat Occurrence: Stinging nettle occurs in waste places and roadsides.

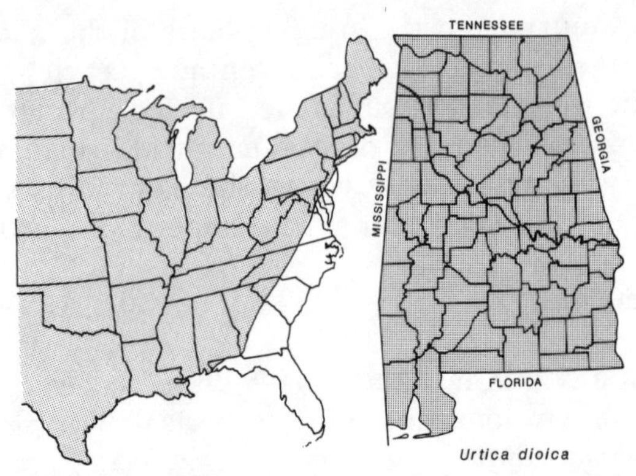

Range in eastern United States   Known range in Alabama

Urtica urens BURNING NETTLE

Species Recognition: Burning nettle is a sprawling or weakly ascending plant with inflorescences that are less than 1.5 cm long but no spherical in shape.

Geographic Distribution: Reported sporadically from Newfoundland to British Columbia, south to California. In Alabama known only from Mobile County.

Habitat Occurrence: Burning nettle occurs on roadsides and in waste places

Family JUGLANDACEAE

JUGLANS WALNUT

Juglans is a genus of large monoecious trees with alternate, pinnately compound leaves, having more than 10 leaflets per leaf. The flowers are im perfect. The fruit is a nut with 4 hardened bracts that surround the fruit and do not naturally peel off. Two species occur in Alabama.

Toxic Properties: *Juglans* stems and leaves contain juglandic acid. Prolonged contact with the juice of the stem has caused dermatitis to some individuals, especially harvesters of black walnuts.

Juglans cinerea BUTTERNUT; WHITE WALNUT

Species Recognition: The butternut tree grows to about 18 meters, with even, pinnately compound leaves; that is, there is no terminal leaflet. The fruit is oval in cross section.

Geographic Distribution: New Brunswick to Minnesota, south to Arkansas and Georgia. Can be expected in the northern one-third of Alabama.

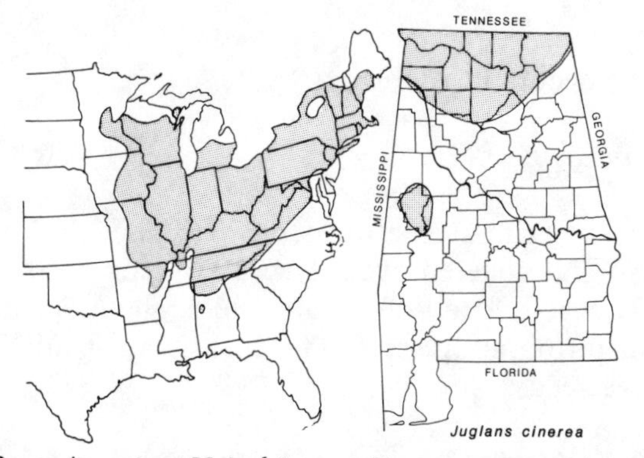

Range in eastern United States   Known range in Alabama

Habitat Occurrence: Butternut occurs in rich woods and river terraces, especially deciduous woods with sandy loam.

***Juglans nigra* (Plate 5)** BLACK WALNUT

Species Recognition: Black walnut trees may reach a height of 23 meters. They have alternate, pinnately compound leaves with a terminal leaflet; therefore, the number of leaflets is always odd. The fruit is terete in cross section.

Geographic Distribution: Western Massachusetts to Minnesota, south to Florida and Texas. Can be expected throughout most of Alabama.

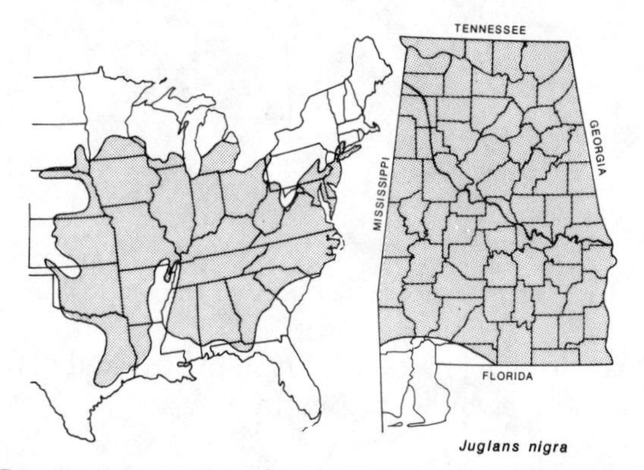

Range in eastern United States   Known range in Alabama

Habitat Occurrence: Black walnut occurs in rich deciduous woods, especially with sandy loam soil.

Family EUPHORBIACEAE

CNIDOSCOLUS STINGING NETTLE; BULL NETTLE

A single species occurs in Alabama.

Cnidoscolus stimulosus STINGING NETTLE

Species Recognition: Stinging nettle is an herb about 0.5 meter tall with milky sap and covered overall with stinging hairs. The leaves are palmately lobed. The flowers are white, about 2 cm across, and arranged in terminal clusters. The fruit is a capsule, usually with 3 fairly large seeds.

Toxic Properties: If Alabama has what might be called a venomous plant, this is it. Stinging nettle causes a painful dermatitis inflicted by stinging hairs that contain an irritant juice. The juice is injected into the victim by the needle-sharp hairs. The dermatitis consists of inflammation, red rash, and itching. The stinging hairs are found on all parts of the plant except the roots.

Geographic Distribution: Southeastern Virginia to the Florida Keys, west to Mississippi and eastern Louisiana. In Alabama can be expected from Calhoun and St. Claire counties south.

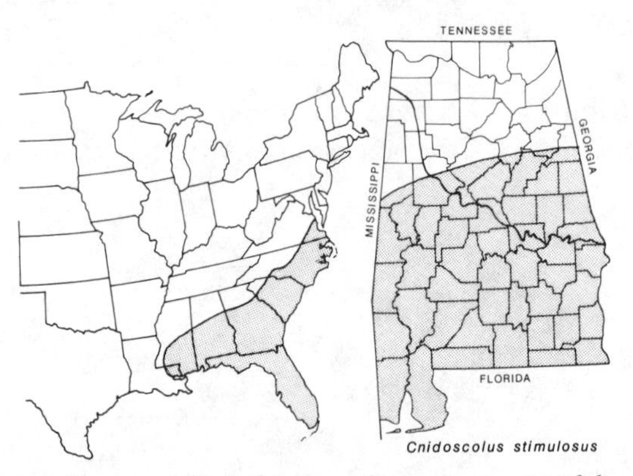

Range in eastern United States Known range in Alabama

Habitat Occurrence: Stinging nettle occurs in the Coastal Plain in dry, often sandy areas, and open deciduous woods.

CROTON CROTON; HOGWORT

Croton is a genus of herbs or large shrubs with alternate leaves and colored or resinous sap that is not milky. The leaves and often the stems are

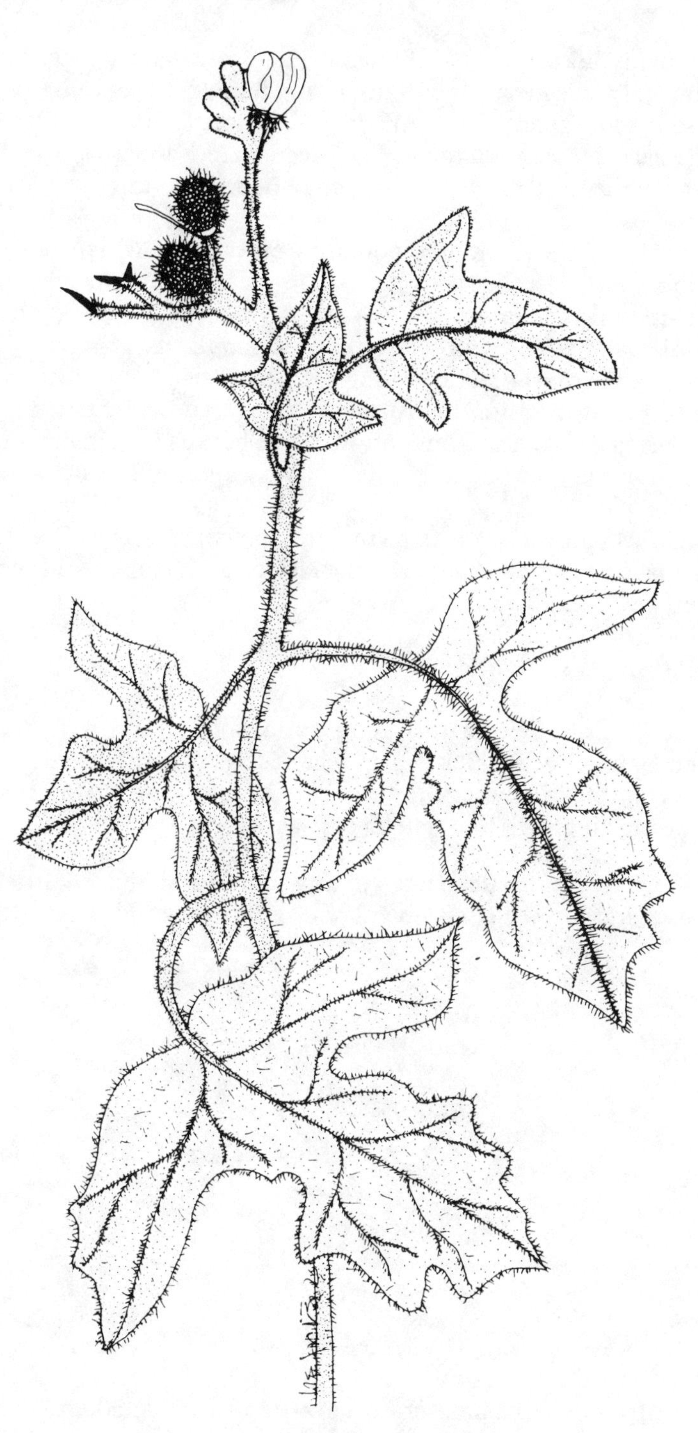

Cnidoscolus stimulosus (Stinging Nettle)

covered by many branched or scaly hairs. The flowers are arranged in terminal clusters and are imperfect. Both male and female flowers are usually produced on the same plant. The fruit is a globose capsule with 3 chambers and 1 large seed in each chamber. Six species are known to occur in Alabama, and a seventh, *Croton texensis*, may be in the state.

Toxic Properties: The toxic compound is croton oil, which is especially concentrated in the seeds but is also present in the leaves and stem. Croton oil is highly toxic. As few as 10 drops of pure oil have been fatal to dogs. This oil is used medicinally as a purgative and is among the most drastic available. Symptoms include intense gastroenteritis. In addition to croton oil, the sap is an irritant to the skin and eyes. Poisoning from croton is rare because the plant is odoriferous and very distasteful; it is seldom eaten by livestock or by humans.

Croton oil is known to be in *Croton capitatus* and *C. texensis* and is suspected in all other crotons native to Alabama. All species contain the irritant sap.

***Croton alabamensis* (Plate 10)** ALABAMA CROTON

Species Recognition: Within this genus, Alabama croton is distinctive as being the only species in the region that is a shrub. Also, it is the only species in which the female flowers have distinctive petals. The shrub grows to a height of 3–4 meters.

Geographic Distribution: Restricted to Tuscaloosa and Bibb counties in Alabama and known from nowhere else in the world.

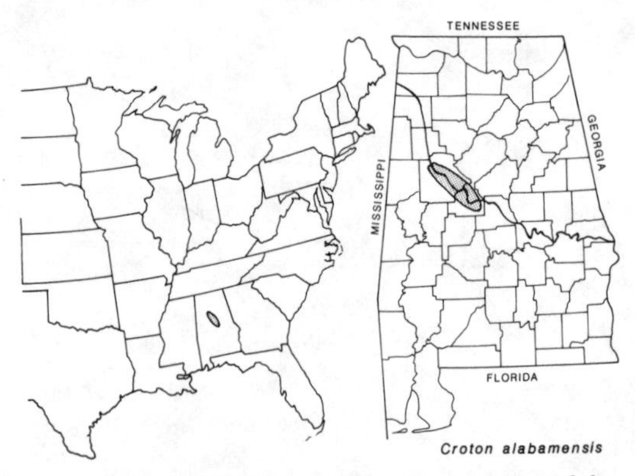

Range in eastern United States   Known range in Alabama

Habitat Occurrence: Alabama croton occurs in rich deciduous woods and in limestone outcrops overlooking the Black Warrior and Cahaba rivers, and Little Schultz and Six Mile creeks.

Croton capitatus WOOLLY CROTON

Species Recognition: Woolly croton is an herb that grows to about 1 meter, with entire leaves that are densely stellate pubescent. The fruits contain 3 brown seeds, each having a reticulate surface.

Geographic Distribution: Southern New York to southern Ohio, Indiana, Illinois, Iowa, and Kansas, south to northern Florida and Texas. Occurs throughout Alabama.

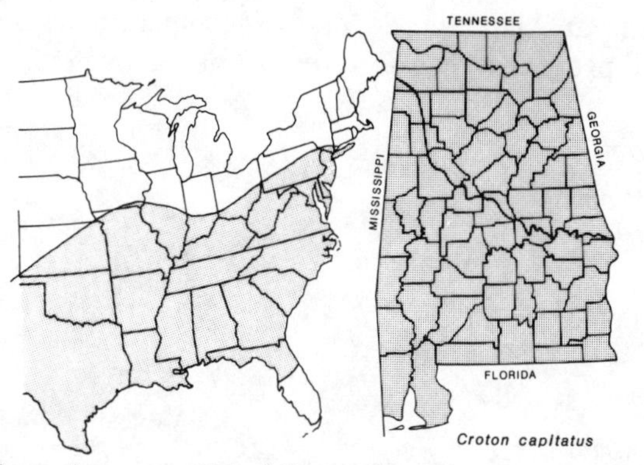

Range in eastern United States Known range in Alabama

Habitat Occurrence: Woolly croton occurs in disturbed sites with dry open soil and in waste places.

Croton glandulosus CROTON

Species Recognition: With its serrate leaves, *Croton glandulosus* is the easiest of the Alabama species of crotons to recognize. It is an erect herb

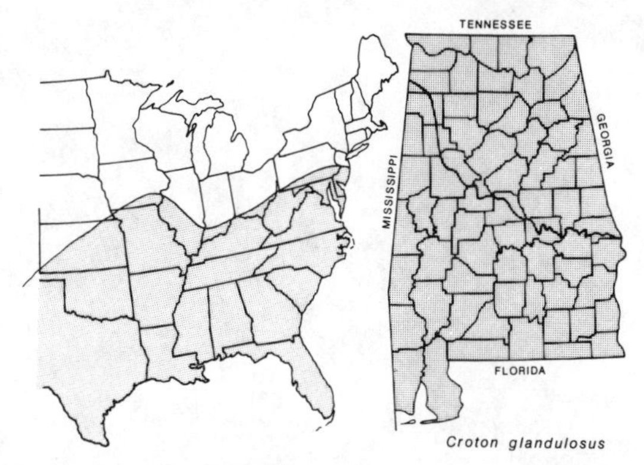

Range in eastern United States Known range in Alabama

that grows to about 60 cm, having fruits with 3 seeds, with a reticulate surface.

Geographic Distribution: New Jersey and Pennsylvania to Indiana, Illinois, Iowa, and Kansas, south to Florida and Texas. Can be expected throughout Alabama.

Habitat Occurrence: *Croton glandulosus* occurs in sandy or dry soil in woods, waste places, or open fields.

Additional Species in Alabama: Several other species with similar appearances and properties as those mentioned above occur in this region.

Croton argyranthemus SILVER CROTON

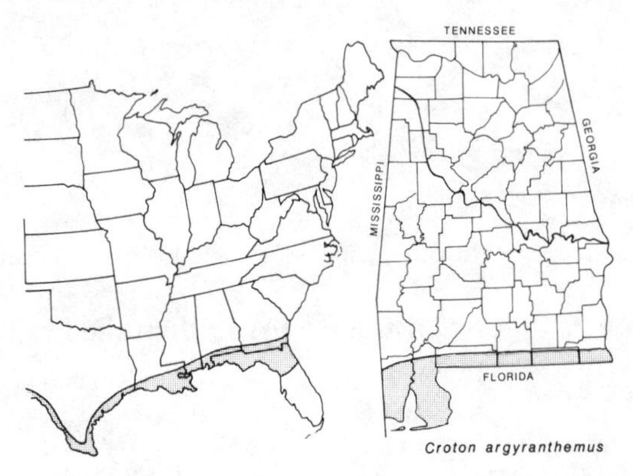

Range in eastern United States Known range in Alabama

Croton monanthogynus PRAIRIE TEA

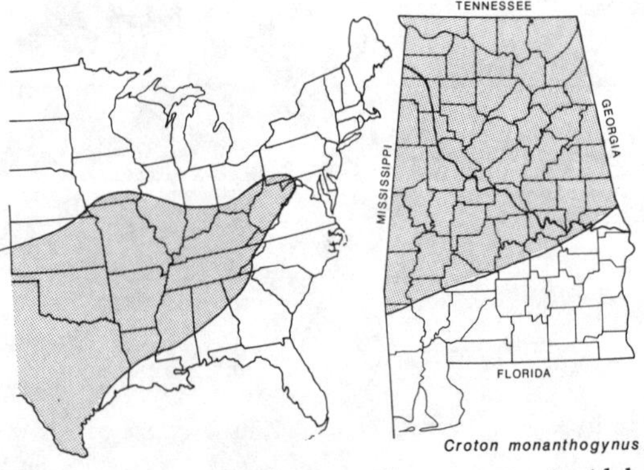

Range in eastern United States Known range in Alabama

Croton punctatus BEACH TEA

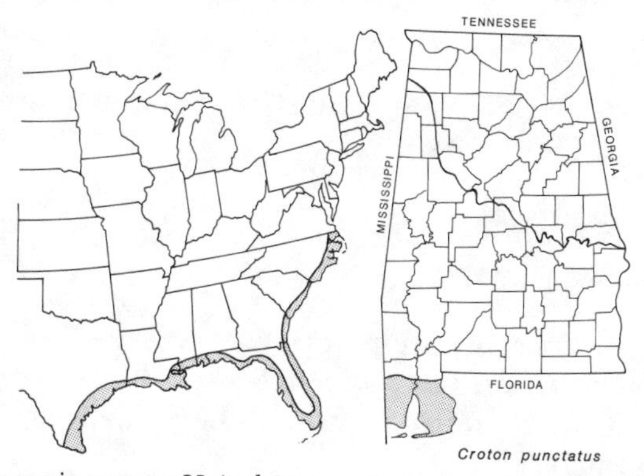

Range in eastern United States Known range in Alabama

Family VITACEAE

AMPELOPSIS PEPPER VINE

Only one species that occurs in Alabama is toxic.

Ampelopsis arborea PEPPER VINE

Species Recognition: Pepper vine is a high-climbing vine with white wood and few tendrils. The leaves are divided into many leaflets. The fruits are dark blue or black spherical berries with 2–5 seeds. The seeds are elongate and about 0.5 cm long.

Toxic Properties: Pepper vine has been known to cause dermatitis in a few people who came in contact with the leaves and berries. The toxic substances, however, are unknown. The leaves and fruits have a peppery taste and are considered to be inedible.

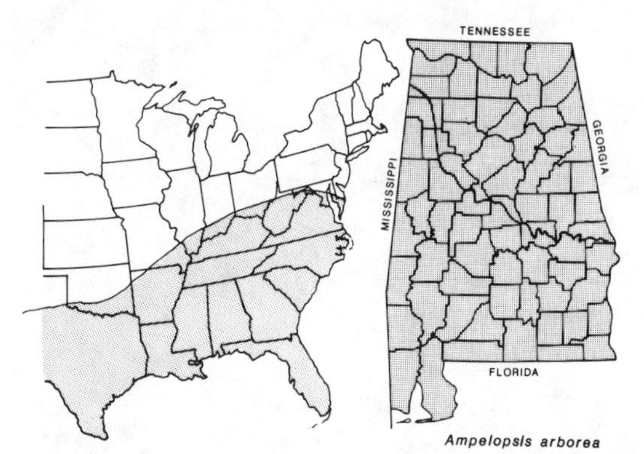

Range in eastern United States Known range in Alabama

Geographic Distribution: Eastern Maryland to southern Illinois, Missouri, and Oklahoma, south to Florida and Texas. Can be expected throughout Alabama.

Habitat Occurrence: Pepper vine is most common in swampy woodlands.

Family ACERACEAE

ACER MAPLES

Acer is a genus of trees or shrubs with watery, sugary sap, opposite leaves that are either simple and palmately lobed or pinnately compound. The flowers are small and imperfect with 2 cells in the ovary. The fruit is a double-winged samara.

Toxic Properties: It is perhaps surprising that a species that produces the sap for maple syrup would be toxic in any way. Yet there are records of 2 forms of toxicity. First, there are 2 cases of death of livestock in West Virginia involving both cattle and horses after eating maple leaves. The toxic compound is not known, and other details are lacking. The second form of toxicity is from the oil of the pollen, which results in a contact dermatitis that causes red skin rash and blisters. Little advice can be offered for avoiding the pollen, except for warning parents to keep their children out of maple trees during early spring. In addressing the first problem, however, the advice is simple: Don't eat maple leaves.

Acer negundo BOX ELDER

Species Recognition: A small tree with opposite pinnately compound leaves and a yellow-green double samara.

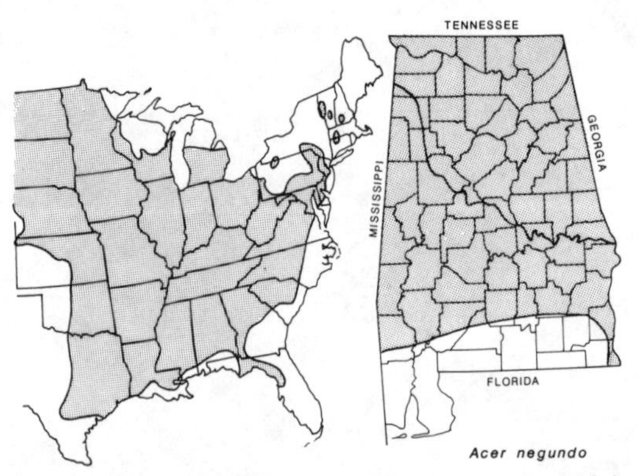

Range in eastern United States Known range in Alabama

Geographic Distribution: New England and southern Ontario to southeastern Minnesota, and south to Florida and Texas. Occurs throughout the northern three-fourths of Alabama.

Habitat Occurrence: Box elder occurs on river floodplains, in woods, waste places, and along fencerows.

***Acer rubrum* (*Plate 11*)** RED MAPLE

Species Recognition: Red maple is a tree with opposite, palmately veined, and lobed leaves with a double samara that is red. The bark is usually gray and smooth.

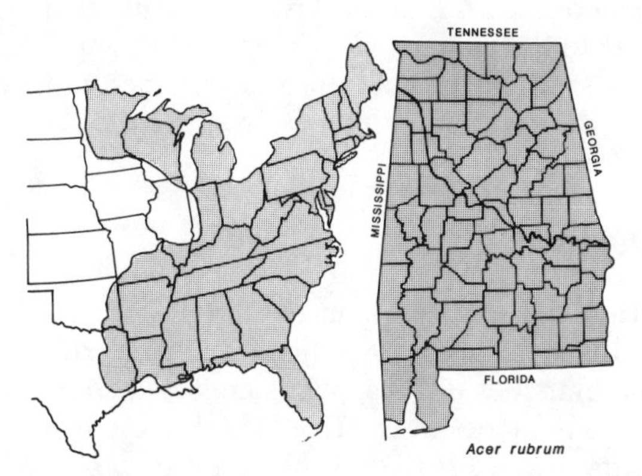

Range in eastern United States Known range in Alabama

Geographic Distribution: Quebec to Manitoba, and south to Florida and Texas. Occurs throughout Alabama.

Habitat Occurrence: Red maple occurs most commonly in swamps, along streams, or in alluvial woods.

Family ANACARDIACEAE

TOXICODENDRON POISON IVY; POISON OAK; POISON SUMAC

Toxic Properties: The active ingredient is an oily sap that contains urushiol, which is a phenol. This sap is present in the ducts of all parts of the plant and is exposed to the surface when the plant is bruised. Urushiol is nonvolatile and, therefore, will not evaporate and diffuse through the air. Thus, one cannot contract dermatitis by getting near the plant. Instead, one must actually touch the chemical. Since the chemical is nonvolatile, it will adhere to anything that contacts the plant, e.g., a pet, gardening tool, gloves, etc., until washed or rubbed off the object. Someone touching the pet or object that has touched the chemical can develop dermatitis.

Dermatitis appears within a few hours to a few days. The rash may last 7–14 days, but recontamination can occur. Excess toxins can be washed away, but the toxin binds with skin proteins quickly and washing does not remove those molecules already bound with the skin. About 25 percent of the human population is not sensitive to the oil, but one's sensitivity can change at any time, especially with repeated exposure. Some persons have eaten small amounts of the leaves and fruits in an attempt to desensitize themselves. This ingestion has occasionally caused gastroenteric upsets and even death; it is not recommended. Livestock and other animals seem to be immune to the plant. It should not be burned, for the nonvolatile oil adheres to smoke particles, which can then cause serious inflammation of the nasal membranes, respiratory tract, and even the lungs.

Toxicodendron radicans POISON IVY

Species Recognition: Poison ivy is an extremely variable species, ranging from small subshrubs to large liana-like vines that seemingly cover trees. The branches of these vines are often so large that they may appear to be branches of the tree. The leaves are compound and are divided into 3 lobed leaflets. The lobes are pointed at their tips, in contrast to those of poison oak, which are rounded. The cluster of flowers or fruits is produced near the top of the stem just below the leaves. The fruits are round to laterally flattened, creamy yellow to tan, and with a hard pit.

Geographic Distribution: Widely distributed in the Western Hemisphere, from southern Canada to western Guatemala, including the eastern

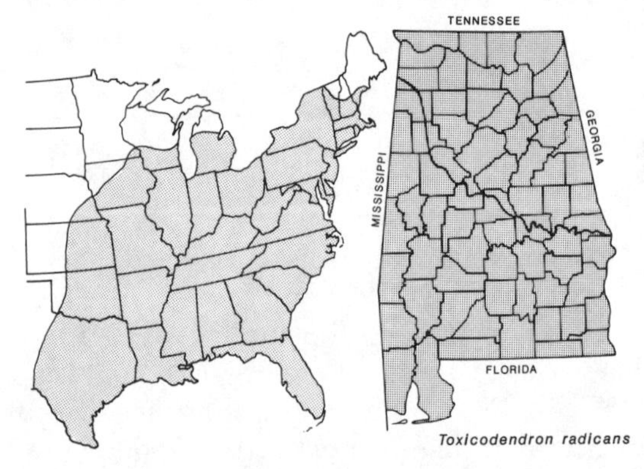

Range in eastern United States Known range in Alabama

third of the United States, throughout Mexico, Bermuda, the Bahamas, and also in eastern Asia. It can be expected in every county of Alabama.

Habitat Occurrence: Poison ivy is characteristically associated with moist, wooded habitats, although it can be found in almost any terrestrial habitat of Alabama. The species is predominantly a scandent vine, although it occasionally is a small subshrub.

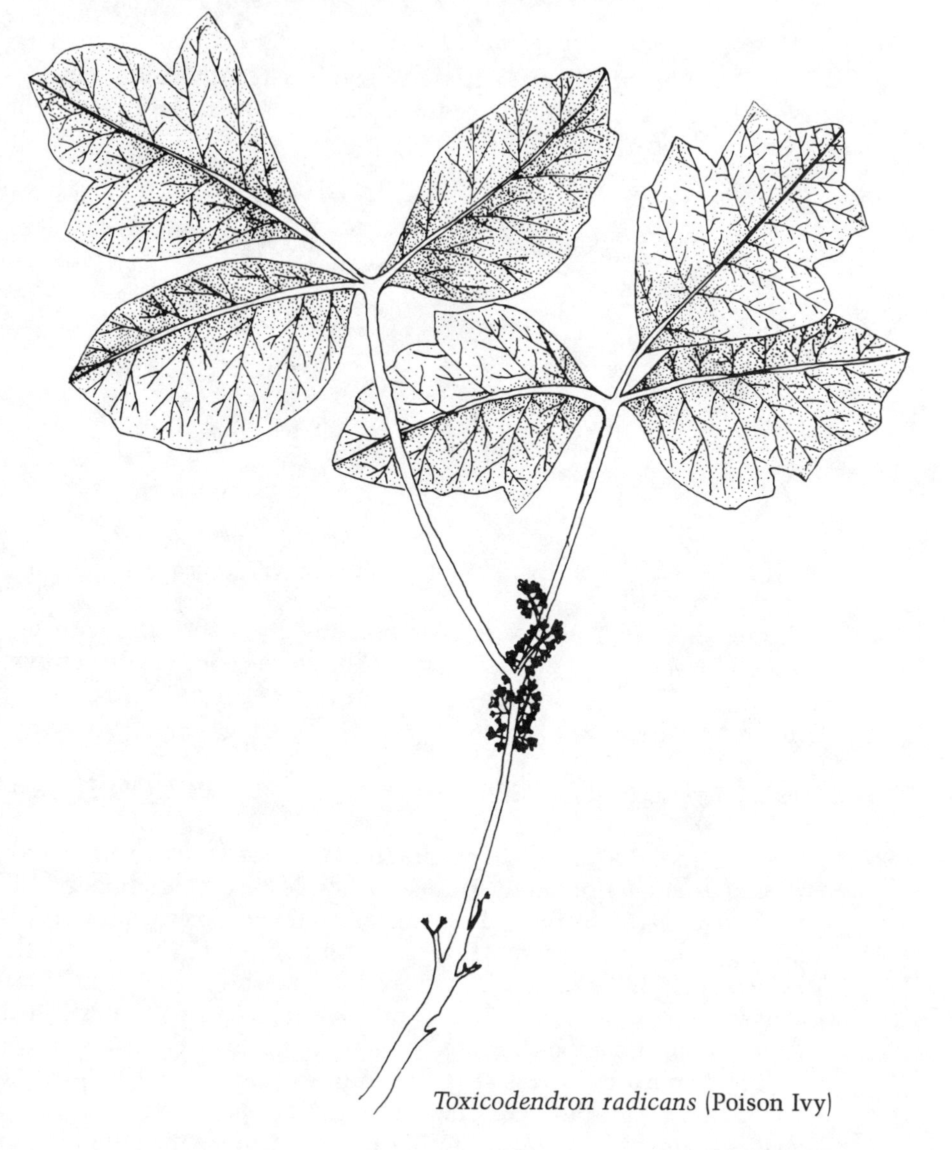

Toxicodendron radicans (Poison Ivy)

Toxicodendron toxicarium POISON OAK

Species Recognition: Poison oak is a small shrub that normally is less than 1 meter tall and has hairy branches. The compound leaves are divided into 3 lobed leaflets. The lobes are rounded at their tips, in contrast to those of poison ivy, which are pointed. The cluster of flowers or fruits is produced near the tip of the stem, just below the leaves. The fruits are round, yellow-brown to tan, and have a hard pit.

Geographic Distribution: Central Texas to southeastern Kansas, southern New Jersey, and northern peninsular Florida. Can be expected in every part of Alabama.

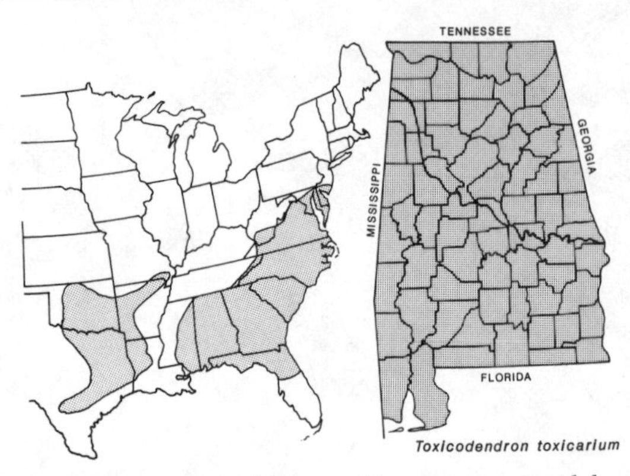

Range in eastern United States Known range in Alabama

Habitat Occurrence: Poison oak is characteristically associated with dry habitats, such as pine-oak sandhills. The species, unlike poison ivy, is not a climbing vine; instead, it grows as an erect understory shrub, usually in sandy soil.

***Toxicodendron vernix* (Plate 12)** POISON SUMAC

Species Recognition: Poison sumac ranges from a large shrub to a small tree, reaching heights of about 7 meters. The leaves are compound, divided into 7–15 leaflets. The cluster of flowers and fruits is produced in the axil of lower leaves, and remains on the stem after the leaves fall. The fruits are round, yellowish white, with a hard pit. Poison sumac superficially resembles the sumacs (genus *Rhus*), from which it can be separated by having fruits as described above rather than fruit clusters at the terminus of the stem. The individual fruits of *Rhus* are red. The leaves of this species are among the first to change colors in the autumn, and persons who gather leaves for decoration often develop severe cases of dermatitis.

Geographic Distribution: Maine to Minnesota, south to eastern Texas and northern peninsular Florida. Can be found throughout Alabama.

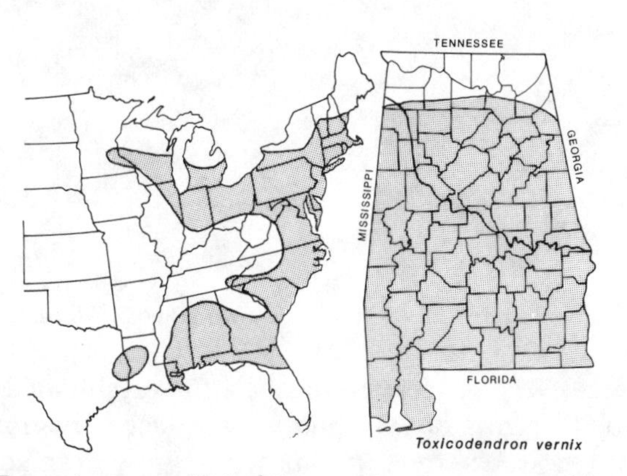

Range in eastern United States   Known range in Alabama

Habitat Occurrence: Poison sumac is characteristic of wet woodlands, such as swamps, pine flatwoods, and bogs. It is especially common along the borders of these woodlands where sunlight is plentiful.

Family SIMAROUBACEAE

AILANTHUS TREE-OF-HEAVEN

A single species occurs in Alabama.

***Ailanthus altissima* (*Plate 12*)** TREE-OF-HEAVEN

Species Recognition: Tree-of-Heaven is a fast-growing softwood tree that grows to 20 meters in height. The leaves are up to 70 cm long and are divided into 11–25 leaflets, each leaflet with a conspicuous gland at its base. The flowers are produced in compact panicles and are not showy. The fruits are oblong membranous samaras, with 1 seed in the middle.

Toxic Properties: Tree-of-Heaven has been known to cause dermatitis to some people when they contact the flowers or leaves. The active ingredient is not known. The species is also suspected to cause gastroenteritis when ingested.

Geographic Distribution: Native to Asia and Australia but naturalized from Massachusetts to southern Ontario and Iowa, then southward to Florida and Texas. Can be expected throughout Alabama.

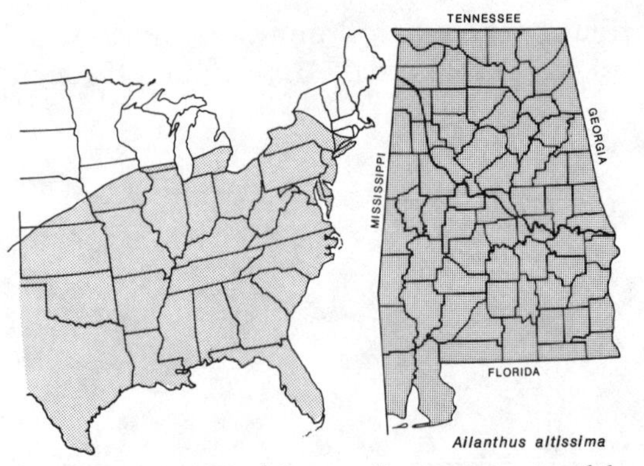

Range in eastern United States   Known range in Alabama

Habitat Occurrence: Tree-of-Heaven has become established most often in mesic woods, rights-of-way, and waste places. In some larger cities of the United States, trees are known to grow from soil that gathers in cracks of buildings. This is the species of *A Tree Grows in Brooklyn*.

Family ARALIACEAE

ARALIA DEVIL'S-WALKING-STICK

A single toxic species occurs in Alabama.

***Aralia spinosa* (Plate 14)** DEVIL'S-WALKING-STICK; HERCULES' CLUB

Species Recognition: Devil's-walking-stick is a small tree that grows to about 8 meters, with prickles on the trunk and large compound leaves that range up to 1 meter in width. The flowers are white, in large open clusters. The fruits are purplish berries that are about 0.5 cm in diameter.

Toxic Properties: Persons handling the bark and roots of this species have developed dermatitis with inflammation and blisters. Ingestion of large amounts of either the roots, seeds, or leaves is said to be harmful. Feeding experiments with seeds have shown them to be lethal to guinea pigs. The active ingredient is apparently unknown but is possibly a volatile oil, a resin, or araliin, which is a glycoside. All occur in the plant.

Geographic Distribution: Western New York to southern Illinois and Iowa, south to Florida and Texas. Occurs throughout Alabama.

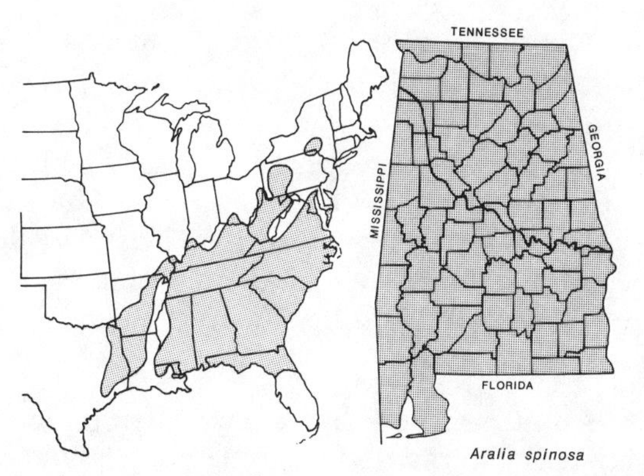

Range in eastern United States   Known range in Alabama

Habitat Occurrence: Devil's-walking-stick occurs in rich deciduous woods, especially along bluffs overlooking rivers.

HEDEREA ENGLISH IVY

Only one species occurs in Alabama.

***Hederea helix* (Plate 14)** ENGLISH IVY

Species Recognition: English ivy is an evergreen creeping or climbing vine with dark green leaves that are 3–5-lobed with lighter markings along the veins. The flowers are in umbels and produce black berries, each having 5 seeds.

Toxic Properties: All parts of the plant, especially the leaves and berries, contain the 2 toxins hederagenin and hederin. These are both saponic glycosides. Although most people do not respond negatively to English ivy, a sensitive person may have severe allergenic dermatitis within 48 hours after contact with the plant. Large quantities taken internally may cause diarrhea, excitement, nervousness, labored respirations, convulsions, coma, and possibly even death.

Geographic Distribution: Widely cultivated throughout the southern United States, in many cases, persisting after cultivation. Can be expected throughout Alabama.

Habitat Occurrence: English ivy is common around old homesites and in woodlands near homesites.

Family BIGNONIACEAE

CAMPSIS TRUMPET CREEPER

A single species occurs in Alabama.

Campsis radicans ***(Plate 18)*** TRUMPET CREEPER

Species Recognition: Trumpet creeper is a woody vine that trails over shrubs and herbs or climbs by aerial roots. The leaves are divided into 7–15 leaflets and have no tendrils. The showy flowers, 4–5 clustered at apex of stem, are 6–8 cm long with a red-to-orange corolla. The fruit is a fusiform capsule, 10–12 cm long, 2–3 cm wide, with numerous winged seeds.

Toxic Properties: Trumpet creeper has been known to cause an inflammation or blistering dermatitis in some people who come in contact with the leaves and flowers. The active ingredient is not known. The species is also thought to be poisonous if ingested.

Geographic Distribution: Connecticut to Michigan and Illinois, south to Florida and eastern Texas. Can be expected throughout Alabama.

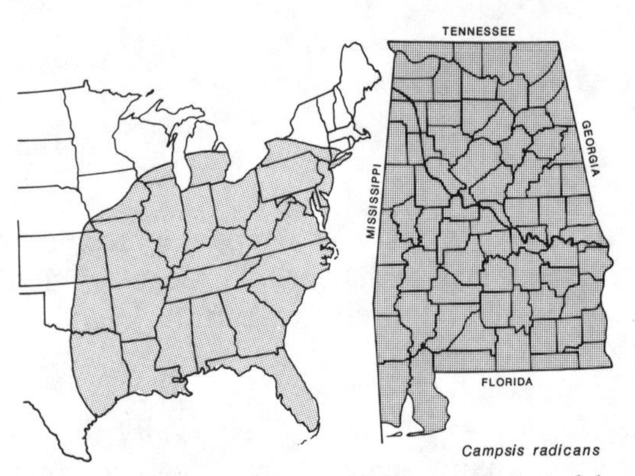

Range in eastern United States Known range in Alabama

Habitat Occurrence: Trumpet creeper occurs in low woods, thickets, waste places, and along fencerows and roadsides.

CATALPA CATAWBA TREE

A single species occurs in Alabama.

Catalpa bignonioides CATAWBA TREE

Species Recognition: Catawba tree is a medium-sized tree up to 20 m tall. The leaves are opposite, cordate at base, and short acuminate at apex. The flowers are showy, 2.5–4 cm long, and white with yellow and purple-brown spots. The fruit is a cylindrical capsule 30 cm or more long and 1–1.5 cm in diameter with numerous flat, winged seeds.

Toxic Properties: Contact with the flowers has been known to cause dermatitis to some persons. The active ingredient, however, is unknown. Smoke from the pods is inhaled as a relief for asthma.

Geographic Distribution: Southern New England to Michigan, and south to Florida and Texas. Can be found throughout Alabama.

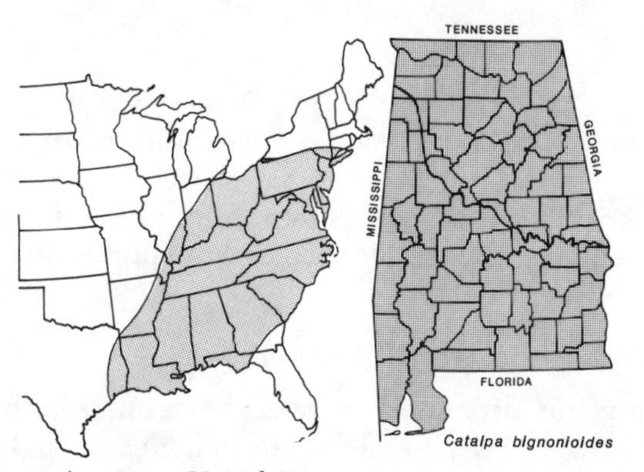

Range in eastern United States   Known range in Alabama

Habitat Occurrence: Catawba trees occur naturally along stream banks and in low woodlands, and are becoming more common along roadsides and fencerows.

Family ASTERACEAE

CONYZA HOGWEED; HORSEWEED

One species occurs in Alabama.

Conyza canadensis CANADA HOGWEED

Species Recognition: Hogweed is an erect plant up to 2 meters tall with narrow linear leaves that are 1.2–5 cm long. The white-to-lavender flowers are up to 5 mm wide and 3 mm high.

Toxic Properties: The leaves contain an irritant called oleoresin that is suspected of causing dermatitis. Ingestion of this species by livestock may also be a problem. It has been demonstrated that if a lamb consumes 3 percent of its weight of a western species of *Conyza* the results can be fatal.

Geographic Distribution: Throughout the United States, including all of Alabama.

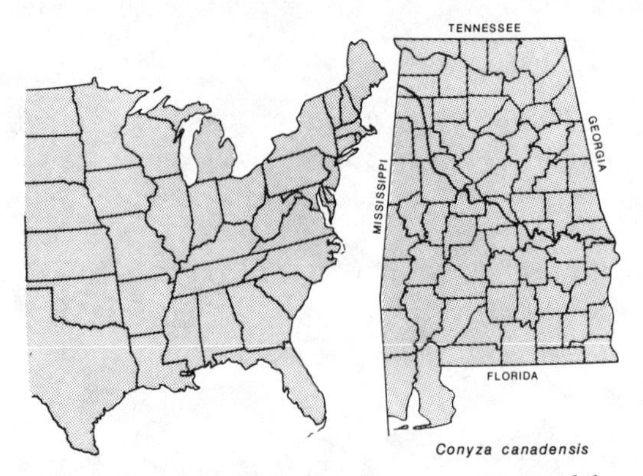

Range in eastern United States   Known range in Alabama

Habitat Occurrence: Hogweed occurs most commonly in heavily disturbed sites. It is one of the earlier species to populate strip-mined areas.

AMBROSIA RAGWEED

Ambrosia is a genus of annual or, rarely, perennial herbs with opposite leaves near the bottom of the stem and alternate leaves higher on the stem. The leaves are lobed or dissected. The flowers are clustered into small heads. All the flowers on 1 head have only 1 sex, either stamens or carpels. The heads are at the terminus of the stem and are in racemes. Two species occur in Alabama.

Toxic Properties: Stems, leaves, and pollen contain oleoresin which, when in contact with skin, can cause dermatitis. The pollen also contains a water soluble protein. Pollen from ragweed is a major cause of hay fever allergies in the United States, a phenomenon often incorrectly blamed on Alabama's former state flower, goldenrod.

Ambrosia artemisiifolia COMMON RAGWEED; ANNUAL RAGWEED

Species Recognition: Common ragweed leaves are divided pinnately; that is, small segments are attached along a central stalk or rachis. The plants grow up to 1 meter.

Geographic Distribution: Throughout the United States and southern Canada. In Alabama can be expected in all counties.

Habitat Occurrence: Common ragweed occurs in waste places, fields, and on roadsides.

Ambrosia trifida (Giant Ragweed)

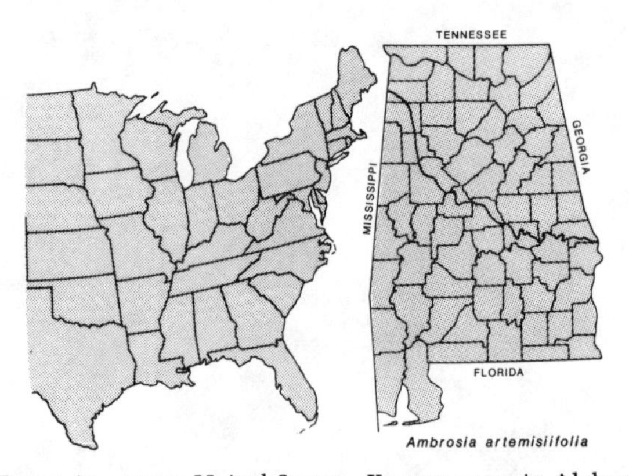

Range in eastern United States   Known range in Alabama

Ambrosia trifida GIANT RAGWEED

Species Recognition: Giant ragweed is a coarse herb that grows 2–5 meters tall with leaves that are 3–5 palmately lobed; that is, 3–5 lobes appear to arise at one point.

Geographic Distribution: Southwestern Quebec to British Columbia, south to Florida, Alabama, Texas, and Arizona. Throughout Alabama.

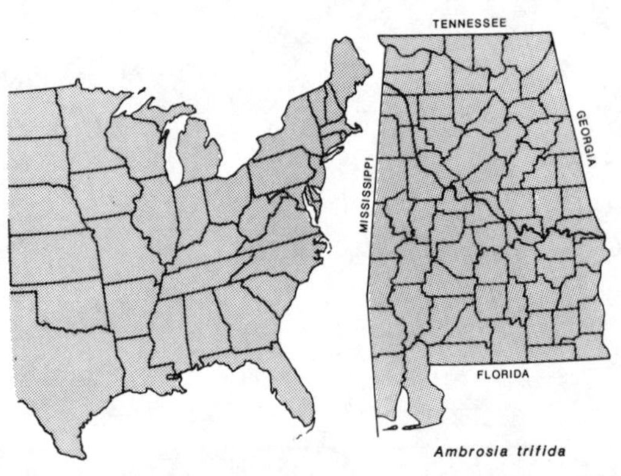

Range in eastern United States   Known range in Alabama

Habitat Occurrence: Giant ragweed occurs in waste places and rich alluvial soil.

Chapter 6

Venomous Invertebrates

Alabama is inhabited by thousands of species of animals that are capable of forcibly injecting a poisonous substance and are thus technically venomous animals. There are more than 2,000 species of the order Hymenoptera (bees, ants, wasps) in this state, and most of them are capable of stinging. Likewise, almost all members of the order Araneae (spiders) can bite. The stings and bites of these organisms are usually backed up by a venom supply.

The good news is that most of these animals are so small or dispense so little venom that they are of little or no concern to humans. For instance, only 2 species of spiders in Alabama—the black widow and the brown recluse—have the potential to cause serious injury. Even the large and formidable-looking wolf spiders, the "tarantulas," are overrated in the minds of most people. They rarely ever bite, and when they do, the effect is usually less severe than the sting of a honeybee. Relative to their total numbers, therefore, only a small number of venomous spiders and Hymenoptera will be discussed here.

Various coastal invertebrates and a variety of aquatic or terrestrial insects will be dealt with in a similar manner. Many are technically venomous but are unlikely to have any serious impact on a human, due to their small size or to the rarity and improbability of encounter. The focus of the present study will be on potentially dangerous species.

Descriptions of a few venomous species of invertebrates that pose little danger are included because their venom is thought by the general public to be dangerous, even though their true effect is usually inconsequential.

Phylum COELENTERATA

Class Scyphozoa — JELLYFISH

Venom Mechanism and Administration: The dangerous parts of a jellyfish are located on the tentacles attached to the harmless blob that constitutes the body. The tentacles are equipped with thousands of structures called nematocysts that cause the problem. When a person swimming in the Gulf of Mexico brushes a tentacle, a reaction by the nematocysts is triggered. A tiny hairlike structure extends to the outside of each nematocyst capsule, and when touched the entire thread that is coiled inside, projects outward. The thread bears spines and releases venom. Although a single nematocyst would be inconsequential, hundreds or thousands may be triggered, causing a unified effect. The tentacles and nematocysts are designed for prey gathering, but the nematocysts of the sightless creature floating in the gulf respond to touch and chemical cues, whether fish or human.

Venom Biochemistry and Symptoms: The primary ingredients of nematocyst venom are believed to be peptide compounds which have a digestive role. The immediate effect on a person is a stinging or burning sensation that may continue for some time. Itching and general irritation on the skin's surface are also characteristic of an encounter with a jellyfish. Scratching or rubbing the site of the sting can cause the barbed spines to penetrate deeper and should be avoided. The application of salves or isopropyl alcohol may relieve the pain, and ammonia has been suggested to deactivate the toxin, but both alcohol and ammonia have been reported to aggravate the problem in some instances.

Cooking salts containing meat tenderizers are recommended by some as a home remedy for minor jellyfish stings. The salt is believed to digest the stinging threads so that they become ineffective. An extensive sting over several parts of the body may require medical attention, based on the response of the patient. According to a study reported in the *Southern Medical Journal* (J. W. Barnett, H. Rubinstein, and G. J. Calton, 1983, "First Aid for Jellyfish Envenomation," 76:870–72), commercial meat tenderizer with papain does dissolve the stinging parts embedded in the skin. However, they noted that when high concentrations of the papain are used, the person's skin may begin to peel. A more effective first-aid treatment recommended was to sprinkle the sting area with baking soda, which also dissolves the nematocysts. In addition, they found that the application of vinegar was effective in reducing the stinging sensation in Portuguese man-of-war stings but that vinegar should not be used for stinging nettles. So it would appear that baking soda is the safest and surest home remedy for jellyfish stings.

Family PHYSALIIDAE

PHYSALIA

Physalia physalia (pelagica) PORTUGUESE MAN-OF-WAR

Species Recognition: This large jellyfish floats on the surface of the water like an attractive pink and blue ship, belying the danger that lies below. The float is bladder-shaped, crinkled at the top, and may be up to 20 cm long. Beneath the surface dangle long tentacles, often more than 10 meters in length. The tentacles are used to grab prey, especially small fishes, which are killed by the stinging cells on the tentacles.

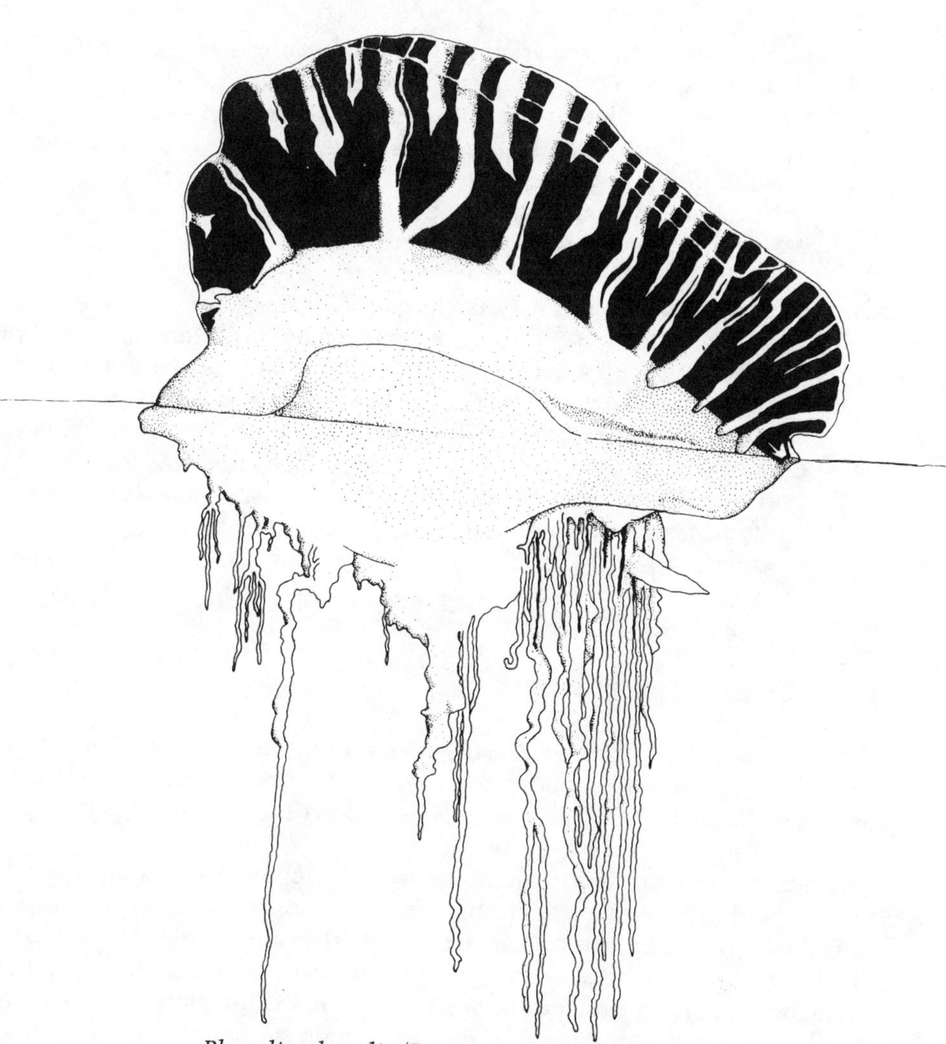

Physalia physalia (Portuguese Man-of-War)

Geographic Distribution: Throughout tropical and some warm temperate seas. In the Gulf of Mexico and along the Alabama coastline.

Habitat Occurrence: Restricted to saltwater habitats but may be close to shore where swimmers are. Occasional specimens are even washed onto beaches, and the nematocysts can still function if stepped on by a barefoot bather.

General Life History and Behavior: The Portuguese man-of-war, like many other jellyfish, does not exhibit any active behavior; it just floats. The part above the surface is a gas-filled float, and the animal moves only by water currents or by the wind blowing against the crest, like a sail. One of the remarkable biological features of the Portuguese man-of-war is that it is a single species, but not a single animal. Instead, the complete organism comprises thousands of individuals. Each of several different functions, such as feeding, digestion, and reproduction, are carried out by different individuals that form this colonial organism. All share in the effort to operate like a single animal and are obviously highly effective.

Other Jellyfish: Various other jellyfish inhabit Alabama's coastal waters, and all carry nematocysts as part of their armament. Many are small or are equipped with small tentacles, and the consequences of encountering them may not be noticeable. Several kinds cause mild stings that make swimming unpleasant. Jellyfish sometimes occur in large enough numbers to drive bathers from the water, and though the Portuguese man-of-war is the worst offender, several jellyfish of the Gulf of Mexico can cause stings that warrant at least home remedy attention or family sympathy.

Other Coastal Invertebrates

Four other phyla of invertebrates have venomous species reported from the Gulf of Mexico. Any could conceivably be found along the Alabama coastline, although their occurrence is rare and the circumstances for a person to be stung would be even less likely.

Within the phylum Echinodermata are certain sea urchins and sea cucumbers with venomous spines that can puncture the skin if the animals are picked up and handled or stepped on, whether alive or dead. Some cone shells in the genus *Conus* of the phylum Mollusca have venomous spines, although there is some uncertainty about whether those along the Alabama coast can cause any problem. Likewise, various species of marine worms of the phyla Platyhelminthes (flatworms) and Annelida (segmented worms) have venomous spines, but whether any are likely to be encountered is questionable. The likelihood of a typical swimmer or fisherman being painfully injured by any of these creatures is small, and the consequences are not likely to be medically serious.

Even if more scientific information were available, a comprehensive account of the venomous marine invertebrates of the Alabama coastline would probably be a short one. Furthermore, accounts would certainly be mild when compared to many tropical regions of the world where venomous marine organisms are more abundant and more potent. Also, it is worth noting that staff members who were contacted at large hospitals in the Mobile area could not locate or recall records of patients admitted for injuries from any of these animals during recent years.

Phylum ARTHROPODA

Class Chilopoda CENTIPEDES

Centipedes are venomous animals characterized by a flat body, a pair of claws on the underside of the head for injecting venom, and many legs that seem to want to keep moving. Centipedes have one pair of legs for each body segment, which number 15 or more in adults. They are carnivorous, preying on smaller invertebrates. The similar millipedes (Class Diplopoda) eat living or dead plant material, have 2 pairs of legs on most segments, and are strictly nonvenomous. Many millipedes have a more rounded or cylindrical body shape than centipedes, though some are flattened to varying degrees. Some centipedes in Alabama are bright red or orange whereas others may have an olive or brown appearance. Although enormous centipedes (almost a third of a meter in length) occur in the Southwest, those in Alabama range from less than 3 cm to about 10 cm in length.

More than 175 species of centipedes, representing more than a dozen families and 4 orders, have been reported from the eastern United States. Many of these occur in Alabama. Most species can be distinguished only by expert taxonomists. Only general information on the class will be provided. As far as can be determined, any centipede with front claws large enough to get a grip on human skin can bite and inject venom if handled. This seldom happens because of their small size and because in most situations they are too busy trying to escape.

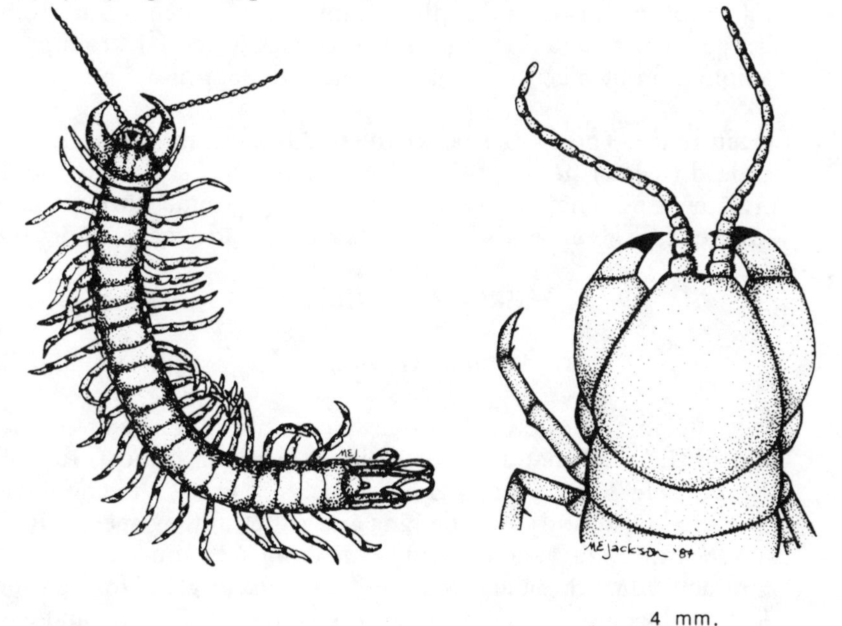

Scolopocryptops sexspinosus (Red Centipede)

Venom Mechanism and Administration: Centipedes inject venom through a pair of fanglike claws located on the anterior segment of the body (see drawing). Each fang has an opening near the end that connects via a duct to the venom gland. The venom is stored in the claw segment and is injected directly into the victim. Although the tail segment of some centipedes is elongate and elaborate in appearance, only the front end of a centipede does the biting.

Venom Biochemistry and Symptoms: Centipede venom is a complex of proteins that probably vary among different families or even species. The effect of Alabama centipede bites on humans is of little consequence from a medical standpoint. Among the common species, the smaller ones can pinch slightly but cause no effect. With some of the larger ones, the points where the fangs enter sting as if acid had been poured on the spot. A pair of tiny red pinpricks appears for a few minutes along with the burning sensation. Although it is conceivable that some people could respond allergically to the bite of a centipede, no documented case of a serious bite has come to our attention. In fact, it is difficult to find people in Alabama other than biologists who have been bitten by centipedes. One reason for this is that they are only apt to be encountered by someone actively digging in litter, removing bark from a tree, or otherwise searching in likely habitats where centipedes should be and most people should not. Another reason is that the organisms are much more eager to escape than to bite, so unless one makes a special attempt, envenomation by a centipede is not likely to happen.

Habitat Occurrence: The species occurring in Alabama are generally found in dead trees or heavy ground litter where they search for smaller invertebrates. They are seldom seen above ground during the day but are often abundant beneath dead leaves, bark, rocks, or decaying logs.

Class Arachnida

Order Araneae SPIDERS

Venom Mechanism and Administration: Spiders have 2 fangs at the front part of the head that are used to inject venom. In most, the fangs are movable, folding into a groove when not in use. A venom gland is located posteriorly to each fang and varies among species in size and location. The venom gland is surrounded by muscle tissue that contracts when the spider bites. Pressure on the gland forces venom down a duct that passes through the center of the fang and ejects through an opening at the tip. The venom is thus injected into the victim in the fashion of a hypodermic syringe and needle. The fangs of most spiders are too small to break human skin, and fewer than 3 dozen of the more than 30,000 species of spiders in the world are considered a serious threat to humans. Only 2 of these live in Alabama.

Venom Biochemistry and Symptoms: As with most venoms, that of spiders consists of an exceedingly complex protein structure, varies significantly among species, and has not been thoroughly analyzed in most instances. A primary component is neurotoxic in some species, but proteolytic enzymes that act to destroy cell tissues are usually present and may be dominant in others.

The serious consequences of a black widow bite are considered to be primarily of neurotoxic origin with respiratory paralysis being a common symptom and ultimate cause of death in cases ending in fatality. Other symptoms include intense muscular pain throughout the body, irregular breathing patterns, and an increase in heartbeat rate and in blood pressure.

The bite of the brown recluse is usually more localized in its effect, and fatalities are rare. However, necrosis (the destruction of living tissue) occurs at the site of the wound, and a large and deep area of dead cells may develop. A permanent scar usually results once the wound has healed. The bite may be initially painful with swelling, but most reports indicate no immediate pain and that 2–8 hours may pass before the bite becomes painful. In fact, many people are apparently unaware when they are bitten, and the ulceration and tissue damage may not occur for several days.

The treatment of a bite from either of these 2 species is best done through professional medical services; however, the application of ice to spider wounds is recommended by many medical authorities. The primary objective after envenomation by a black widow or brown recluse should be to get the victim to a doctor or hospital as rapidly as possible.

Family LOXOSCELIDAE

LOXOSCELES

Loxosceles reclusa BROWN RECLUSE

Species Recognition: The brown recluse spider is light gray to chestnut brown with a darker-colored fiddle-shaped mark on the upper surface between the head and abdomen. The narrow portion of the "fiddle" points toward the abdomen. The brown recluse has 6 eyes rather than the 8 of many other species. Individuals can be up to 1 cm in body length.

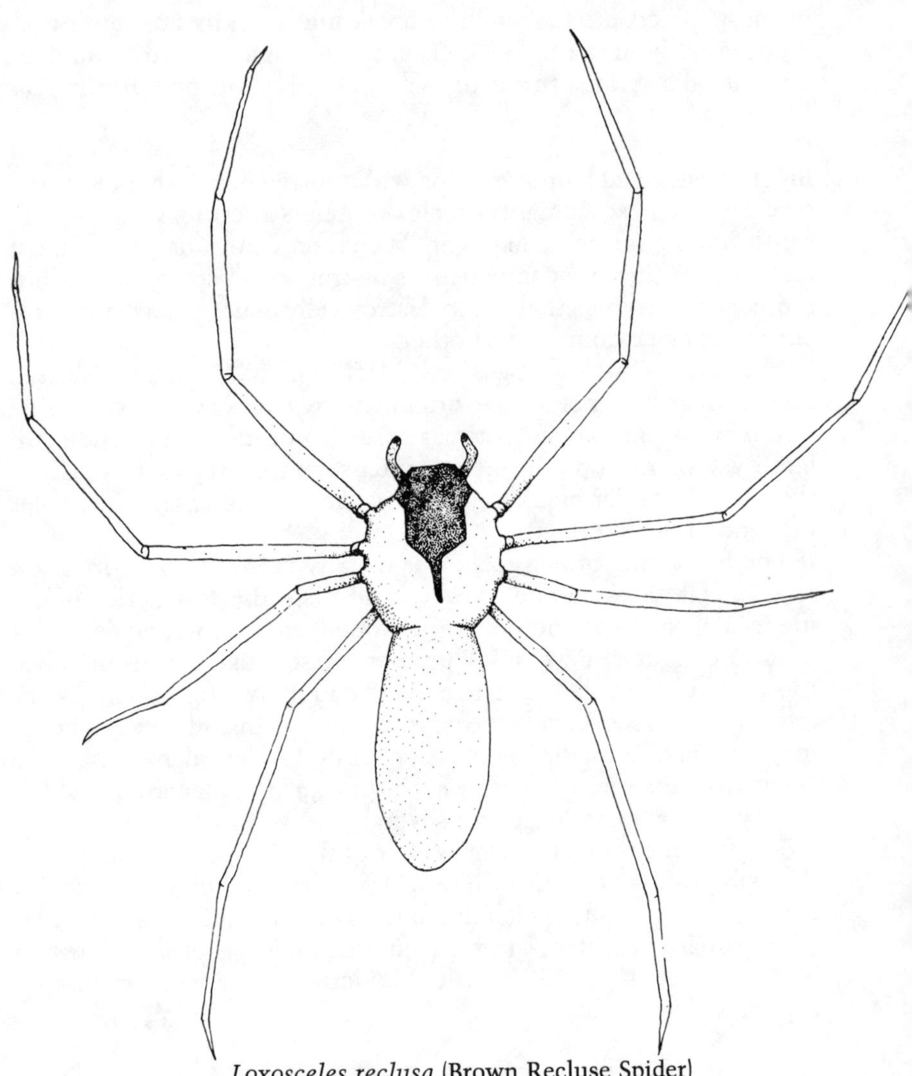

Loxosceles reclusa (Brown Recluse Spider)

Geographic Distribution: Southeastern Gulf Coast states to Tennessee an some midwestern states to the Southwest. The species has recentl been reported from various localities along the East Coast, presum ably accidentally introduced by humans. The brown recluse can b found throughout Alabama, but because of the current uncertaint of its exact geographic distribution in the eastern United States, range map is not given.

Habitat Occurrence: Although at home in sheltered and protected natura habitats, such as beneath fallen trees or dense brush, the brown rec luse is noted for its tendency to live in dark areas of barns, outbuildings of various kinds, as well as in homes. Its frequent occurrence among clothing stored in closets, basements, or attics is a major cause of bites.

General Life History and Behavior: This secretive spider spins its web in dark and secluded areas. An interesting phenomenon based on at least 3 incidents, one in Oklahoma, one in South Carolina, and one in Alabama (reported by Dr. Earl Cross, Department of Biology, University of Alabama), is the occurrence of numerous brown recluses inside a seldom-used dwelling. In each case, dozens of adult spiders were under insulation or were crawling around on the floor, tabletops, and other furniture. No explanation has been given for such massive outbreaks, but let us all hope that they remain a rare event.

Family THERIDIIDAE

LATRODECTUS

Latrodectus mactans BLACK WIDOW SPIDER

Species Recognition: The shiny black appearance and large bulbous abdomen with the bright red hourglass figure on the underside distinguish the black widow spider from others in its range. Females may be more than 1 cm in body length, whereas males are generally less than one-half the size of the females The northern widow (*Latrodectus variolus*), on which the hourglass is separated into 2 parts, is recognized by some authorities as a distinct species that is closely related to the black widow and possibly could be found in the northern parts of Alabama.

Geographic Distribution: Widespread, throughout most of the United States and all of Alabama. Uncertainty of taxonomy of the black widow—whether more than one species is actually represented—confounds the presentation of a distribution map for the species.

Habitat Occurrence: Black widows are most common in areas of sandy soil, beneath debris and litter, or in dark, concealed places. Old boards, tin cans, cardboard, or other solid items are favored hiding places and should be picked up with caution in areas where black widows are common. Outhouses were a common habitat of both humans and black widows before the mid-1900s and were the sites of a major portion of the bites in the United States.

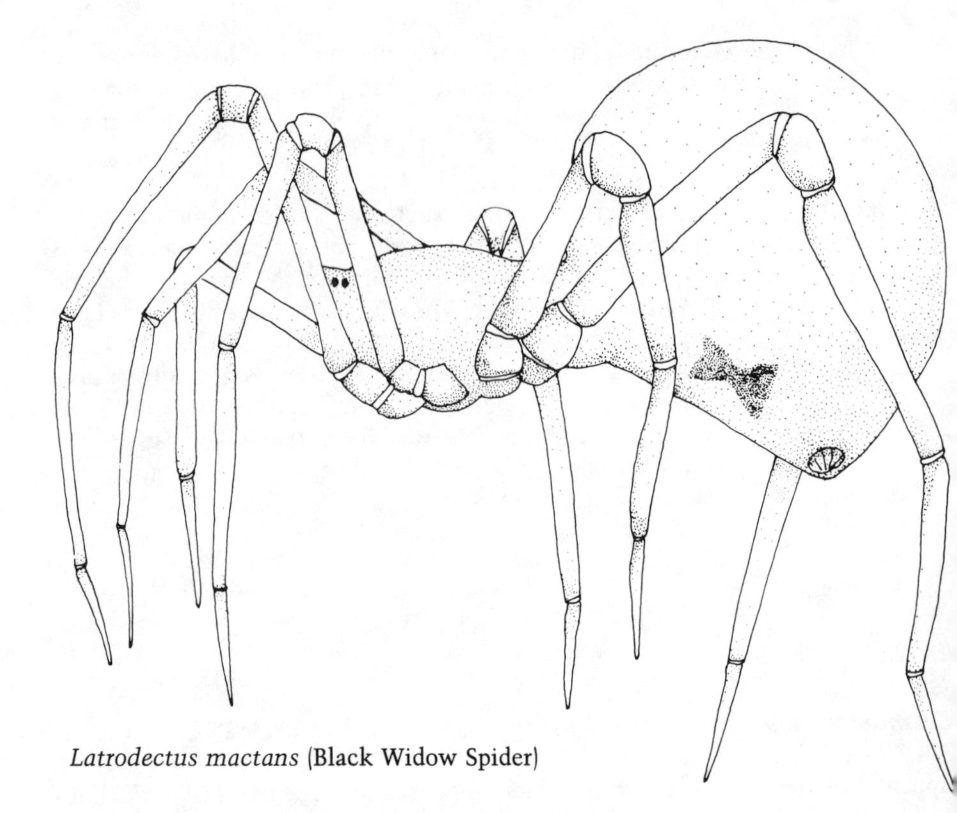

Latrodectus mactans (Black Widow Spider)

General Life History and Behavior: According to some reports, one of the most fascinating events in the life history of this species follows mating whereupon the female in some instances eats the smalle male. If this is true, it is an extreme example of the appetite cravin by pregnant females. Because of her larger size, the bite of the femal is far more serious to a human than that of the male. In fact, there is some question whether the male has ever bitten a human.

Family LYCOSIDAE

Although the black widow and brown recluse are the only Alabama spi ders that can cause serious bites under normal circumstances, the wolf spi ders bear mentioning because of their size. These are the large, ground dwelling spiders that can be gray, brown, or black. They may reach a bod length of more than 2 cm.

They live in leaf litter and in soil burrows and emerge to attack prey with out the use of a web. The name *wolf spider* comes from their reputatio as active hunters. Serious bites to humans from the wolf spider are rare c unknown, but the 2 enormous fangs at the front of the head give one cause to consider what the effect might be. In the case of Julian Harrison of the College of Charleston, who was bitten on the hand, the wound was painful for several minutes, and swelling ensued within a short while. Wolf spiders, however, usually want to escape rather than bite. Most of those picked up will quickly scamper out of one's hands and head for a hiding place.

All other Alabama spiders are assumed to be innocuous, although if one wishes to work at it, some of the larger ones can presumably be made to bite. As a whole, spiders should be appreciated and enjoyed as a fascinating group of top-level carnivores whose primary use of venom is to capture prey.

Order Scorpionida SCORPIONS

Although most people have a cautious respect for scorpions, aware that they can sting, the mechanism of the sting and the levels of danger are not usually understood. The facts are simple. Scorpions have a stinger on the tip of the tail, a weapon that is used primarily for killing prey, mostly insects. It is used less frequently for stinging bothersome animals, including people, who might do them harm. The pinching claws at the ends of the front legs have no venom associated with them. They are for holding crickets, grasshoppers, or other small animals while the scorpion first stings and then eats them. Some scorpions, some as nearby as the southwestern deserts of the United States, can deliver a lethal sting to humans. In Alabama the single species is not deadly, but it can and will sting.

Venom Mechanism and Administration: Scorpions have a hollow stinger on the tip of a long tail. A poison gland filled with venom is located inside the last segment of the tail and connects to the stinger by means of a duct. The venom is injected beneath the skin through openings in the tip of the stinger.

Venom Biochemistry and Symptoms: Scorpion toxins are biochemically complex proteins that vary among species. The amino acid sequences of some kinds have been analyzed, especially those that can be lethal to humans. The venom of the Alabama species, the southern unstriped scorpion, has not been thoroughly analyzed, but it presumably comprises peptide compounds and possibly enzymes.

Scorpion stings can affect both the nervous and cardiovascular systems of humans, although the manifestation of a sting varies dramatically with different species. A sting from the southern unstriped scorpion can result in painful swelling around the wound. Unless there are unusual symptoms, medical attention is not required, but a cold compress and sympathy will help.

Family VAEJOVIDAE

VAEJOVIS

Vaejovis carolinianus SOUTHERN UNSTRIPED SCORPION

Species Recognition: The southern unstriped scorpion varies from light brown or reddish to slate gray or almost black. The largest speci-mens are about 5 cm long.

Geographic Distribution: Found in scattered localities from central Ken tucky, eastern Tennessee, western parts of Virginia and the Caroli nas, and through Georgia and Alabama, mostly above the Coasta Plain. Isolated records are also available for eastern Mississippi an the border area of southwestern Mississippi and Louisiana. Foun in several counties above the Fall Line in Alabama and in a fev below, as far south as Dallas County.

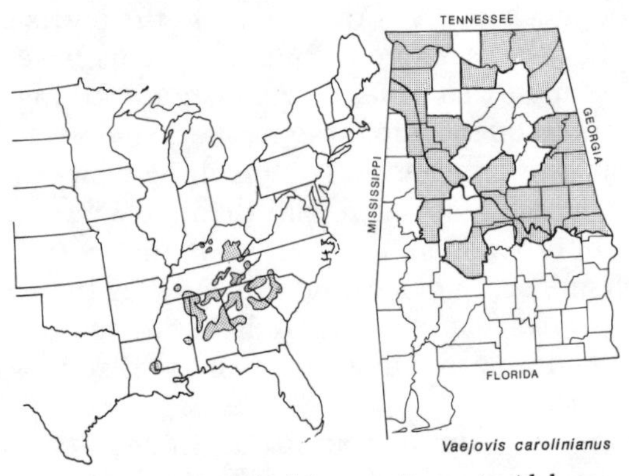

Range in eastern United States Range in Alabama.

Habitat Occurrence: The southern unstriped scorpion is generally re stricted to moist woodland habitats where it lives beneath leaves logs, and other litter.

General Life History and Behavior: This secretive little creature is probabl more common than casual excursions in the woods would sugges It feeds on small invertebrates that also live beneath rocks or decay ing wood. Scorpions use their front pinchers to grab their prey an then insert the stinger into a soft spot to kill the victim. The south ern unstriped scorpion gives live birth to young in the fall, and a many as 26 babies have been recorded. Most scorpions are activ from early spring until late fall and probably retreat deep into th ground during winter.

11 mm.

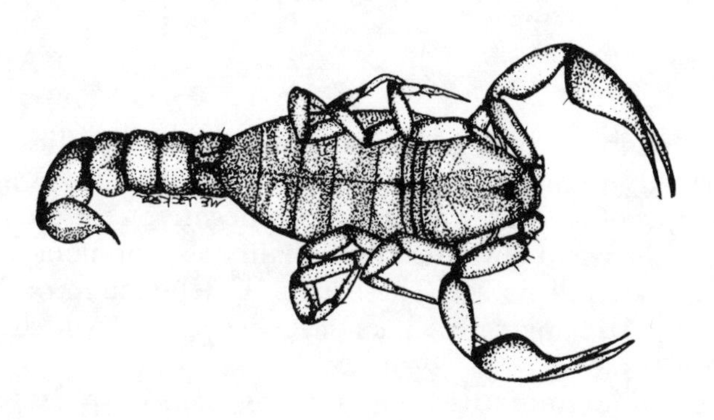

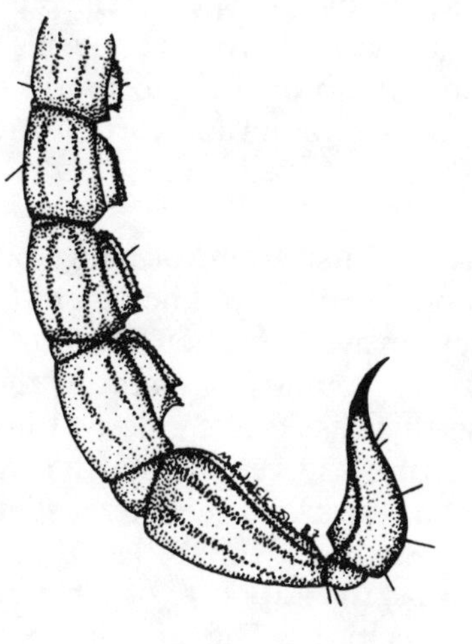

Vaejovis carolinianus (Southern Unstriped Scorpion)

Class Insecta

Order Lepidoptera

STINGING CATERPILLARS

Most people think only pleasant thoughts about butterflies. Some moths can be startlingly beautiful, and except for the peskiness of a candlefly or the holes in a wool blanket, most evoke no negative sentiments. Nonetheless, many moths and a few butterflies qualify as venomous animals when they are in the caterpillar stage. These juvenile delinquents of the order Lepidoptera are the stinging caterpillars, a few species of which are found in Alabama.

The danger comes from urticating (stinging) hairs or spines covering parts of the body and carrying a venom supply. Stinging caterpillars are a worldwide phenomenon, and more than 40 species have been reported in North America. Two genera of moths actually have species that are venomous as adults. However, only one (the brown-tailed moth) of the several species that are venomous as adults is found in North America, and it does not occur in Alabama.

Venom Mechanism and Administration: Stinging caterpillars do not seek out or attack people in any way. They merely mind their own business while munching away on their favorite vegetation, being well protected from many would-be enemies by the arsenal of poisonous hairs or spines on their backs and sides. Although the exact mechanism of hair structure and venom transfer from a caterpillar to its victim varies among species, the general structure and process are similar. In fact, in the many caterpillars that are hairy or spiny and completely harmless, the structure of the hair or spine is similar to that of the venomous forms. The only difference is that in the stinging caterpillars the cell at the base of the spine contains a reservoir of venom. Some have hairs that detach from the caterpillar, and some have hairs or spines that break off in the skin. In some species, the venom reservoir connects with a hollow spine, and the venom is actually ejected onto or into the victim as the spine breaks. Those having detachable hairs carry some venom inside the hair so that when the structure leaves the caterpillar and enters the skin, venom is released.

Venom Biochemistry and Symptoms: The biochemical makeup of caterpillar venom is even less thoroughly understood than that of many other venomous animals. One reason for the lack of laboratory analyses has been the difficulty of obtaining a sample for analysis. The venom is contained within individual cells and must be extracted separately from other fluids if it is to be analyzed precisely. Histamines, serotonin, various proteolytic enzymes, and proteins have been identified in the venom of several species, and formic acid has been notably absent in most, but the exact biochemical structure of any caterpillar venom is yet to be determined.

Initial symptoms, immediately after an encounter with a stinging caterpillar, are burning, itching, and a general irritation at the site of contact. Scratching is a normal response, but should be avoided because additional hairs may be broken and the venom released. With Alabama species, the effect is primarily a localized one that includes redness and swelling. The site of the wound may develop a noticeable wheal and be painful for several minutes, hours, or even days in severe cases. Fatalities have actually occurred from stinging caterpillars in some regions of the world, and serious reactions such as fever, vomiting, and muscle cramps have been reported. Although serious cases from the species that occur in the Southeast are rare, someone with severe allergies could conceivably suffer a medically significant reaction, and many a child has spent an unhappy day or two as a consequence of something as innocuous looking as a saddleback caterpillar. The unfortunate event of getting stinging hairs in the eyes has been known to result in serious visual complications, and inhalation can cause respiratory problems.

Treatment: For most people, a confrontation with a stinging caterpillar is a short-term experience that one wishes had not happened. Try not to scratch. Run cold water over the affected area to wash off any stray hairs that have not been embedded. One home remedy is to place adhesive tape over the wound and then yank it off to remove any hairs that may be sticking out of the skin. You'll probably get a few of your own hairs, too, but it could be worth it. The application of antihistamine salve can help relieve the itching to some degree, but the salve is not absorbed effectively through the skin and may actually make the skin more sensitive. A more suitable remedy may be hydrocortisone cream (0.5 percent), which is usually available without a prescription.

Be prepared not to enjoy life as much for the next hour or possibly several, and in some instances the effects of the sting linger for a day or two. If serious complications of any sort develop, see a physician. Or, if the stinging hairs get into the mouth, nose, and particularly the eyes, one should seek medical assistance. Normally, stinging caterpillars cause only a minor pain that can be lived with, but the possibility exists for an allergic reaction that deserves medical attention.

Family LIMACODIDAE

PARASA

Parasa (= Latoia) indetermina STINGING ROSE CATERPILLAR MOTH

Species Recognition: The front half of the moth's body is green; the rear half is yellowish. The front wings have a broad green area bordered by a brown band; the back wings are yellowish. The caterpillar is light in color with several dark stripes down the center of the back. Half a dozen or so protuberances bearing spines extend outward from the upper sides.

Geographic Distribution: Most of eastern United States from New York to Florida, and west to the Plains and Texas. Throughout Alabama.

Habitat Occurrence: The stinging rose caterpillar feeds on oak, hickory, maple, and dogwood trees, as well as rosebushes.

PHOBETRON

Phobetron pithecium HAGMOTH; MONKEY SLUG

Species Recognition: This is a not-so-attractive black-bodied moth with a wingspan of about 2.5 cm. Males are smaller than females and have translucent wings with black markings on the rear half of each wing and around the wing margins. The back wings of the female are dark, and the front wings are yellowish and brown.

The hagmoth caterpillar is light brown and is sluglike in appearance, usually with 9 pairs of hairy, yellowish-tipped, leglike extensions. The stinging hairs can cause intense localized pain, itching, and swelling.

Geographic Distribution: Eastern United States from New England to Florida, and west to Nebraska and Mississippi. Found throughout Alabama.

Habitat Occurrence: These caterpillars feed primarily on the leaves of trees, including oaks, willows, and dogwoods.

SIBINE

Sibine stimulea SADDLEBACK CATERPILLAR MOTH

Species Recognition: The saddleback caterpillar moth is a dark brown-to-black, heavy-bodied species with a wingspan of 2.5–3.5 cm. The front wings are a darker brown with a single white spot near the body and 1 or a few more near the front wing tips.

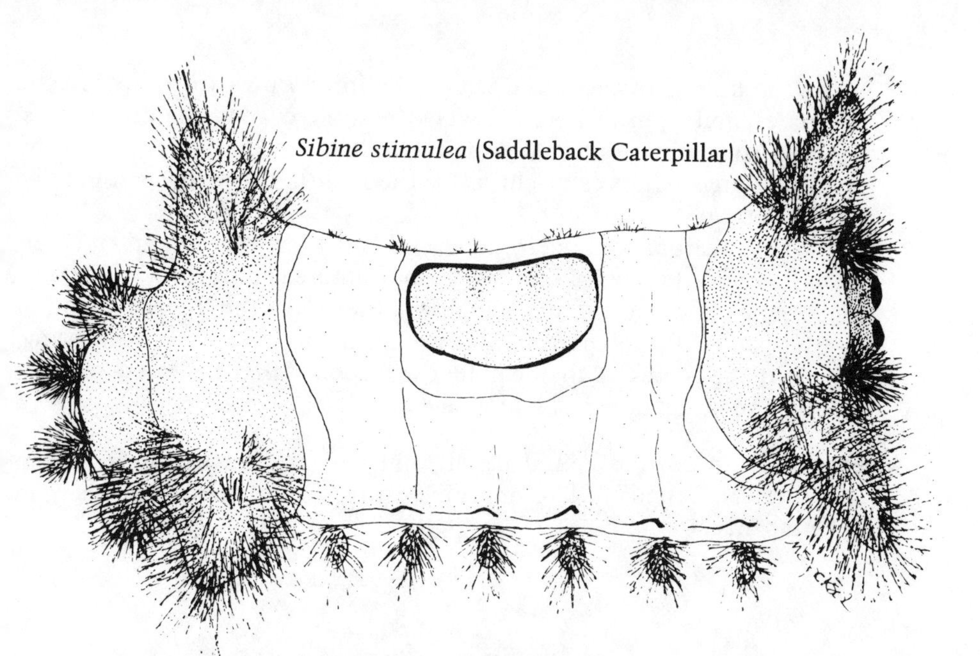
Sibine stimulea (Saddleback Caterpillar)

The fact that the species is named after the caterpillar stage speaks to its influence on people. This little chocolate brown denizen with pointed antennae-like extensions on each end looks like it has been outfitted with a bright green riding blanket with a white-bordered brown oval in the center. Stinging hairs line the sides, and an encounter with the caterpillar can result in a very painful and memorable sting that can cause a major welt. The aftereffects are localized at the wound, but may last for hours or even days.

Geographic Distribution: Lower New England to Florida, and west to eastern Plains states and Texas. Found throughout Alabama.

Habitat Occurrence: Saddleback caterpillars can subsist on a variety of plants such as apple trees, blueberries, corn, and plum bushes. The species is common in many areas and is frequently encountered in parks and recreation areas.

Family SATURNIIDAE

AUTOMERIS

***Automeris io* (Plate 23)** IO MOTH

Species Recognition: Both the adult io moth and the stinging caterpillar stage are common and easily recognized. The harmless and attractive moth has a large, circular blue-black spot with a white center in the middle of each hind wing. The front wings are yellow in the males, and a pinkish wash covers the inner edge of the hind wings. The front wings of females are brownish, and the pink area on the hind wings is not as bright as on the males. The wingspan may be more than 7 cm.

The creature of concern, however, is the large, bright green caterpillar with its longitudinal red stripe bordered below by a white one. A series of green tufts of stinging spines lines the upper portion of the body. The io caterpillar is not likely to cause a serious sting, but if one is picked up, the holder will soon realize his mistake. Itching and stinging may last for several hours.

Geographic Distribution: Found throughout eastern North America from southern Canada to Florida, and west to Texas. Occurs anywhere in Alabama.

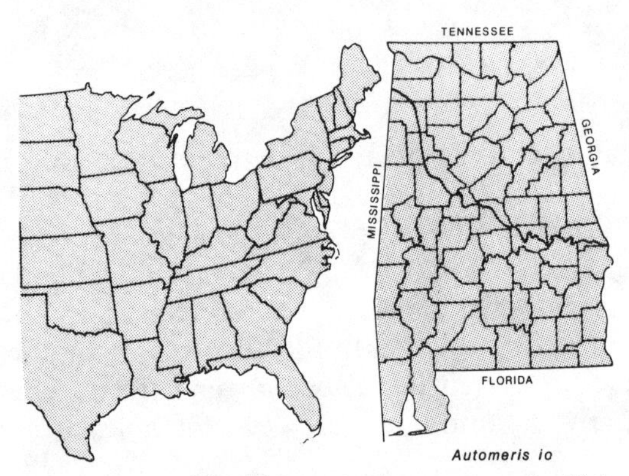

Range in eastern United States   Known range in Alabama

Habitat Occurrence: Io caterpillars eat the leaves of a wide variety of trees and other plants including oaks, maples, corn, and clover.

HEMILEUCA

Hemileuca maia BUCK MOTH

Species Recognition: The adult buck moth has a short, fat body and a wing span of 5–8 cm. The wings are dark with a white center band from front to back. Each of the 4 wings has a small, kidney-shaped eye spot. The terminal tufts of the male's body are red; in the female they are black.

The caterpillars are variable in color but are generally brown or yellowish brown with short tufts of spines on the dorsal surface. The tufts of spines on the sides are longer than those dorsally because of the presence of a central shaft that extends outward.

Geographic Distribution: Eastern United States from Maine to Wisconsin and Texas to Florida, and throughout Alabama.

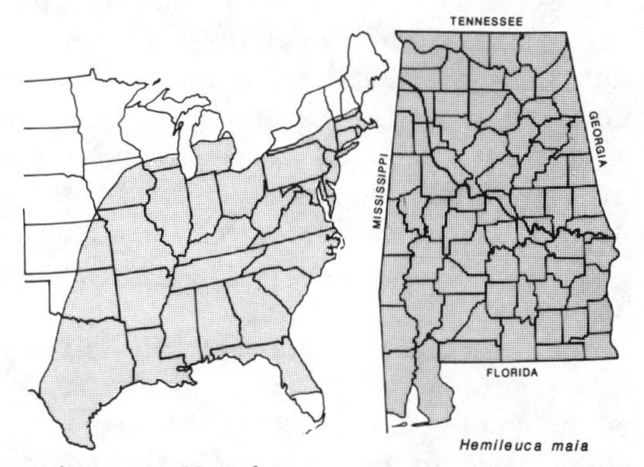

Range in eastern United States Known range in Alabama

Habitat Occurrence: Buck moth caterpillars are partial to oak trees and can be particularly abundant in scrub oak forests. An interesting case was reported in Maryland where buck moth caterpillars migrated from a scrub oak habitat to a strawberry field, occurring in such large numbers that the strawberry pickers had to stop operations because of the stings.

Other Stinging Caterpillars: Other species of Alabama moths have caterpillars that can cause irritation to the skin to varying degrees. Among the more notable are the members of the family Megalopygidae, including the black-waved flannel moth (*Lagoa crispata*), the southern flannel moth (*Megalopyge opercularis*), and the white flannel moth (*Norape ovina*).

Order Hymenoptera

WASPS, BEES, AND ANTS

Wasp is the general name given to a host of flying or crawling venomous insects that include paper wasps, hornets, yellow jackets, and velvet ants. Besides the large and obvious stinging kinds, the term encompasses a wide variety of small parasitic and solitary species. All female wasps have a stinger on the end of the tail, but most are harmless to humans.

Of the more than 4,000 species of wasps in the United States, hundreds of kinds live in Alabama. Practically all go about their life's business without contact with humans and are too small or otherwise inoffensive to have any effect on humans whatsoever. However, a few species deserve attention as truly venomous animals. Most belong to the family Vespidae.

Family VESPIDAE

This family includes primarily social wasps that live in colonies. The responsibilities of the colony are divided among 3 castes: a fertilized female or queen, sterile female workers, and fertile males. Well-armed with stingers and venom, most will unhesitatingly attack any intruder to protect their nests. The general reproductive pattern begins with a solitary queen building a nest in the spring. She lays eggs and then obtains food for the growing larvae. The prey is primarily other insects that the queen catches and kills. The larvae develop into female workers, with stingers, that begin adding to and protecting the nest.

The workers forage for nectar that they eat themselves. They also bring back insects or spiders for feeding the other larvae that develop into additional workers for the colony. Colony size continues to grow through the summer as the worker population increases. During late summer and early fall, the queen begins to lay eggs that develop into larvae that turn into males or become females that will be future queens. The workers begin to die during the cooler months of autumn, and the young queens are fertilized by the males. Eventually all colony members die except for the fertilized young queens. The queens leave their nests and seek a safe place to hibernate until spring when the cycle begins again. Nests are not used a second year.

Venom Mechanism and Administration: The stinger is a modified ovipositor located at the tip of the abdomen and thus is present only in the females. The stinger, associated with a venom gland, serves to inject the poison directly into the victim. The stinger itself is short and is normally contained inside the abdominal cavity at the end of the body. A poison sac is located deeper in the abdomen and connects to the stinger shaft by a poison duct. The poison sac is surrounded by muscle which contracts to inject the venom when the animal stings. The penetrating part of the stinger is composed of 2 barbed shafts that lie parallel but operate independently. Thus, during the stinging process one of the shafts, which are called lancets, is pushed into the victim. Although the stinging act appears to be one of jabbing the stinger into a person, the process is actually one in which the 2 lancets are alternately driven deeper and deeper. The poison duct opens directly between them and the venom can then be injected beneath the skin of the victim.

Venom Biochemistry and Symptoms: The venom of wasps is extremely complex with the most active constituents that have been identified being histamine, serotonin, dopamine, and noradrenalin. The enzymes phospholipase A, phospholipase B, and hyaluronidase are also present.

A sting from a member of this family can be extremely serious and is nearly always intensely painful. Physiological effects include the contraction of smooth muscles in the area, an increase of the permeability of blood capillaries, and a dilation of the blood vessels that can result in a drop in blood pressure. Many deaths have been documented from wasps in this family, and the causes include anaphylactic shock, respiratory problems, and bacterial infections. Most people experience only a great pain from a wasp sting, but for those who are allergic to the venom, a single sting can produce serious problems. Although the pain is immediate and serious medical complications are usually apparent within a few minutes or hours, there have been cases of actual death from wasp stings as long as 4 days after the sting itself.

Treatment: The application of ice or cold compresses to the sting generally relieves the pain and presumably prevents the venom from disseminating over a wider area. The application of meat tenderizer or a variety of salves that can be purchased from the drugstore are also effective for some persons. Obviously, if a serious sting appears in the offing, the victim should be taken immediately for medical treatment. Some medical doctors recommend that persons with a known allergy to wasp stings carry an emergency sting kit that includes syringes with epinephrine. Such a kit should be used only under the direction of a medical doctor.

POLISTES

Polistes annularis PAPER WASP

Species Recognition: This is the common bronze-colored or brownish wasp that lives in open-faced nests. It approaches lengths of 3 cm. The gray paper nest is attached by a single pedicel and may be more than 26 cm in diameter.

Geographic Distribution: Found throughout Alabama.

Habitat Occurrence: Nests of this species are common in bushes and vegetation alongside streams, lakes, and other bodies of water, as well as in barns and other buildings in rural areas.

General Life History and Behavior: New nests are begun in early spring, and the colonies increase in size during the summer, often consisting of more than 100 individuals. They are generally slow-moving wasps whose chief prey items are caterpillars. During winter, paper wasps hibernate under banks, in stumps, or in holes away from the nest and may be active on warm winter days.

Polistes exclamans GUINEA WASP

Species Recognition: This species is more brightly colored than the bronze paper wasp, with yellow bands around the abdomen. Body length is about 1–2 cm. The completed nest is grayish and usually less than 10 cm in diameter.

Habitat Occurrence: The guinea wasp is more common around houses, barns, and other man-made structures than in natural wooded areas.

General Life History and Behavior: *Polistes exclamans* is similar to *P. annularis* in regard to general ecology, except that it is less active during winter.

DOLICHOVESPULA

Dolichovespula maculata BALD-FACED HORNET

Species Recognition: This large wasp may be almost 2 cm long with a black and white pattern. The nest is generally a large gray oval structure of paper, and the largest ones may be almost 0.6 meters in length and over 0.3 meters in diameter. The upper portion is usually the largest in diameter, tapering somewhat toward the lower end.

Geographic Distribution: Found throughout almost all of the eastern United States and Canada except southern Florida. Bald-faced hornets occur westward into Texas and the Plains states, through Canada to Alaska, and south through most of coastal California and into the southwest except desert areas. Found anywhere in Alabama.

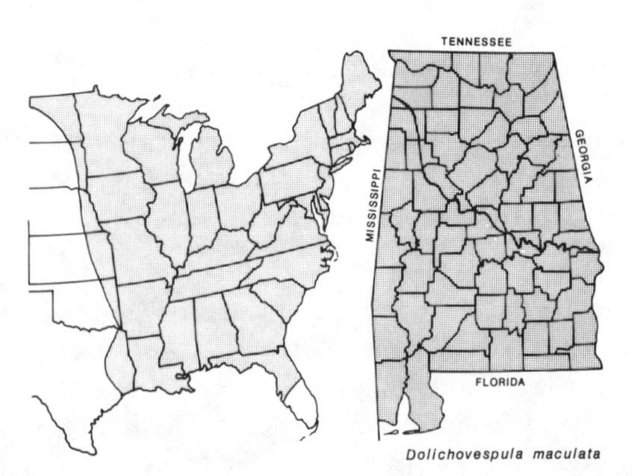

Range in eastern United States Known range in Alabama

Habitat Occurrence: This species does well in forested areas or swamps, but nests are frequently found in suburban areas.

General Life History and Behavior: The bald-faced hornet is one of the more impressive of the social insects in the United States. The workers will vigorously attack anyone molesting a colony, but will not sting unless they are intentionally or unintentionally disturbed. Fortunately, many nests are built high in trees. This reduces the chances of encounters with people, so that only someone looking for trouble is likely to have it with bald-faced hornets.

Related Species: Four other species of hornets in this genus are found in North America, and the geographic ranges of 3 of these closely approach Alabama at the northeast corner along the Appalachians. These are *Dolichovespula arctica*, *D. arenaria*, and *D. norvegicoides*. Although any of these species might be found in the extreme northern tier of counties in Alabama, the most likely one would be *D. arenaria*. Although it is a true hornet, its size and color pattern are those of a yellow jacket. But, unlike the yellow jackets, its nests are located in shrubs and trees and look very much like those of the bald-faced hornet. Hence it is called the aerial yellow jacket.

VESPA

Vespa crabro EUROPEAN HORNET

Species Recognition: This is a large hornet, more than 2 cm long, with a dark brown body and mostly yellow abdomen. The nest is normally inside hollow trees or other enclosed areas and consists of a series of brown paper combs one above the other. The only other wasp of similar size in this region is the cicada killer, which nests in the soil and is sometimes mistaken for the European hornet.

Geographic Distribution: Introduced from Europe into New York before 1860, and now found throughout most of the eastern United States. Potentially, this species can occur anywhere in Alabama, but few colonies have been discovered.

Habitat Occurrence: Most colonies are in heavily forested areas.

General Life History and Behavior: The worker hornets leave the nest in search of prey, which consists of other insects as well as hymenopterans such as honeybees and smaller yellow jackets, and have also been known to be attracted to lights at night.

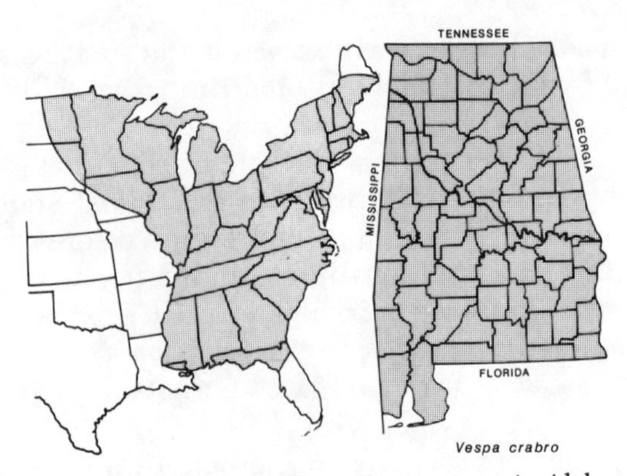

Range in eastern United States   Known range in Alabama

VESPULA

Vespula consobrina BLACK JACKET

Species Recognition: This is a small black and white wasp that nests in underground cavities (frequently rodent burrows) or beneath logs or rocks.

Geographic Distribution: Found throughout the southern half of Canada and down the Appalachians to Alabama. Also found along the western coastal areas and through western Montana, eastern Idaho to Colorado, and in the Great Lakes region. May be present in the extreme northeastern portion of Alabama.

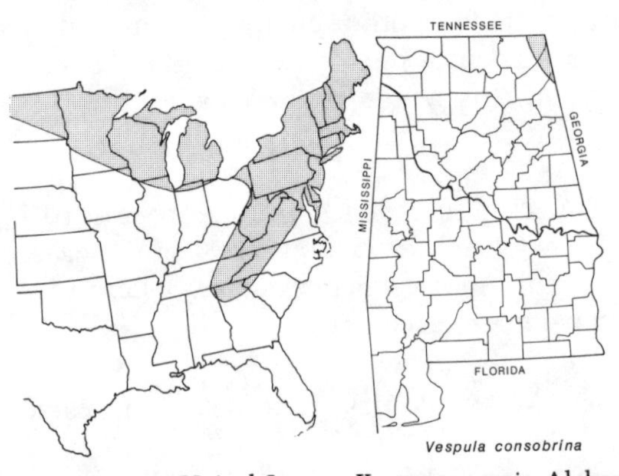

Range in eastern United States   Known range in Alabama

Habitat Occurrence: The black jacket is found primarily in forested areas, often at high elevations.

General Life History and Behavior: This species is infrequently encountered by humans because of its association with heavily forested areas, and the relatively small size of the colony.

Vespula maculifrons EASTERN YELLOW JACKET

Species Recognition: Similar in appearance but slightly smaller than the southern yellow jacket, this species also builds its nest underground or, less commonly, within structures.

Geographic Distribution: Entire eastern United States and southern edge of Canada, as far west as Montana, Wyoming, Colorado, and New Mexico. Found throughout Alabama.

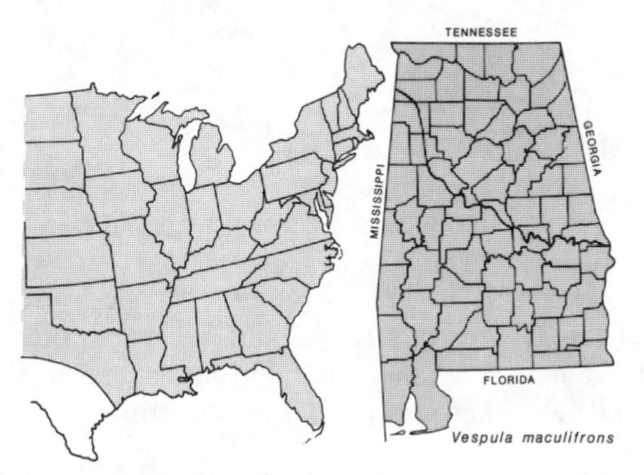

Range in eastern United States Known range in Alabama

Habitat Occurrence: The eastern yellow jacket builds its nest along creek banks and in hardwood forests as well as in urban areas where it may display a preference for the insides of house walls.

General Life History and Behavior: The eastern and the southern yellow jackets are the primary problem species of wasps in Alabama because of their ubiquity in parks, urban areas, and recreational sites. Individuals of either species are quick to defend their disturbed nests and unhesitatingly sting any person or pet that is within striking distance.

Vespula squamosa SOUTHERN YELLOW JACKET

Species Recognition: This species has the characteristic yellow and black banding pattern on the abdomen that typifies a "yellow jacket." Adults have 2 lengthwise yellow stripes between the wings, a feature that positively identifies the southern yellow jacket. All other yellow jackets are completely black between the wings. The paper nests are usually underground and may reach 37 cm in diameter.

Geographic Distribution: Throughout the eastern United States south of New York, Michigan, and Wisconsin, through eastern Texas and Mexico. Found throughout Alabama.

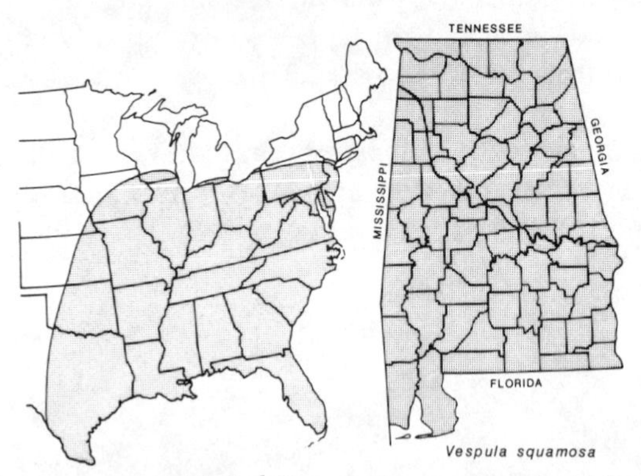

Range in eastern United States   Known range in Alabama

Habitat Occurrence: Southern yellow jackets can be found in almost any habitat throughout their range and are particularly abundant in many urban areas, parks, and recreational sites. This species and the eastern yellow jacket are the 2 most common species in Alabama.

General Life History and Behavior: This species is intensely protective of a nest that has been disturbed. A chemical present in the venom is highly attractive to other southern yellow jackets and thus elicits aggressive group stinging. One of the fascinating biological aspects of the species is that it is a social parasite of other yellow jackets such as *Vespa vidua*, which occurs in the northeastern section of the country, and *V. maculifrons*, which occurs in Alabama. The queen of the southern yellow jacket takes over the nest of the other species and begins laying her own eggs. The workers of the host species, apparently unaware of being usurped, care for the invader's eggs and young. Eventually the colony is populated only by the southern yellow jackets.

Family SPHECIDAE

This family includes the mud daubers, the attractive, even graceful, blue-black or black and yellow wasps that build nests of mud or clay beneath bridges or under the eaves of houses. Mud daubers bring spiders they have stung and paralyzed to the nest for the larvae to eat. They will fly away if disturbed, rather than defend their nests as yellow jackets do. A female mud dauber will sting if handled, but the minor pain subsides shortly.

This family also includes one of Alabama's largest wasps, the cicada killer (*Sphecius speciosus*). This enormous (more than 2.5 cm long) black and yellow wasp paralyzes full-grown cicadas and carries them to an underground nest. The female will sting if caught or stepped on, but the sting is mild considering the insect's size. An interesting phenomenon with cicada killers is that the females are shy and retiring. Like the mud daubers, females do not defend the nest, their main thrust in life seeming to be the acquisition of cicadas. The males, on the other hand, are aggressive toward humans, buzzing around the face and darting back and forth as if to protect their territory. You can run if you like, and the male cicada killer will join in the sport by chasing you, but the animal has no stinger and is absolutely harmless. Cicada killers are gregarious, many having nests in the same area, such as a lawn, but they are essentially harmless to humans. Because of their large size, they are occasionally confused with the European hornet.

Family MUTILLIDAE

The members of this family of wasps are called velvet ants because of their velvety body covering and because the females are wingless and spend their lives crawling on the ground. The flying males look formidable but are quite harmless.

Although the females are walkers, not flyers, and appear antlike, velvet ants are not true ants (which belong to the family Formicidae). Many species are brightly colored, often scarlet or yellow. The largest species in Alabama, the so-called cow killer (genus *Dasymutilla*), is among the most beautiful of the wasps, being bright orange or red. It reaches lengths of more than 2 cm and has a black stinger that is almost half the length of the body. As anyone knows who has picked up one of these wasps, the sting is powerful. Velvet ants are solitary wasps that parasitize other insects, primarily other wasps and bees, by laying their eggs in the nest. The velvet ant larvae feed on the host larvae.

Velvet ants are common in the southern and southwestern United States and occur widely in other parts of the world. In Europe, bumblebee nests are parasitized, and in Africa the tsetse fly is the host species. Velvet ants can be found throughout Alabama with cow killers being particularly common in areas of sandy soil.

The sting from a velvet ant hurts in an uncompromising manner for a short while, but the pain and swelling normally subside within a few hours. The application of ice to the wound usually reduces both the pain and swelling, but some persons may react allergically to the sting; unusual symptoms should be watched for and medical attention sought if necessary. The venom mechanism and characteristics of the sting are similar to those of the other wasps.

Velvet Ant (Family Mutillidae)

Family APIDAE
HONEYBEES AND BUMBLEBEES

Bees, although venomous, are one group generally credited with being worth the occasional discomfort of a sting. Although they provide honey, it is a trivial contribution to humans compared to the value of honeybees and bumblebees in the role of pollination. Fruit production that most people take for granted would be dramatically reduced if the world were suddenly without bees. Alabama has numerous species of bumblebees and one honeybee, all of which can deliver a sting that will not go unnoticed.

APIS

Apis mellifera HONEYBEE

Species Recognition: Among all the venomous species of animals in North America, the common honeybee is most generally recognized by everyone. There may be some variability in the shades of brown and yellow coloration, but only 1 species of honeybee occurs in the country. The many other species of bees are different enough in size, color, or behavior to make identification of the honeybee easy. Anyone who is not sure should ask the nearest grammar school child.

Geographic Range: Originally introduced from Europe and now found throughout North America, including Alabama. Honeybees have been introduced into most parts of the world and are now found wild or, more commonly, in man-made hives on all continents and major islands.

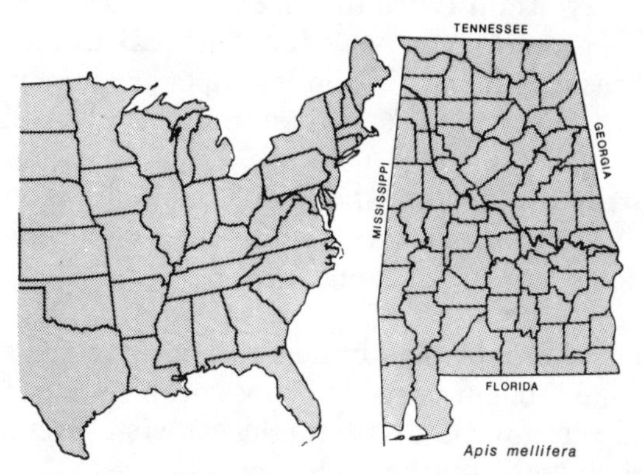

Range in eastern United States   Known range in Alabama

Habitat Occurrence: Honeybees are associated with man-made hives, but wild colonies will nest in hollow trees or other protected places.

General Life History and Behavior: The most fascinating behavior of honeybees is the individual worker's communication with others in the hive. The "language of the bees" was well documented by a European scientist, Karl von Frisch, who demonstrated that an intricate communication system exists in which body movements (dances) of one bee are used to pinpoint the location of pollen sources for others.

Honeybees will sting to protect the hive from intruders. Stings are also the defensive response of a bee that is picked up, stepped on, or that gets pressed against one's skin by clothing. The victory for the bee is, of course, a hollow one, since it leaves not only the stinger but the rear portion of its abdomen, and subsequently dies. But the animal makes its point that the victim should not bother honeybees. Incidentally, the fact that the stinger remains stuck in the skin identifies the honeybee. Wasps and yellow jackets have smooth stingers and can penetrate the skin repeatedly whereas the barbed stinger of the honeybee cannot be withdrawn. The honeybee, thus, gives its own life in defense of the hive.

Honeybees maintain their colonies throughout all seasons and form new ones by swarming when the colony becomes too large. The queen mates in flight with as many as a dozen different drones (stingless males) and then forms a new colony.

BOMBUS

Bombus sp. BUMBLEBEE

Species Recognition: Bumblebees have stout, hairy bodies that are black and yellow. They attain twice the size of honeybees. Because the similarity in appearance and behavior of several species of southeastern bumblebees defy identification attempts by most biologists, it is not particularly meaningful to differentiate between the species.

Geographic Distribution: Throughout the United States and Alabama. *Bombus griseocollis* is a common species of bumblebee, but several others also occur in Alabama.

Habitat Occurrence: Bumblebees build their nests in the ground, typically in abandoned rodent nests, but most people's experience with them is around the flowers that bumblebees visit in enormous numbers beginning in early spring each year.

General Life History and Behavior: The bumblebee is no more of a threat to humans than the honeybee. Ordinarily, to be stung, one must pick up or accidentally press a bee against one's skin in some way. An exception is the unfortunate event of unknowingly disturbing a bumblebee nest. Workers from an antagonized bumblebee colony will indeed attack, so be prepared for a sprint through the woods with several of them after you if you disturb a nest.

Bumblebees are more like wasps than honeybees in their social structure. The colony lives through the warmer months, but only the fertilized queen survives the winter and starts a new colony in the spring.

Family FORMICIDAE

ANTS

Ants are among the most abundant animals on earth. All ants are colonial, and some colonies may have more than a million residents. The basic colonies, like those of wasps or bees, consist of a queen, stingless males, and female workers that sting to defend themselves or their colony. Some accounts indicate that there are as many as 10,000 different species of which more than 600 have been described from North America. More than 200 of these may be found in Alabama.

Although ants are generally thought of as wingless, the queens and males have wings, and many species actually mate in flight. When ants have wings, they can be distinguished from other hymenopterans by their antennae which are not straight, but instead are elbowed at a sharp angle.

The envenomation by ants is primarily by means of a stinging apparatus in the tail, not unlike that of the wasps. Not all ant species sting. For example, common carpenter ants belong to a major group of stingless ants. However, some do bite with large front jaws, and many spray formic acid from a gland in the tail. Formic acid is a powerful toxin that is a major ingredient of ant venom although, like other venoms, the complete makeup is exceedingly complex biochemically. Ant species that spray formic acid, rather than inject through the stinger, can poison by spraying venom into the bite.

As anyone living in Alabama knows, ants can be found almost anywhere, indoors as well as out. Many people are bitten or stung by ants each year, usually without serious consequences. The most dangerous ant to most people in Alabama, and to an ever-increasing part of the southeastern United States, is the imported fire ant (*Solenopsis invicta*). Deaths in livestock and, rarely, people have been reported as a result of this species.

Mobile is credited with being the importation site of the fire ant from South America in the early 1900s. The species has now spread along the southeastern Coastal Plain to the Carolinas and has moved westward to Texas. Imported fire ants occur throughout most of Alabama, being in greatest concentrations in the southern third of the state. Fire ants are small, but their mounds of granulated earth can be more than 30 cm in diameter. Stepping into the soft structure results in an immediate invasion of any part of the foot or leg that touches the mound. Fire ants bite and sting, clinging tenaciously so that they are difficult to brush off, and some must be given special attention to remove them. The effect of a single ant sting can cause a severe systemic response that leads to shortness of breath, dizziness, and even anaphylactic shock. The sensitivity of some people increases over time with the accumulation of fire ant stings. The local effect is swelling, pain, itching, and the eventual formation of a pus-filled area that may persist for several days or even weeks. The immediate application of ice to the wound for up to an hour may alleviate pain and other symptoms. The use of antihistamine, as a salve or taken internally, is also helpful for some victims, and the application of hydrocortisone cream (0.5 percent) is considered to be of value in some instances.

Fire ants are a menace and should be an issue of concern to parents of small children who might wander into a mound. Check with a county agent or local farm supply store for chemicals to eliminate nests in the yard. Stings of other ants vary in their potency and can be a source of irritation in some instances. Ice and antihistamines are effective treatments for stings by most species.

BITING AND SUCKING BUGS

The gradient of effects of poisonous plants and venomous animals already mentioned ranges from mild to lethal. The effect depends on the chemical composition of the poison or venom, the physiological condition of the victim, and other factors. We have defined venom as a substance that is injected for the purpose of subduing prey or protecting the creature from a presumed enemy. However, many insect species that cause pain or discomfort are not truly venomous. Detailed accounts of these various species will not be given, but some merit mentioning because of the obvious and immediate impact they may have on the victim.

Many flies and mosquitoes (order Diptera) are blood-sucking animals that like to get their meals from humans. The saliva from some species is an anticoagulate that makes the victim's blood flow more freely into the mouth tube that has been injected under the skin. This saliva is what causes the itching from mosquito bites. In one sense it could be considered a venom that has been injected into the prey animal. Deerflies, horseflies, and the tiny biting midges known as "no-seeums" also have their effect in a similar manner.

Another group of insects that is not truly venomous includes members of the order Hemiptera, the true bugs. Two common kinds in Alabama that warrant serious attention are the giant water bugs (family Belastomatidae) and backswimmers (family Notonectidae). Giant water bugs are primarily aquatic but will fly to lighted areas at night where they might be picked up by someone interested in what looks superficially like a large beetle. The bite is extremely painful but is a mechanical, not chemical, infliction, which is caused by pinching. Bites from backswimmers frequently occur when a person gathering aquatic vegetation inadvertently picks up the animal. Students sorting through material in a dipnet or seine are often victims. The pain is intense but short-lived, and 5 minutes later there may be no evidence of the bite, only a memory. There is no venom associated with the bite of either of these bugs.

Some other insects such as assassin bugs, fire beetles (these black beetles with red heads actually spray an acid onto the skin), and fleas also deal with humans in ways that are not well received. But none is venomous in the sense we have used the word.

Chapter 7
Venomous Vertebrates

Far fewer vertebrates than invertebrates of Alabama are technically venomous, yet the vertebrates have a commanding psychological lead in human fear because of one group—the poisonous snakes. The term *poisonous* is used because of its common usage, not because *venomous* would be less correct. None of Alabama's venomous snakes has been documented to be poisonous to eat, and in fact at least 3 species are passably good when properly cooked. Caution: It has been suggested that the coral snake may taste bad. However, advice not to eat coral snakes will probably be perceived by most to be of minimal value.

Only 7 species of vertebrates native to Alabama, or anywhere in the entire Southeast, use oral envenomation for both food-getting and protection. These are the 5 species in the family of pit vipers [cottonmouth moccasin, copperhead, eastern diamondback rattlesnake, canebrake (or timber) rattlesnake, and pygmy rattlesnake], the coral snake of the cobra family, and perhaps a surprise to most, the short-tailed shrew. This species is included not because of its threat to the populace of Alabama, but because of its near-unique place in the animal world as a venomous mammal. We will probably wait a long time for the first human death from a shrew. No birds or amphibians are venomous anywhere in the world, and Alabama has no venomous lizards (only 2 species—the gila monster and Mexican beaded lizard—are extant today, and they are confined to the southwestern United States and Mexico).

No fish that inhabits fresh water or that is common in the near-shore coastal systems of Alabama is venomous orally. However, the injection from spines on certain species of fresh- and saltwater catfish can sometimes cause serious injury, although the consequences under normal circumstances are only painful, not life-threatening. Some of the stingrays, however, can cause an extremely serious wound, and deaths have been reported, including in gulf coastal waters. Many marine fishes have venomous spines, and in other parts of the world injuries from some species result in a high incidence of death. Some of Alabama's coastal fishes are venomous, and although being finned may be extremely painful, the chance of losing one's life as a result is unlikely. In addition, the likelihood is low that the average coastal visitor or inhabitant will encounter most of the venomous saltwater fishes.

Emphasis will be on those fishes likely to be encountered by the average person doing normal things. For example, anyone who comes upon a spinycheek scorpionfish, which lives in the gulf at depths of 30–100 fathoms, would presumably not fault the state of Alabama for harboring this venomous species. Therefore, such unlikely candidates as the spinycheek scorpionfish will not be discussed.

Each species will be treated in a separate account, but common aspects, such as treatment for venomous snakebite, will be considered as a unit.

Class Chondorichthyes

Order Myliobatiformes — Rays

Rays are mostly marine animals that are closely related to the sharks. They are included in this book because 4 of the species in Alabama can inflict painful and potentially dangerous injuries with a venomous spine in the tail. Any of these species can be found along the Alabama coastline and occur throughout most of the Gulf Coast region of the United States; however, only 2 species are likely to be encountered.

Venom Mechanism and Administration: Stingrays inject a venom into the body when the barbed spine penetrates the skin. The barbs point backward so that removal results in torn flesh and a physically painful wound. The venom is located along the spine beneath tissue that releases it into the body upon impact or following attempts to remove the spine.

Venom Biochemistry and Symptoms: The pain from a stingray spine is extreme in the area of contact and is definitely cause for alarm. The venom is a complex protein that has not been thoroughly analyzed.

Treatment: The victim of a stingray attack usually requires medical attention for shock and for the treatment of potential consequences of the venom itself. Antibiotics should probably be administered. Because the venom is extremely sensitive to warm temperatures, bathing the wound in hot water will neutralize it, but in view of the large size of the spines of some stingrays, medical attention may be required for treatment of physical damage.

Family DASYATIDAE

DASYATIS

Dasyatis sabina — ATLANTIC STINGRAY

Species Recognition: This is the more abundant of the 2 species of stingrays common on the Alabama coastline. The Atlantic stingray has a somewhat triangular snout. The color is sandy brown and the size is up to 50 cm across.

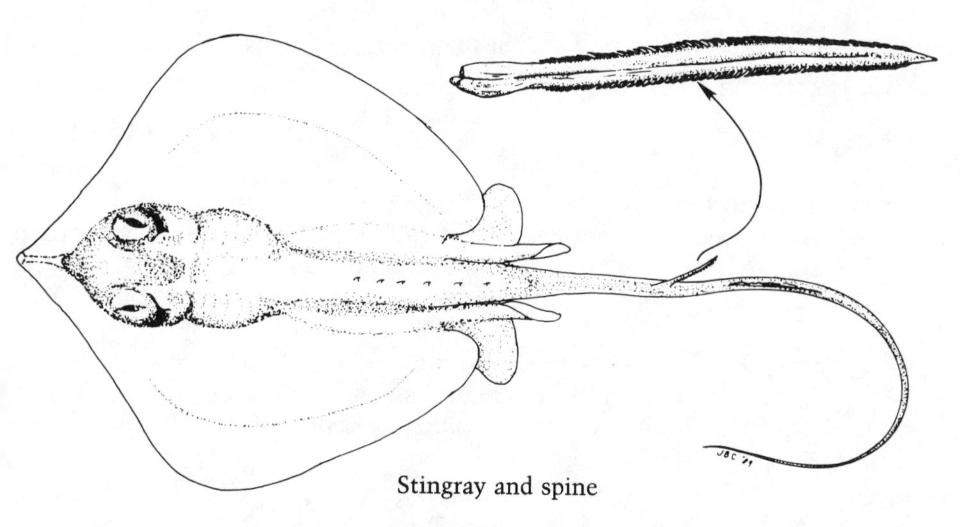
Stingray and spine

Geographic Distribution: Along the Atlantic coast from Chesapeake Bay to Florida and the entire Gulf Coast region.

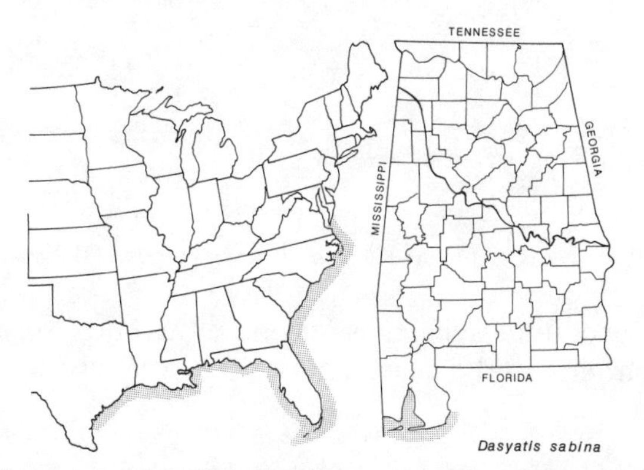

Range in eastern United States Known range in Alabama

Habitat Occurrence: This species is common in shallow-water areas where the sand or silt bottom matches its basic color. It is most frequently seen in estuary shallows and will sometimes enter waters that are only slightly brackish or even fresh. Many Atlantic stingrays are caught on fishing lines throughout the gulf and southern Atlantic. Most injuries occur to swimmers or waders in shallow waters, although occasionally fishermen who carelessly handle a stingray in a boat or on a dock are wounded. In the wintertime stingrays move from the shallow estuaries into the slightly deeper offshore waters.

General Life History and Behavior: Atlantic stingrays eat a wide variety of crustaceans and marine worms and are also considered to be scavengers of animal remains. Stingrays are live-bearers that have several young at a time.

Dasyatis sayi BLUNTNOSE STINGRAY

Species Recognition: The bluntnose stingray is similar in appearance to the Atlantic stingray but has a more rounded front edge. It is larger than the Atlantic stingray, with most adults being more than 60 cm across and some reaching widths of more than 1 meter.

Geographic Distribution: Ranges throughout the Atlantic coastal region from New England to Brazil and all of the Gulf of Mexico. Can be expected along any portion of the coast of Alabama.

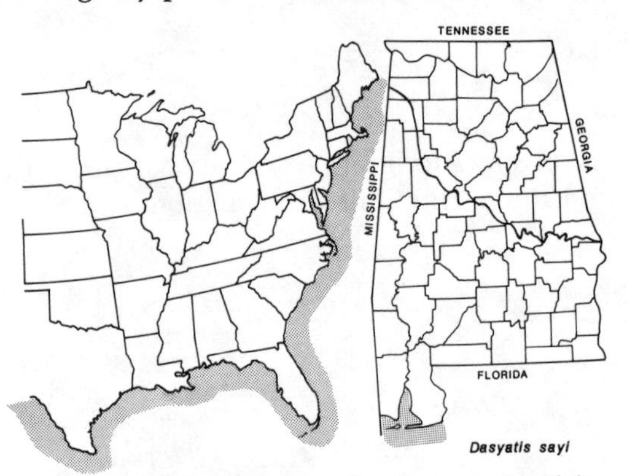

Range in eastern United States Known range in Alabama

Habitat Occurrence: This species is less likely to enter fresh water and is more common in the coastal ocean environment than is the Atlantic stingray.

General Life History and Behavior: Little is known about the ecology and behavior of the bluntnose stingray.

Family GYMNURIDAE

GYMNURA

Gymnura micrura SMOOTH BUTTERFLY RAY (NONVENOMOUS)

Species Recognition: Butterfly rays are easily distinguished from the 2 stingrays by having a very short tail and a much shorter body relative to the width of the "wings." The tips of the wing may be more than a meter apart, but because of the shorter body length, this species does not reach the large size of the stingrays. Butterfly rays have no spine and therefore are not dangerous like the stingrays. The species is included only because it looks similar to the dangerous forms.

Geographic Distribution: Throughout the gulf and Atlantic regions, including warmer sections of the eastern Atlantic. Uncommon, but may be encountered anywhere along the Alabama coastline.

Habitat Occurrence: Butterfly rays are found on the bottom, but are less likely to be found in estuary areas because of a preference for salty water.

General Life History and Behavior: Little is known of the behavior and ecology of this species.

Family MYLIOBATIDAE

AETOBATUS

Aetobatus narinari SPOTTED EAGLE RAY

Species Recognition: This strikingly colored ray has a white belly and a brown or gray back with numerous equally spaced white spots. The head is set out forward of the wings, and the tail is almost twice as long as the body. The spine is near the base of the tail, and therefore is not used effectively on humans in most instances. This is the largest ray along the Alabama Gulf Coast and may be more than 2 meters in width and weigh more than 180 kg.

Geographic Distribution: Worldwide in all temperate or tropical seas. Found throughout the gulf and may be expected, though rarely, anywhere along the coast of Alabama.

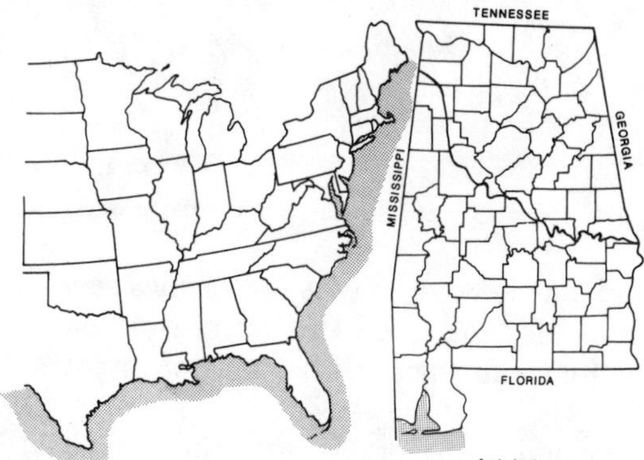

Range in eastern United States Known range in Alabama

Habitat Occurrence: This species inhabits open water and is unlikely to be seen in the nearshore coastal areas.

General Life History and Behavior: The eagle rays eat mussels and other shellfish, which they crush with powerful teeth. This species is not considered common in most areas, although schools of many dozens have been reported in the ocean. Eagle rays are live-bearers and normally have a half dozen or more young.

RHINOPTERA

Rhinoptera bonasus COWNOSE RAY

Species Recognition: Cownose rays are similar in appearance to butterfly rays, except that the wings are more pointed and the head is extended forward from the center of the body. A deep groove passes between the eyes down the middle of the skull. Most cownose rays are about 60 centimeters in width, and large ones may be close to 1 meter. The tail is long and whiplike, but the spine is near the base, so there is less chance of its hitting a swimmer.

Geographic Distribution: Found throughout the Atlantic Ocean from New England to Brazil and throughout the Gulf of Mexico. It is possible that this form occurs in the Indian and Pacific oceans.

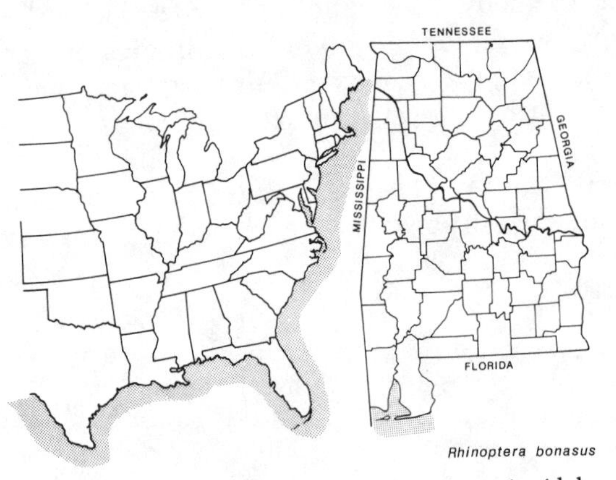

Range in eastern United States Known range in Alabama

Habitat Occurrence: The cownose ray is sometimes seen in schools in open waters, but is also found on the bottom in coastal areas. It is less likely to be found in the shallow areas of estuaries.

General Life History and Behavior: As with the other rays, little is known of the general ecology and life history of this species. The schooling behavior is believed to be associated in some way with reproduction

Class Osteichthyes

Order Siluriformes Catfishes

The only species of freshwater fishes in Alabama that are considered truly venomous are certain catfishes. Although venom can be injected and the outcome is temporarily painful, these fishes do not constitute a serious danger. Most fishermen learn quickly to avoid the spines of a catfish through careful handling. In addition to the freshwater catfishes, Alabama also has 2 venomous saltwater species that are common on the coast.

Venom Mechanism and Administration: The 2 pectoral spines and the single dorsal spine are the mechanisms of envenomation. On a large specimen the stab alone would be painful. However, many species carry venom in a thin sheath surrounding the spine so that upon injection of the spine, the poison is released beneath the skin. The spines are movable and can actually be moved into an upright position by the catfish. Most catfish stings are a result of careless handling.

Venom Biochemistry and Symptoms: The chemical makeup of catfish venom has not been analyzed for most species. Presumably, it is a complex of proteins that varies among species to the extent that those in some parts of the world (not in North America) can cause a lethal sting whereas others can cause a sting equivalent to that of a wasp or even of less consequence. Alabama species belong in these last 2 categories.

Even a tiny madtom catfish can give a jolt to the person who picks it up. The effect is clearly a sting, not just a jab from a sharp object. The burning and stinging sensation, numbness, and swelling can last from several minutes to hours from larger specimens. Under most circumstances, no Alabama catfish is likely to give someone a medically serious sting that is attributable to the venom, but the spine of a large catfish can cause a puncture wound that in its own right requires medical attention.

Ordinarily, the pain from a catfish sting should be alleviated by application of ice, and the wound itself should be treated as any other such injury.

Family ICTALURIDAE

Three genera and 18 species of freshwater catfishes are found in Alabama. They vary in level of toxicity, but most have not been thoroughly enough tested to permit a valid comparison of the effect of the venom on humans. In one test of mosquitofish response to venom from various species, the

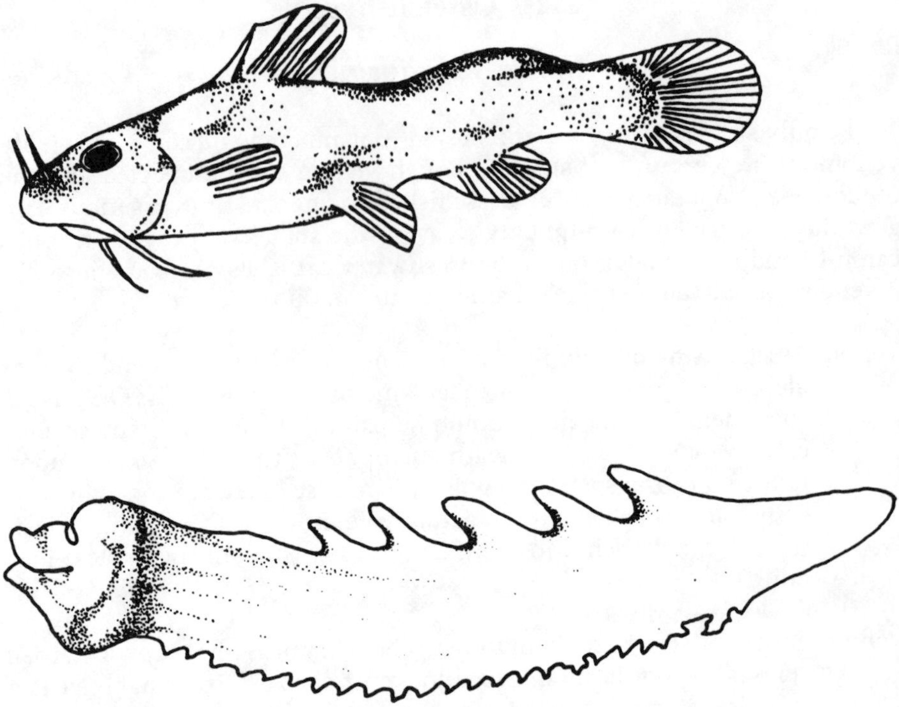

Catfish and Spines (*Noturus* sp.)

black bullhead (*Ictalurus melas*) and the slender madtom (*Noturus exilis*) were most toxic. The speckled madtom (*N. leptacanthus*) and flathead catfish (*Pylodictus olivaris*) were considered to be nontoxic. However, the response of another fish may not be a valid indication of how mammals might respond, so it is best to consider all catfishes guilty of being venomous until proved innocent.

ICTALURUS

Ictalurus punctatus CHANNEL CATFISH

Species Recognition: This is considered the premier catfish of North America by many fishermen, fish eaters, and catfish farmers. Like other catfishes, channel cats have several fleshy barbels near the mouth. The fish is grayish in color, has a forked tail, and may be more than 1 meter in length. Blue (*I. furcatus*) and white (*I. catus*) catfishes also have forked tails, whereas bullheads do not.

Geographic Distribution: Throughout the central United States, extending east into New York, and south to Florida. The channel catfish has now been introduced into many far western localities. Found throughout Alabama.

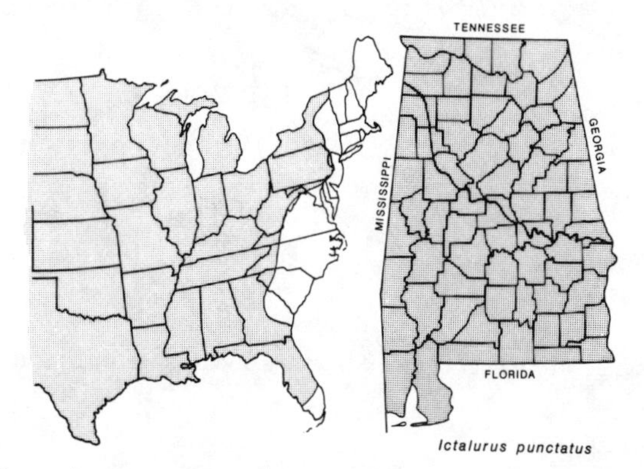

Range in eastern United States   Known range in Alabama

Habitat Occurrence: The channel catfish historically occurred in large clean rivers but now can be found in streams, lakes, and reservoirs because of introductions. Specimens are occasionally found in brackish waters.

Life History and Behavior: The female channel catfish lays eggs during late spring and early summer beneath the bank or on the bottom in a spot cleared by the male. The male serves as a guard for the developing eggs. Young channel catfishes eat a variety of invertebrate prey whereas larger specimens eat fish or other animals.

Additional Species in Alabama: Several other species similar to the channel catfish also occur in Alabama waters.

Ictalurus brunneus SNAIL BULLHEAD

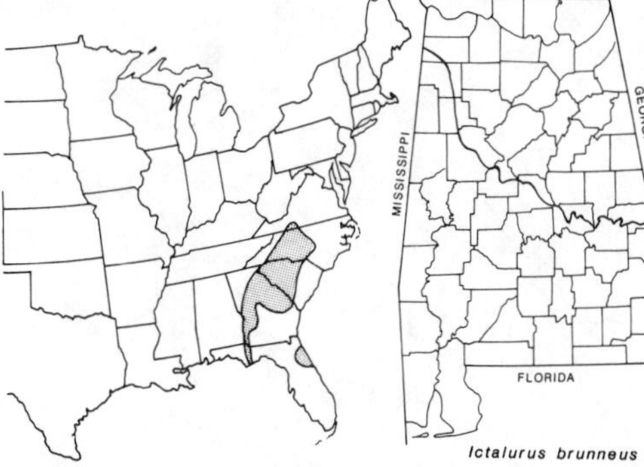

Range in eastern United States   Known range in Alabama

Ictalurus catus WHITE CATFISH

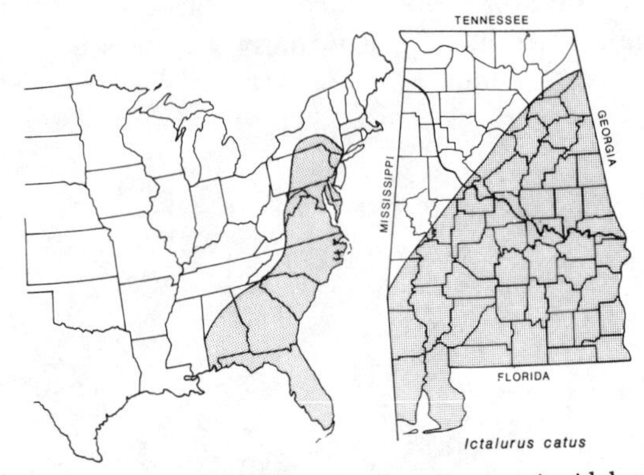

Range in eastern United States Known range in Alabama

Ictalurus furcatus BLUE CATFISH

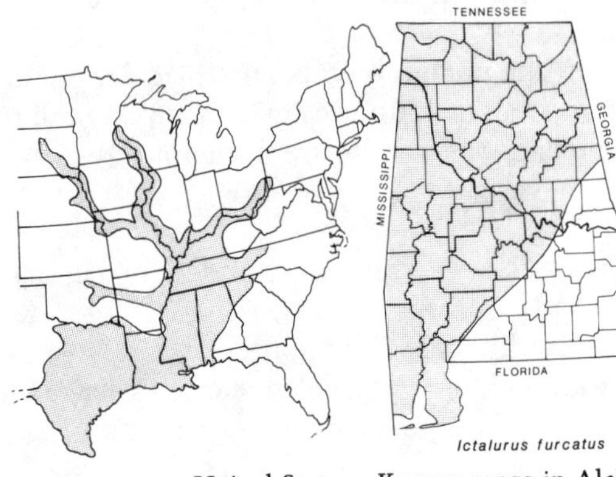

Range in eastern United States Known range in Alabama

Ictalurus melas BLACK BULLHEAD

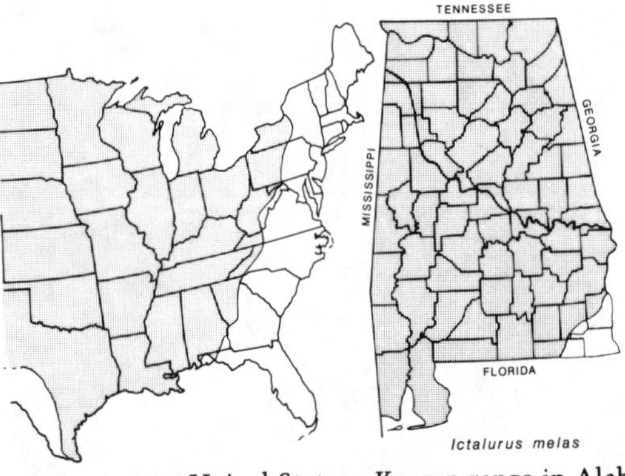

Range in eastern United States Known range in Alabama

Ictalurus natalis YELLOW BULLHEAD

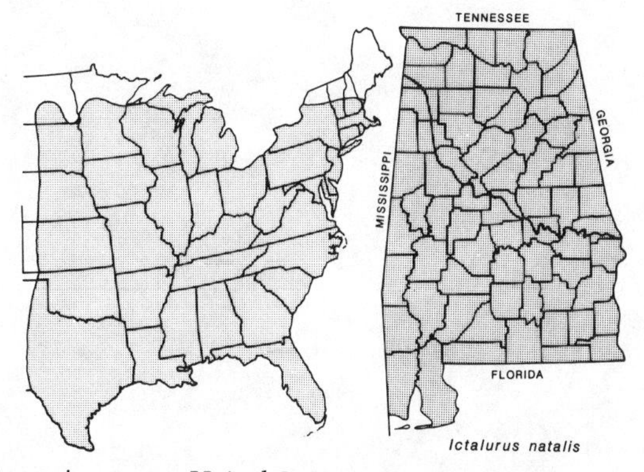

Range in eastern United States Known range in Alabama

Ictalurus nebulosus BROWN BULLHEAD

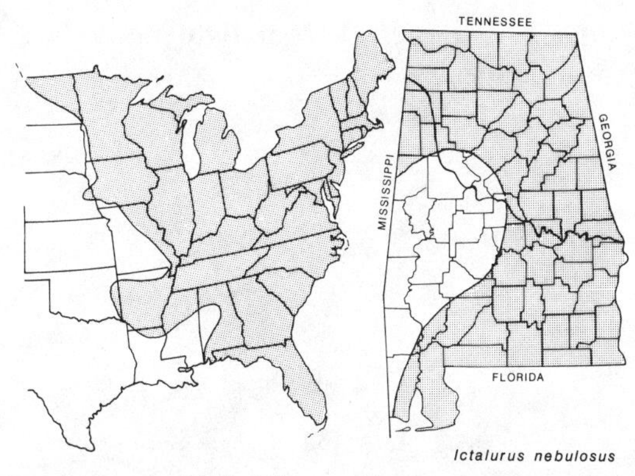

Range in eastern United States Known range in Alabama

Ictalurus serracanthus SPOTTED BULLHEAD

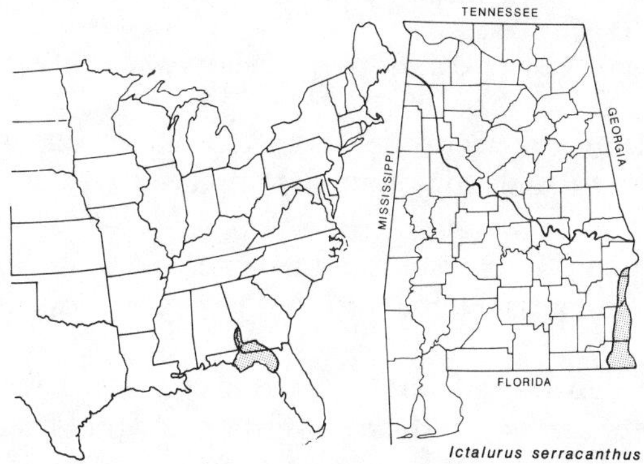

Range in eastern United States Known range in Alabama

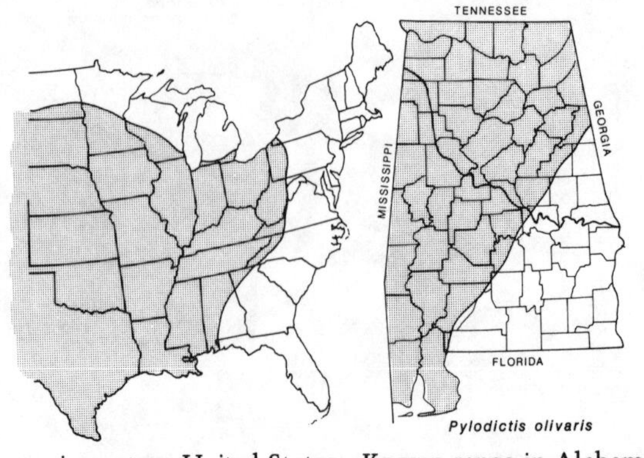

Range in eastern United States Known range in Alabama

NOTURUS MADTOM CATFISHES

Most of the madtom catfishes are small (usually less than 12 cm in length), stream-dwelling species.

Noturus nocturnus FRECKLED MADTOM

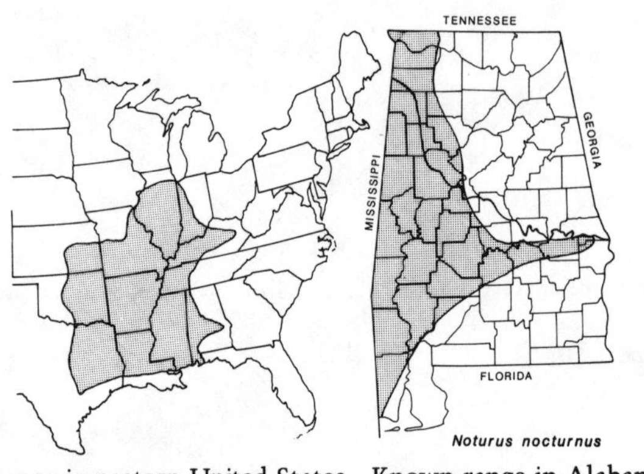

Range in eastern United States Known range in Alabama

Geographic Distribution: Central Mississippi River Basin to Gulf Coast from Texas to Alabama. Found throughout western Alabama.

Habitat Occurrence: The freckled madtom lives in moderate-to-large streams and is associated with both deeper pools and riffle areas.

Life History and Behavior: Female madtoms lay eggs in a protected site (empty beer cans seem to be a common and preferred egg-laying hab itat in many Alabama streams) during spring or summer. The male guards the egg mass from potential predators, and the young hatch in about a week's time. Juvenile madtoms are generally no more than 2–3 cm in length a month or so after hatching, but the venomous spines are already functional.

Additional Species in Alabama: Several other species with similar appearances and properties as the freckled madtom also occur in this region.

Noturus elegans ELEGANT MADTOM

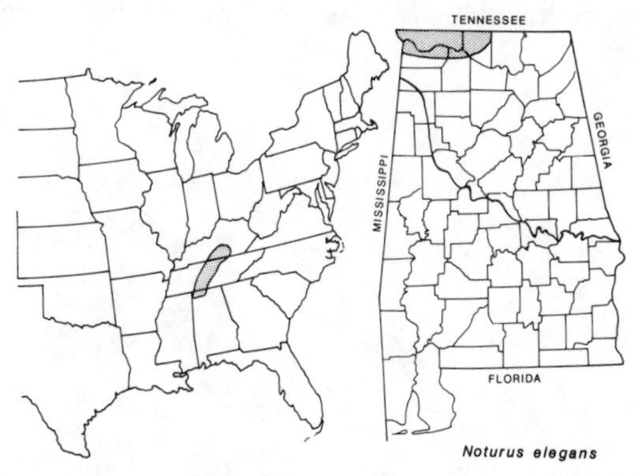

Range in eastern United States Known range in Alabama

Noturus exilis SLENDER MADTOM

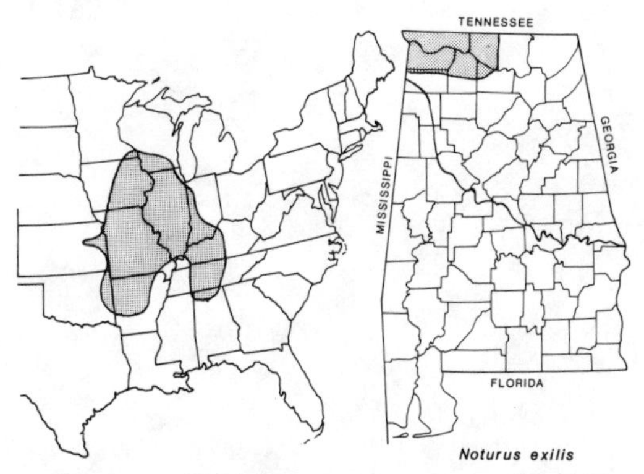

Range in eastern United States Known range in Alabama

Noturus flavus STONECAT

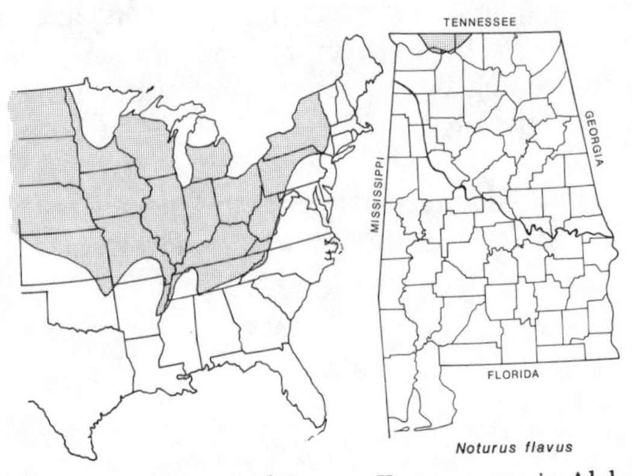

Range in eastern United States Known range in Alabama

Noturus funebris BLACK MADTOM

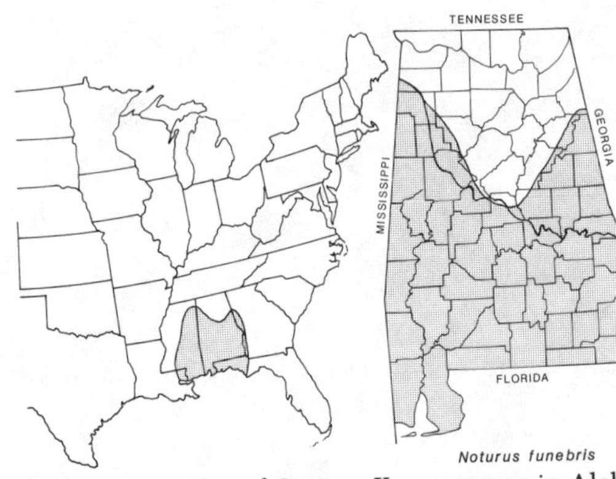

Range in eastern United States Known range in Alabama

Noturus gyrinus TADPOLE MADTOM

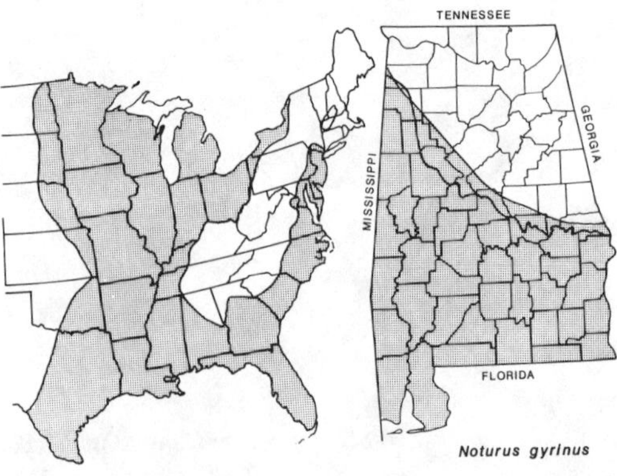

Range in eastern United States Known range in Alabama

Noturus leptacanthus SPECKLED MADTOM

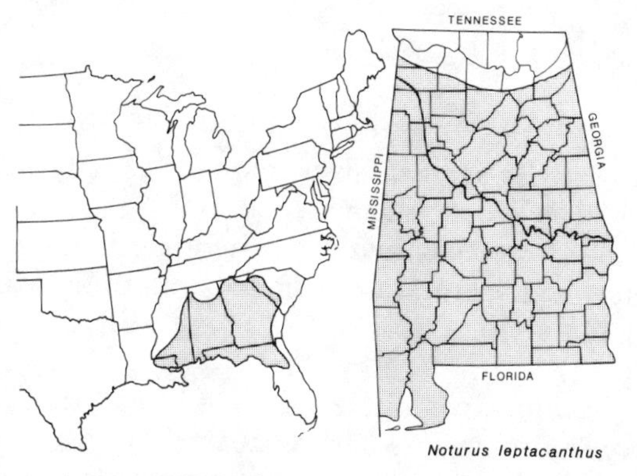

Range in eastern United States Known range in Alabama

Noturus miurus BRINDLED MADTOM

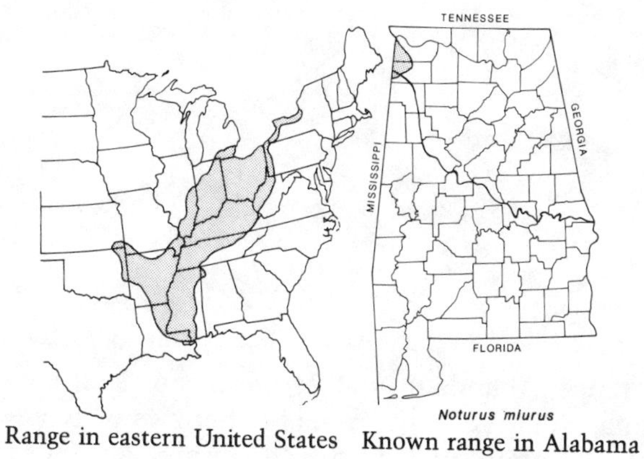

Range in eastern United States Known range in Alabama

Noturus munitus FRECKLEBELLY MADTOM

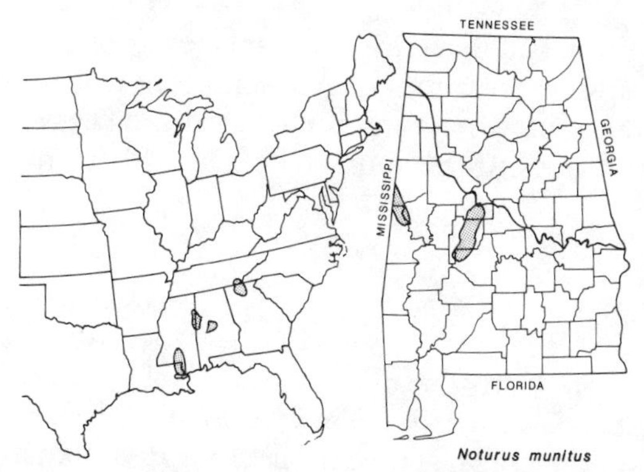

Range in eastern United States Known range in Alabama

Family ARIIDAE

ARIUS

Arius felis SEA CATFISH

Species Recognition: The sea catfish is a small (usually less than 30 cm), olive-colored species with an erect dorsal spine. Six barbels are present, 4 on the lower jaw and 2 on the upper. The venomous spines of the dorsal and pectoral fins can cause a mildly painful injury that is usually not serious.

Geographic Distribution: Coastal New England to Florida and coastal Gulf of Mexico. Common in Alabama coastal areas.

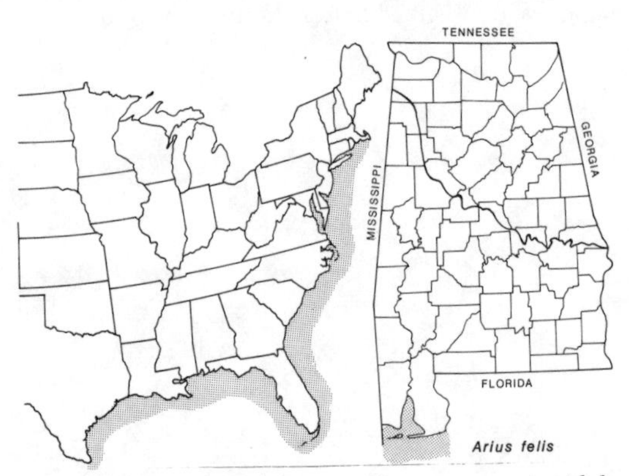

Range in eastern United States Known range in Alabama

Habitat Occurrence: Abundant in all estuarine and shallow coastal areas of Alabama, where it is the bane of fishermen who are trying to catch almost any other kind of fish. The disappointment of catching one is more than doubled when the grumbling fisherman is speared by one of the spines.

General Life History and Behavior: This species is an omnivorous scavenger that is present year-round. A fascinating feature of the sea catfish is that it is a mouth brooder. The fertilized eggs are held by the male in his mouth until the young hatch, at which time they are released.

BAGRE

Bagre marinus GAFF-TOPSAIL CATFISH

Species Recognition: Gaff-topsail catfishes are often larger (often more than 60 cm) than sea catfishes. They have only 2 barbels on the lower jaw. The tip of the dorsal spine has an elongated filament extending from it. The dorsal and pectoral spines are venomous.

Geographic Distribution: Along Atlantic coast from New England to Florida and South America. Found throughout the northern Gulf of Mexico. Common in coastal waters of Alabama.

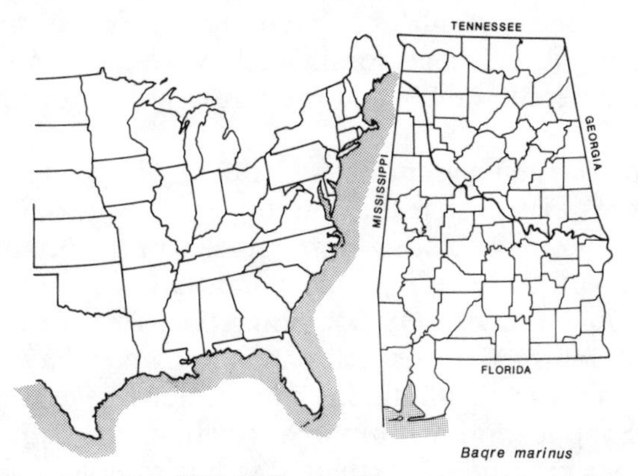

Range in eastern United States   Known range in Alabama

Habitat Occurrence: This species is found in estuaries, sounds, and open ocean waters, as well as near shore.

General Life History and Behavior: The gaff-topsail catfish is also a mouth brooder. As with the sea catfishes, this species is unlikely to be encountered except by fishermen. The primary danger is in handling specimens dangling from a fishing line.

Order Perciformes

Family SCORPAENIDAE

SCORPAENA SCORPIONFISH

Scorpionfishes are closely related to stonefishes, which are considered to be among the most dangerously venomous fishes in the world. Fortunately, stonefishes do not occur anywhere near the Gulf of Mexico. However, scorpionfishes do, and although the dangers are not of equal significance, 3 species can be encountered in Alabama's coastal waters. These are the barbfish (*Scorpaena brasiliensis*), the spotted scorpionfish (*S. plumieri*), and the smoothhead scorpionfish (*S. calcarata*). All are small fishes, the spotted scorpionfish reaching about 30 cm, the barbfish 20 cm, and the smoothhead only 10 cm in length. Some scorpionfishes are relatively colorful, the barbfish being mostly dull red and mottled. The most common species is the spotted scorpionfish which has dark and light brown and gray mottling.

Venom Mechanism and Administration: The dorsal, pelvic, and anal spines of scorpionfishes consist of a distinct venom gland that lies alongside each spine and is covered by a sheath. When the spine penetrates the skin, the venom is released.

Envenomation by scorpionfishes in Alabama's coastal waters would most likely result from handling. It is conceivable that a scorpionfish might be stepped on in shallow water; some are known to extend their spines and stand their ground when threatened.

Venom Biochemistry and Symptoms: Little can be said about scorpionfish venoms except that they are highly complex proteins.

The symptoms of a sting from an Alabama scorpionfish could be painful, result in swelling, and potentially lead to serious consequences. According to Dr. Bob Shipp (Department of Biology, University of South Alabama), no one is known to have died from the scorpionfishes in the Gulf, but amputations have resulted. He suggests the application of ammonia to relieve the pain of minor injuries, but soaking in water that is as hot as is tolerable is believed to destroy the components of scorpionfish venom and to relieve the pain as with stingray wounds. Obviously, medical attention should be sought for a presumably serious sting.

Other Marine Fishes

More than 300 species of fishes in the world have been documented to have venomous spines. In addition to the catfishes and scorpionfishes, several venomous species live in the Gulf of Mexico and may occasionally wander inshore along the Alabama coastline. For many, the questions of whether they qualify as venomous and what the potential effects of envenomation might be on humans are unresolved. For example, some books declare that stargazers have mild venom; others discussing the southern stargazer (*Astroscopus y-graecum*) mention their capacity to create an electric shock but say nothing about their being venomous. A simple, sensible rule is to avoid handling marine fishes with which one is not familiar. Although serious consequences are unlikely, the possibilities of pain and discomfort from physical or chemical injury are real.

Class Reptilia

Order Squamata LIZARDS AND SNAKES

Family CROTALIDAE PIT VIPERS

This family of snakes includes all the pit vipers, so named because of the heat-sensitive pit between the eye and nostril. Any of the 5 species occurring in Alabama can give a serious bite, and in large specimens of at least 3 of the species the consequences sometimes can be lethal.

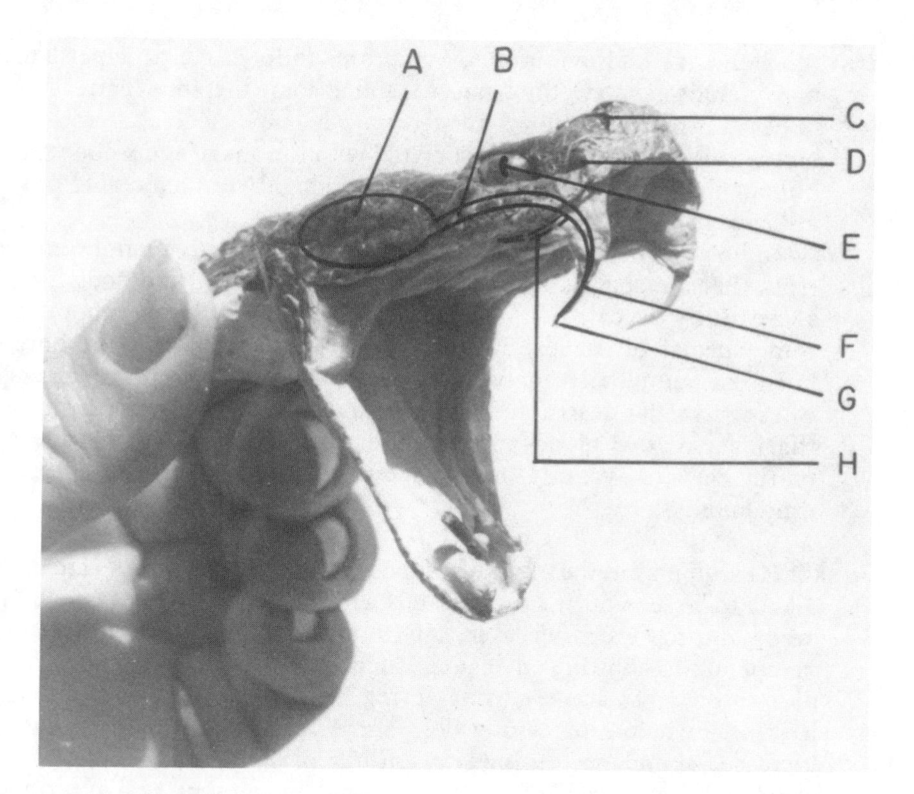

Head and Poison Apparatus of a Pit Viper (Canebrake Rattlesnake, *Crotalus horridus*)

Venom Mechanism and Administration: All Alabama pit vipers inject venom into the victim through 2 hollow fangs located on the front of the upper jaw.

The head of a canebrake rattlesnake can be used to demonstrate the mechanism by which venom is administered to a victim (see illustration). Poison glands (A) are located on either side of the head. On a large rattlesnake, these may contain several cubic centimeters of venom at any one time. Impact and pressure force the venom from the sacs through venom ducts (B) into the hollow fangs (F). An opening (G) in the tip of the fangs allows the system to operate in the same manner as a hypodermic syringe. The dotted line (H) indicates the position of the movable fangs when the mouth is closed. Pit vipers usually have a series of replacement fangs which move into position, one at a time on either side, if a fang is damaged or removed. In contrast, the fangs of coral snakes are considerably smaller and more or less permanently fixed in position in the front of the mouth.

The heat-sensitive "pit" (D) of a pit viper is located slightly below a line drawn from eye (E) to nostril (C).

Venom Biochemistry and Symptoms: Symptoms following a pit viper bite may include pain, swelling, nausea, and in some instances can result in death if the victim does not receive adequate medical attention. Numerous variables, some associated with the snake and some with human physiology, can interact to make a highly unpredictable situation. Many factors work in favor of the victim.

The venoms of Alabama pit vipers include highly complex proteins that are species specific. Modern biochemical methods have successfully revealed the amino acid sequences of the venom in some species of snakes, but chemical makeup and structure have not been completely analyzed in any species. The primary effects can involve the destruction of cells and tissues and cause hemorrhaging and other blood- and tissue-related problems. A direct effect on the nervous system is not characteristic of pit vipers in the eastern United States.

Seasonality: Pit vipers are warm-weather species, as are all snakes, so are unlikely to be seen during cold periods. However, they may be encountered during warm days in any month of the year. In warm temperate-to-subtropical regions such as Alabama, specimens appear to be most active during spring, perhaps seeking mates and looking for food after winter dormancy. There is also activity and increased abundance during late summer or early fall for some species, probably because of their overland movement to wintering sites and the arrival of newborn young. Wherever they occur, pit vipers should be watched for on any warm day.

AGKISTRODON

***Agkistrodon contortrix* (Plate 23)** COPPERHEAD

Species Recognition: Adults are generally 0.6–1 meter long. Their overall appearance is light brown or pinkish with darker saddle-shaped crossbands. The head is solid brown.

Geographic Distribution: Eastern United States from the Florida Panhandle to Massachusetts, southern New York, Pennsylvania, and midwestern states, to eastern Kansas and Oklahoma and western Texas. Found throughout Alabama.

Habitat Occurrence: This species is usually found in wet wooded areas or on the high ground of swamps in the southern part of Alabama and is common in rocky, wooded areas in the northern portion. Copperheads occur in most habitats including hardwood, swamp forest, and mixed pine-hardwood areas. They are primarily terrestrial but are often found in the vicinity of water.

General Life History and Behavior: Although less venomous than a similar-sized rattlesnake or cottonmouth, the copperhead is quick to bite, once a person moves within striking distance. Also, the camouflage potential of a brown and copper snake coiled among the leaves of late fall can often result in an outdoorsman's unwitting entry into the copperhead's strike range. Needless to say, where copperheads are known to be common, the first order of safety is to watch carefully what one steps on, sits on, or picks up.

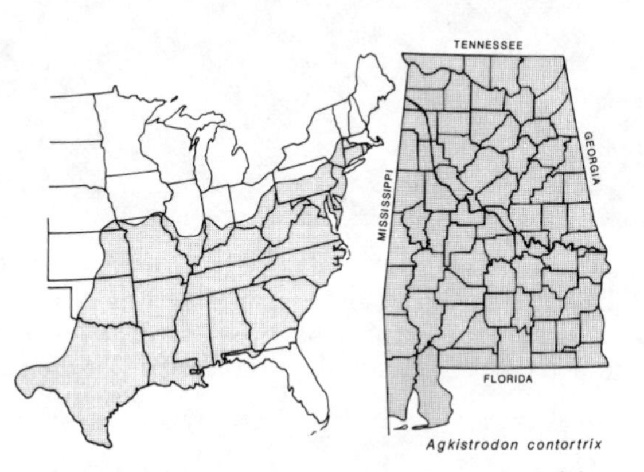

Range in eastern United States Known range in Alabama

***Agkistrodon piscivorus* (Plate 23)** COTTONMOUTH MOCCASIN

Species Recognition: Adults reach lengths of 1–1.5 meters and are generally heavy-bodied. The color pattern is variable, but dorsal colors of adults are usually drab brown or olive with darker crossbands; the belly is a combination of dull yellow and brown, and the underside of the tail is usually black. The young have brown-to-reddish crossbands and are often mistaken for copperheads.

Geographic Distribution: Southeastern Coastal Plain, Mississippi Valley to southern Illinois, and west to central Oklahoma and Texas. Restricted primarily to and most abundant in habitats below the Fall Line in Alabama, but present in most regions of the state.

Habitat Occurrence: This species is found in association with every type of wet habitat including estuaries, tidal creeks, and salt marshes; although aquatic areas are generally nearby, it often wanders overland.

Copperhead (*Agkistrodon contortrix*)
Copperheads can be found in almost any type of habitat in Alabama, especially hardwood areas. They are usually the most abundant venomous species in the mountainous rocky areas in the northern part of the state.

Cottonmouth Moccasin (*Agkistrodon piscivorus*)
Cottonmouths are almost invariably associated with aquatic habitats, although they may wander some distance from water. This is the most common venomous snake in wetland habitats below the Fall Line.

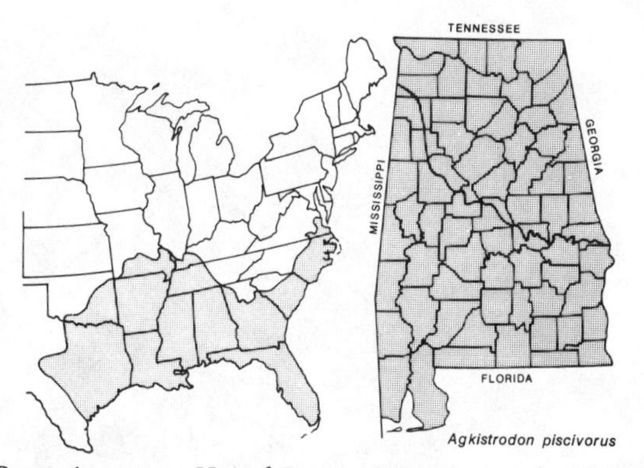

Range in eastern United States   Known range in Alabama

General Life History and Behavior: Within its habitat type, this is the most common venomous snake throughout the southern half of Alabama. Its swamp-colored camouflage and often sluggish, slow-to-retreat manner make it easy to step upon for the incautious person. Cottonmouths are primarily active at night and are readily apparent in the beam of a flashlight. Some are active during the day, but most hide beneath debris or coil inconspicuously almost anywhere in the habitat—alongside a cypress knee, on a log, in a tree, or in the grass. If disturbed, the cottonmouth will often stand its ground and give an open-mouthed threat display, allowing the human trespasser to be well warned of danger.

Several nonpoisonous varieties of water snakes inhabit the environments of cottonmouths. But all of them, no matter how formidable looking or how similar in appearance to the cottonmouth, will immediately retreat to the water when encountered. Cottonmouths occasionally enter the water when startled but, in contrast to water snakes, will often swim on the surface rather than submerge.

CROTALUS

***Crotalus adamanteus* (Plate 24)** EASTERN DIAMONDBACK RATTLESNAKE

Species Recognition: Adult diamondbacks are usually 1–1.5 meters, frequently more than 1.8 meters, and sometimes over 2.1. For many years the Ross Allen Reptile Institute in Florida offered a $1,000 prize (in preinflation money) for anyone bringing in an 8-foot diamondback rattler. No one ever collected, despite the lore about giant rattlers. The basic color is light-to-dark-brown with a distinct combination of brown and yellow diamonds. The tip of the tail is solid black with rattles.

Eastern Diamondback Rattlesnake (*Crotalus adamanteus*)
The most formidable of all southeastern snakes, the eastern diamondback rattlesnake is primarily a coastal region species in Alabama and has not been found north of Choctaw or Barbour counties.

Geographic Distribution: Found throughout Florida and in lower Coastal Plain of the southern states from eastern Louisiana to northeastern North Carolina. In Alabama the species is restricted to the southern tier of counties and is no longer common in most localities.

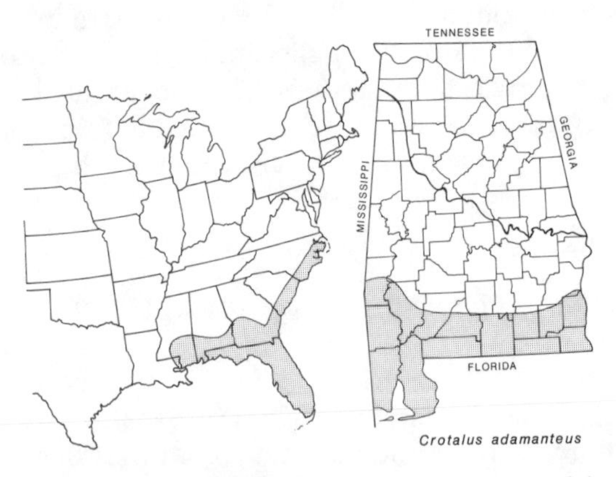

Range in eastern United States   Known range in Alabama

Habitat Occurrence: Diamondbacks are characteristically associated with palmetto stands or with dry terrestrial habitats such as sandhills with pine and scrub oak. However, specimens are occasionally encountered in other habitats within the geographic range of the species, including swamp margins, farmlands, and near beach dune areas.

General Life History and Behavior: All pit vipers of the United States are live-bearers, rather than egg layers, and the diamondback has an average litter of about a dozen. The babies are about 35 cm, look like the adult, and can deliver a venomous bite at birth. Eastern diamondbacks eat warm-blooded prey, primarily small mammals such as mice, rats, and rabbits. Diamondbacks present the most fearsome threat display of any eastern poisonous snake. When disturbed, they rattle and arch the head and neck above the body in a strike posture. Contrary to some beliefs, rattlesnakes need not be coiled to strike, and their strike range can be more than half their body length in certain circumstances. A distance of 4 feet from a threatening 7-foot eastern diamondback rattler could well be within the danger zone.

Because of its size, irritability, and toxicity (see chart), the eastern diamondback is generally regarded as the most dangerous snake in the eastern United States. The victim of an effective bite by this species could be in serious trouble.

Crotalus horridus ***(Plate 23)*** CANEBRAKE (TIMBER) RATTLESNAKE

Species Recognition: Heavy-bodied adults are usually 0.9–1.2 meters and occasionally 1.5 meters long. The basic color is gray with black crossbands that are usually chevron-shaped. The last few centimeters of the tail are solid black with rattles.

Geographic Distribution: Found throughout most of the eastern United States except for upper New England, Michigan, northern portions of midwestern states, and peninsular Florida. Known as the canebrake rattler in southern Alabama and the timber rattler farther north, the species is found throughout the state.

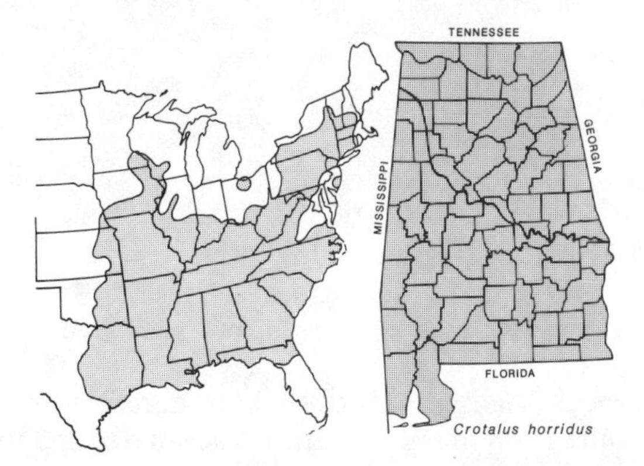

Range in eastern United States Known range in Alabama

Habitat Occurrence: This species occupies a wide diversity of terrestrial habitats. In southern Alabama it is found most frequently in deciduous forests and on high ground in swamps. In the northern part of the state, it occupies mountainous habitats as well as low-lying areas.

General Life History and Behavior: This formidable-looking pit viper may indeed be the most docile in behavior. Canebrake rattlesnakes of this region may not rattle or strike if approached in a slow, easy manner. If restrained, however, canebrakes will strike at their offender, and the whirring rattles of a large one can be heard several yards away. Despite its ubiquity, the canebrake rattler is seldom abundant in a localized area at any given time of the year. Hibernation dens where large numbers congregate conspicuously are rarely found in the nonmountainous terrain of the Alabama Coastal Plain.

SISTRURUS

***Sistrurus miliarius* (Plate 24)** PYGMY RATTLESNAKE

Species Recognition: Pygmy rattlesnakes are heavy-bodied, but usually only slightly more than 30 cm long. The general color is dull gray with dark gray or brown blotches on back and sides; small rattles occur on the end of the tail.

Geographic Distribution: Atlantic and Gulf Coastal Plain and Piedmont from North Carolina to eastern Texas, Oklahoma, and southern Missouri. Found throughout the southern two-thirds and western edge of Alabama.

Range in eastern United States Known range in Alabama

Habitat Occurrence: This species occurs in wet areas of wooded habitats or swamps and is often encountered in scrub oak and longleaf pine forest habitats or other wooded sites in noncoastal areas.

Canebrake (Timber) Rattlesnake (*Crotalus horridus*)
The largest species of poisonous snake found throughout Alabama is at home in a variety of habitats, including swamp margins, rocky areas, and pine forests.

Pygmy Rattlesnake (*Sistrurus miliarius*)
Despite its small size, this species can be very dangerous because of its irritability, tendency to bite when disturbed, and potency of venom. It is found throughout Alabama.

General Life History and Behavior: Next to the coral snake, the pygmy rattler is probably the most infrequently seen poisonous snake in Alabama. Isolated specimens are occasionally found, mostly in mixed pine-hardwood areas. The small size diminishes the ultimate threat of these frequently ill-tempered snakes, but serious bites have occasionally been reported. The tiny rattles, proportionately smaller than those of a young canebrake, are ineffectual as a warning signal in most instances. The small fangs would not likely penetrate a typical hunting boot.

Family ELAPIDAE

This family of snakes, represented by only one species in the eastern United States, includes not only the coral snakes of the New World but also the kraits and cobras of the Old.

Venom Mechanism and Administration: In contrast to pit vipers with their large, movable fangs, coral snakes have fangs that are small and remain in a fixed position in the front of the mouth. The venom is injected in the same manner, via a duct from a poison sac through hollow fangs. These snakes can move very quickly and can bite rapidly when handled, but to ensure an effective bite and administer an adequate dose of venom in a large animal such as a human, they generally must hold on to the victim long enough to inject the venom beneath the skin. This trait alone minimizes the probability of any clear-headed adult sustaining a serious bite. However, a serious bite could be life-threatening, so anyone bitten by a coral snake should consider emergency measures, such as contacting a poison center immediately.

Venom Biochemistry and Symptoms: The venom of snakes in the family Elapidae is generally more neurotoxic than that of most pit vipers. The chemical components and structure of coral snake venom are complex and have not been thoroughly analyzed. The primary effects on human victims are directly to the nervous system and result in paralysis of the respiratory centers. A problem reported for many coral snake bites is that little pain is experienced and minimal swelling occurs. One might not recognize that a victim, particularly a child, has been bitten. More obvious effects that are likely to occur before respiratory problems are dizziness, vomiting, and spastic reflexes. Treatment for coral snake bites is most effectively done at medical facilities, with the use of special antivenin.

Micrurus fulvius (Plate 24) EASTERN CORAL SNAKE

Species Recognition: Adults are usually about 60 cm in length, but are occasionally more than 1 meter. Red, yellow, and black rings encircle the body. The narrow yellow rings alternate with red and black rings. The front end of the head is always black, followed by a wide yellow band.

Geographic Distribution: Atlantic and Gulf Coastal Plain from North Carolina to central Arkansas, Texas, and northern Mexico. Restricted primarily to areas below the Fall Line in Alabama.

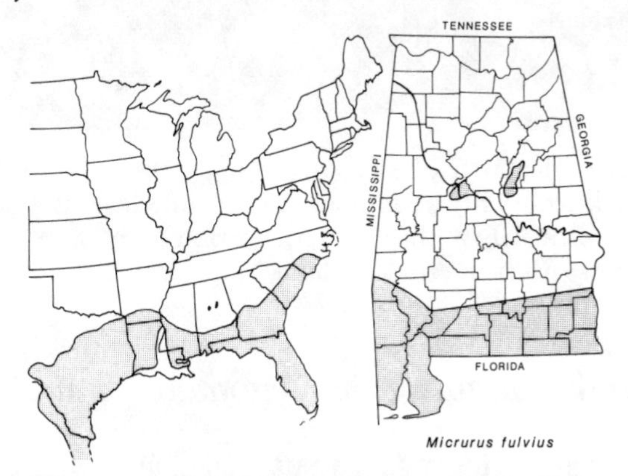

Range in eastern United States Known range in Alabama

Habitat Occurrence: This species has been found in a wide variety of terrestrial habitats including woods, fields, and margins of aquatic areas. Like other species of reptiles, coral snakes are active during warm periods of the year.

General Life History and Behavior: Rare throughout Alabama, the coral snake deserves mention because of its uniqueness among the venomous snakes and its potential danger. As the only native member of the cobra family east of the Mississippi River, this species is, drop for drop, in a venom class all its own. But, because of small size, infrequent occurrence, and perhaps timidity, the coral snake's record hardly ranks with those of other dangerous snakes of the eastern United States. Few deaths from coral snake bites have been recorded in Alabama. Yet, the potential seriousness of a bite from this species warrants a universal warning by parents to their children never to pick up a snake, no matter how pretty, without a knowledgeable adult's assurance of its safety.

Eastern Coral Snake (*Micrurus fulvius*)
The coral snake, the only venomous species in Alabama that is not a pit viper, is found primarily in the southern portion of the state. It is terrestrial and spends most of its life underground.

Factors Involved in Venomous Snakebite

The chances of receiving a dangerous snakebite in the United States are not great. But the chances always exist, particularly in Alabama, where 6 venomous species occur. Five probabilities affect the likelihood of serious snakebite. Careful consideration of each may help dispel myth and rumor and will indicate points at which a person has some control over his fate.

I. Probability of Encounter

Most habitats in nonurban areas of Alabama have at least a few commonly seen species of snakes, and it is a near certainty that people who take frequent trips to wooded areas, lakes, streams, or rivers during the warmer months will eventually encounter some.

Many different environmental factors influence the activity levels of different species. Temperature is the single most important factor affecting snakes and most other reptiles. For this reason, a warm spell in February can bring out several Alabama snake species, whereas a cool period in late April can inhibit their activity. Time of day is also important, although its exact relationship with aboveground activity is not clearly understood.

Rainfall probably stimulates activity in some situations, but its influence, too, is difficult to assess, and no hard and fast rule can be set forth. Unfortunately, our limited knowledge and understanding of the habits and general ecology of most snake species provide little predictability about how a given species will respond to conditions of season, weather, and location.

II. Probability of the Snake Being Poisonous

Six of the 40 species of snakes in Alabama are venomous; however, the different species of both types vary in their prevalence. Because of the high number of uncontrolled variables involved, there is no way of guessing the chance of a trod-upon snake being poisonous. A rough rule to remember is that 10–20 percent of the snakes encountered in Alabama will be venomous. This proportion is high enough to invoke caution.

III. Probability of Being Bitten

Most snakes are seen in time to avoid the bite. The first step in dealing with any snake is to regard it as venomous until identification is confirmed. If you are not positive of the species, leave it alone completely, for killing one does little to reduce the snake population and only increases your chance of being bitten. Many snakebites result from a lack of respect for a small snake. Frequently the victims are children, but many are adults who pick up specimens thinking they are dead or nonpoisonous. Accidentally being bitten by an unseen snake is possible, but to be bitten after you have seen the animal is usually inexcusable. All things considered, a thinking adult has a minimal chance of being bitten by a venomous snake in Alabama.

IV. Probability of the Bite Being Dangerous

One of the least predictable facts of venomous snakebite is the seriousness of a bite immediately after the victim is struck. Table 5 shows the range of effects that have been reported for actual cases. The first category indicates that one-fourth of venomous snakebites are, for all practical purposes, harmless and require no medical treatment. In fact, approximately two-thirds of victims in North America show little or no effect from being bitten. At the other end of the spectrum, more than 10 percent of cases could result in serious consequences, such as loss of feeling in a limb, amputation, or even death if proper treatment is not administered. Although it is difficult to determine the seriousness immediately, it is encouraging that in recent years fewer than 1 in 500 snakebite victims in this country have died as a consequence.

Some variables are more dependable than others in estimating the seriousness of a bite. Obviously, a large eastern diamondback rattlesnake is to be more feared than a small copperhead since few, if any, deaths from a copperhead bite have been recorded in the United States in the last 25 years. But, suppose the rattlesnake had recently killed 2 rabbits for a meal so that its venom supply was greatly reduced. Or, if only 1 fang penetrated the skin, the venom dose would be only one-half. On the other hand, if a copperhead sank both fangs right into the center of the jugular vein of a child, the effects would be immediate and severe, possibly lethal. Unfortunately, information of this type is seldom available when the snake bites its victim. Furthermore, a person is not likely to have a firm knowledge of his own physiological condition, which would determine his body's response to an injection of venom. Uncertainties such as these always leave the seriousness of the bite in doubt.

Table 5. An indication of the probability of an Alabama snakebite being serious is given below. A collection of more than 1,300 case histories from mostly southern states reveals that most pit viper bites result in minimal or no effect to the victim.*

Level of venom introduction into victim	Symptoms	Recommended medical treatment**	Percentage of sample
None	Fang marks, but little or no pain	None	25
Slight	Slight swelling and pain	Minimal; usually no antivenin required	38
Moderate	Pain, swelling, possible nausea, symptoms of shock	Medical attention necessary with administration of antivenin	22
Severe	Increased intensity of pain, swelling, possible unconsciousness in later stages, and other symptoms.	Medical attention essential, high levels of antivenin	14

*Table adapted from (1) H. M. Parrish, J. C. Goldner, and S. I. Silberg, 1966. Poisonous snakebites causing no venenation. Postgraduate Medicine 39(3):265–69 and from (2) L. H. S. Van Mierop, 1976. Poisonous snakebite: A review. 2. Symptomatology and treatment.

**See Table 6 for controversial aspects of first-aid and medical treatment.

V. Probability of Receiving Improper Treatment

The treatment of poisonous snakebite can be of great significance. Improper treatment of the bite of a small pit viper can easily lead to more serious medical complications than undertreatment of the bite of a moderate-sized one. The issue of proper treatment is debated among physicians, yet some medical facts make certain recommendations appear unquestionable (Table 5). Most medical authorities agree categorically on the following points.

DO NOT drink or eat anything, including alcoholic beverages, stimulants, or medicine.
DO NOT run or engage in unnecessary physical exertion.

DO remove rings, bracelets, watchband or any other constricting object since swelling may occur shortly after the bite.
DO get the victim to a hospital, poison center, or local medical doctor as expeditiously as possible.
What constitutes proper first aid and medical treatment is not agreed upon by different medical authorities or even herpetologists with snakebite experience. Aspects of treatment must be considered, particularly by the victim who must make a decision about incision and suction, use of constriction band, artificial cooling, and use of antivenin in the field. Each of these can be best assessed through the cost/benefit model indicated in Table 6 on page 318.

Class Mammalia

Order Insectivora

Family SORICIDAE

BLARINA

Blarina brevicauda SHORT-TAILED SHREW

Two of the world's mammals have been documented to be venomous. One is the duckbill platypus of Australia in which the male has on its rear feet a venomous spine with which it defends itself against predators, including humans. The other is the short-tailed shrew, a small species of the eastern United States, whose saliva is toxic. The venomous saliva in the mouth cavity enters the wound when the shrew bites another animal, usually prey it has attacked to satisfy its voracious, insatiable appetite.
No one is going to be bitten by a short-tailed shrew without a special ef-

Table 6. Controversial first-aid treatments for North American venomous snakebites. Proper first-aid treatment of snakebite is highly controversial among physicians and herpetologists. One reason that a clear, unequivocal medical statement cannot be made is that the effects of the bite can be so variable (see Table 5). The recommendations below are based on the impact that an "average" snakebite would have on a healthy person.

First-Aid Treatment	Possible Benefits	Possible Cost
Incision and suction	Removal of venom	Infection, excessive bleeding, cutting of critical vein, artery, nerve, or tendon
Constriction band*	Retards movement of poison through body	Concentrates poison in unfavorable local area: restriction of circulation
Cooling	Reduces pain and swelling: neutralizes some poisons: retards movement of poison through body	Concentrates poison in unfavorable local area: cell damage from continued low temperatures
Antivenin** administration (self-administered in field without medical guidance)	Neutralizes poison	Infection; anaphylactic shock, other reactions

*The use of a *constriction band* is recommended under certain circumstances. This is certainly a controversial issue. However, if a constriction band is used, it should be just tight enough to restrict lymphatic flow without obstruction of veins returning blood to the heart. Obstruction of veins would promote further swelling and possibly further local tissue damage. Some describe this as a band that is applied over 2 fingers. Also, if a constriction band is applied correctly, it should never be removed until the victim is in an environment ready and able to cope with sudden cardiovascular collapse, because the removal of the band may suddenly release a large amount of venom all at once.

**The use of *antivenin* also is rather controversial. Many medical doctors do not advocate the use of antivenin, which is produced from horse serum, unless symptoms sufficient to indicate that moderate or severe envenomation has occurred. This reluctance to use antivenin in mild cases is because of the high incidence of serum sickness following its use (more than 35 percent).

Professional Medical Assistance Available Within 1 Hour	Victim to be Isolated from Medical Services for Extended Time Period
No action	Careful cutting with blade in vicinity of bite within minutes after being bitten; begin sterile suction immediately for up to ½ hour
No action	Light constriction, frequently removed
No action	Limited use to lessen pain
No action	Careful administration under sterile conditions

fort to pick up one bare-handed. Even then, serious harm to the person is unlikely although the sharp front teeth may cause a nasty little bite, and the venom can result in a burning sensation and subsequent minor swelling. The best treatment is to wash the wound thoroughly and treat it as any other cut. One can likely get some attention later by showing others the injury from the only venomous mammal of the six largest continents.

Species Recognition: All shrews are small with pointed noses and tiny eyes. The short-tailed shrew is the color of pencil lead, reaches a length of 10 cm (large by shrew standards), and has a shorter tail than most species.

Toxic Properties: The chemical composition of the venom is unknown.

Geographic Distribution: Found throughout the eastern United States, from southern Florida to southern Canada, and throughout Alabama.

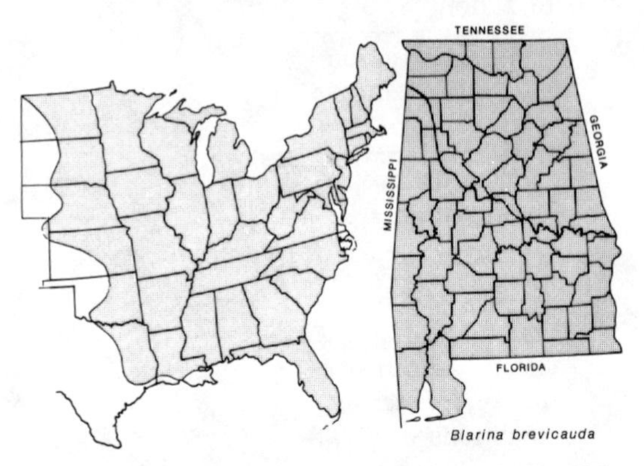

Range in eastern United States Known range in Alabama

Habitat Occurrence: This shrew is most abundant in wooded areas but can be found in almost any terrestrial habitat in its range.

General Life History and Behavior: The short-tailed shrew is at home beneath the leaf litter where it makes an endless search for small invertebrates. It sometimes eats plant material as well. A half dozen or more young are born during the warmer months, a female often producing 2 or more litters during the year. Although technically a venomous animal, the consequences of its bite can be considered primarily an item of conversation.

Chapter 8

Harmless Animals That Appear Dangerous

Many plants and animals are protected by similarity in color, structure, or behavior to poisonous or otherwise dangerous species. This can be the key to survival for a harmless plant or animal. An example is the monarch butterfly, which is able to float with impunity around a flock of insect-eating birds with its bold display of orange and black coloration. Because the caterpillars of monarch butterflies feed on milkweed sap, which is poisonous to some animals, the adult monarch is poisonous to them also. Birds somehow know this, and few will attempt to attack a monarch once, and almost never twice. Viceroy butterflies, on the other hand, would be a good meal; but because they also have the orange and black coloration and look almost identical to a monarch, birds leave them alone, too. Thus, the similarity of appearance serves as a protective measure for the viceroy.

The bright red warning of a cardinal flower in an otherwise dark swamp or of a velvet ant walking boldly across an open field is a signal that they should not be eaten or perhaps even touched. Bright red is often a danger signal in nature, although the reason for concern is not always apparent. It is interesting that red is used for traffic lights and with today's stop signs to give a similar message. Even the flags of many of the most powerful countries of the world use bright red as a dominant color. Because a showy appearance often indicates a hazard, and some species use mimicry as a protective device, it is understandable that humans and other animals would often be cautious even of harmless organisms.

Because of these phenomena, many of Alabama's harmless native species, particularly certain animals, are viewed with suspicion and frequently indicted and convicted without a trial. Perhaps the most universal example in Alabama and the rest of the world is the snake. Most people in the South no longer believe that all snakes are venomous, although the widespread fear and anxiety caused by any snake suggest that a high level of ignorance still exists. One of the most common examples of unnecessary fears about snakes is directed toward the several species of water snakes common to the aquatic habitats throughout Alabama. For example, the harmless diamondback water snake that is so abundant in Alabama's rivers and in many other freshwater habitats resembles a cottonmouth moccasin in many aspects. They are brown and black, have enlarged heads with bulging jaw muscles that can be mistaken for venom sacs, and will actually display by expanding the head further to resemble the triangular shape of a pit viper. They also bite. We have picked up hundreds of these snakes, been bitten by dozens, and have never suffered any ill consequences. Diamondback water snakes are not venomous and are, in fact, important components of these aquatic ecosystems. Yet untold numbers are killed annually in the belief that cottonmouth moccasins are being removed from the face of the earth.

Although no venomous lizards live east of the Mississippi River, some species in Alabama are frequently considered to be quite dangerous. These are the blue-tailed skinks, small black lizards, which have yellow stripes and bright metallic-blue tails when they are young. They are often referred to as scorpions, and it is easy to find people in Alabama who know someone who has been stung by one of these harmless creatures. Of course, one can never identify the person someone else knows who was poisoned by a blue-tailed skink; that person does not exist. Blue-tailed skinks defend themselves in 2 ways that could possibly affect a human, and stinging is not one of them. First, they do bite, but even the largest ones, the males, which are brownish with a bright red head in the springtime and reach a length of up to 30 centimeters, can do no more than pinch. Their teeth are too tiny to break the skin and their grip is no stronger than a clothespin. Their other threat is something most humans would never have to worry about. Skinks taste bad, very bad. Some biologists even think that the reason for their bright blue tails is a warning to predators to leave them alone. Veterinarians in rural areas often report of pets, particularly cats, that have eaten skinks and become ill or even died. So, don't eat skinks and you should be all right; definitely do not worry about them as venomous animals.

A variety of other snake species acquire protection through color pattern or behavior. The 2 species of hognose snakes (*Heterodon platyrhinos* and *H. simus*) of Alabama go through elaborate behavioral displays of expanding the area behind the head so that they look like cobras (they are sometimes called spreading adders) and hissing. The culmination of this threat display sometimes ends with the animal rolling over on its back, opening the mouth so that its tongue hangs out, giving the appearance of a dying snake. Some specimens will even bleed from inside the mouth, adding even more credibility to the act. Although the threatening behavior might make a predator think twice before attacking such a creature, no one has demonstrated categorically why the death-feigning act is effective. Nonetheless, hognose snakes do it. Scarlet king snakes with their beautiful red, yellow, and black bands that encircle the body have been said to gain protection from some predators by being mimics of venomous coral snakes, which occur in the same region in Alabama.

Even the amphibians, none of which is venomous anywhere in the world, have some members that strike unnecessary fear in the hearts of humans. The hellbender that inhabits certain stream systems in northern Alabama is a horrendous-looking beast that can reach a length of more than 2 feet. Some people think hellbenders are venomous, but the species is far more innocuous than its appearance would have one believe. Incidentally, one group of amphibians in Alabama that does have poison glands, but no means of injecting it as venom, is the toad genus (*Bufo*). The large, granular glands located on top and either side of the head contain a powerful toxin that has been a source of distress for more than one Alabama yard dog that has tried to bite a toad. Presumably anyone reading this book already knows that toads do not cause warts.

Almost anyone in Alabama knows to avoid the dorsal spines of a large fish; no one likes being jabbed with needles. But one coastal species has a reputation for being venomous when it is not. This is the toadfish (*Opsanus beta*), which could win a contest for the ugliest little fish around. The dark brown, broad-headed monster is aptly named and often hooked on fishing lines in marine waters along the coast. The spines have no venom, and the teeth are small and not particularly sharp. Toadfish can bite hard, but they do not have the venomous qualities of which they are often accused.

Another coastal marine species that natives know is harmless, but which can alarm visitors, is the horseshoe crab (genus *Limulus*). This armored tank of the invertebrate world, with its many spines and long pointed tail, looks formidable but is as harmless as a swallowtail butterfly.

Certain terrestrial arthropods that are viewed with suspicion are the nonvenomous millipedes (class Diplopoda). Superficially these plant-eating animals with their many legs resemble centipedes, which, of course, have venomous front claws. Millipedes, however, only bite the vegetation or dead organic matter on which they feed, and can be picked up without fear. Do not think they are not protected, however, because some can secrete a form of cyanide, presumably a deterrent to animals that might otherwise find them an attractive meal.

Misconceptions and fears are generally greater about animals than plants. Animals are both active and aggressive in some cases, whereas plants are passive. A plant cannot poison you unless it is eaten or touched.

Glossary

The intent is to present information about plants and animals in a style and format that is readily understandable to persons who have no botanical or zoological training. However, certain terms are essential for expressing some features and characteristics. The use of scientific terminology in descriptions in the book is minimized, but in some instances terms in common usage among biologists have been chosen to provide the most succinct description of a trait. In addition to the definitions in the glossary, the line drawings will aid in recognition of particular structures and in identification of certain species. The following terms are used one or more times in the text.

Biological Terms

achene A small, hard, dry, nonsplitting fruit with one seed.

acuminate Refers to a pointed leaf in which the sides below the point are concave.

adventive Non-native species that has been introduced into an area and is becoming established.

alluvial Pertaining to soil, sand, gravel, or similar detrital material deposited by running water.

annual A plant that completes its life cycle in one year or less.

annulus The lower remnant of the veil that covers the gills before the cap opens in the genus *Amanita* and certain other mushrooms.

anther The upper, enlarged part of a stamen that contains the pollen.

axil The upper angle where the leaf joins the stem.

barbed Having rigid points or short bristles.

barbels The fleshy, threadlike extensions around the mouth region of many species of catfishes.

beaked Ending in a beak or prolonged tip.

biennial A plant that completes its life cycle in two years, blooming the second year.

bisexual flower A flower that contains both stamens and pistils. This is often referred to as a perfect flower.

blade The flat, expanded part of a leaf.

bract A leaflike structure associated with a single flower or group of flowers, differing in size, shape, color, or any combination of these from the other leaves of the plant.

bristle A stiff hair or similar outgrowth.

bulb An underground stem bearing numerous fleshy or scaly leaves, as in an onion.

bulbil A bulblike body, especially one borne on a stem or the flowering part of a plant.

calyx The outermost cycle of flower parts, composed of sepals, usually green and leaflike but sometimes like petals; the sepals may be separate or united.

cap The flat or rounded upper portion of a mushroom.

capsule A dry, multiple-seeded fruit

that opens or splits along two or more lines when mature.

carpel One part of the pistil or one of several separate pistils.

ciliate Hairy along the margin.

clawed Narrowed at the base into a thin, stalklike arm.

cleft Cut at least halfway or more from the margin to the midrib, or from the apex to the base.

cleistogamous flower A flower that never opens, and is thus self-pollinating.

compound Made up of two or more similar parts.

cordate Heart-shaped.

corm An enlarged, fleshy, more or less spherical underground base of a stem.

corolla The inner part of the flower, made up of petals. The petals may be separated or united and may be of any color.

creeping Growing flat on or beneath the ground and rooting.

crested Bearing an elevated appendage like a crest.

deciduous Falling away at the end of the growing period or shedding leaves at the end of the growing period; not evergreen.

decumbent Reclining, but with the tips ascending.

decurrent Extending down from and growing attached to the stem.

deflexed Bent downward from the tip.

dehisce To open, as a seedpod.

dentate Toothed, especially with outwardly projecting teeth.

dicotyledon (dicot) Plant having 2 seed leaves during embryonic development.

dioecious Having unisexual flowers, with male and female flowers on separate plants.

disc In the Asteraceae or sunflower family, the central part of the flower head, bearing many small tubular flowers. See Inflorescences, Figure I.

disc flower One of the tubular flowers in the disc of a flower head of the Asteraceae or sunflower family.

dissected Divided many times; cut into numerous segments.

dorsal spine The single, solid projection at the front edge of the fin on the back of some fishes.

downy Covered with short, fine hairs.

elliptic In leaves, shaped like an ellipse, oblong with regularly rounded ends.

entire Not interrupted by lobes or cuts; continuous, as a leaf margin without teeth, lobes, or divisions.

equitant Overlapping in two layers

eyespot A pigmented spot on an animal that has the appearance of an eye but that may not be associated with vision.

fang A long tooth, usually in the front of the mouth and often hollow for transporting venom.

fascicle A bunched cluster.

fertile Capable of producing seed or viable pollen.

filament The stalk of the stamen that supports the anther.

filiform Elongate and slender.

fruit The seed-bearing product of a plant

fusiform Shaped like a spindle, broad in the middle and tapered at the ends.

gills Fleshy plates radiating from the center of the stalk on the underside of a mushroom cap.

glabrous Smooth; not hairy.

glandular Bearing glands or exuding a sticky substance.

glaucous Covered with a fine, white, powdery coating.

globose Globe-shaped; somewhat spherical.

halberd-shaped Like an arrowhead, but with basal lobes diverging hastate. See Figure II for illustration of a hastate leaf.

halophyte A plant that persists or thrives in salty soil.

haustoria Extensions of a parasitic

plant that penetrate the cells of a host plant and extract nutrients.

head A dense cluster of flowers without stalks or on very short stems.

herb A plant with no persistent woody stem above ground.

hilum The scar left on the seed at the point of attachment.

hirsute With firm or stiff hairs.

imperfect flower Flower without both pistils (carpels) and stamens.

incomplete flower A flower with one or more kinds of floral parts missing. The missing parts may be sepals, petals, stamens, or pistils.

indehiscent Not normally opening to release seeds.

inferior ovary The ovary of a flower in which the sepals are fused to the ovary for more than half its length.

inflated Distended or swollen by or as if by air.

inflorescence The flower cluster of a plant.

involucre A circle or collection of bracts surrounding a flower, especially in the sunflower family.

irregular Applied to a flower in which one or more of the organs of the same series are unlike the rest. Bilaterally symmetrical.

keel A ridge like the keel of a boat. Also, the structure formed by the 2 joined lower petals of a sweet-pea-type flower. See Flower Types, Figure I.

labiate With lips, as the two-lipped corolla of many mints.

lanceolate In a leaf blade, long and narrow and tapering to a point; broadest near the base.

lancet One of the 2 barbed projections that form the stinger in certain species of wasps.

larva(e) The newly hatched form of an animal that differs from the adult form in appearance or habits (e.g., caterpillars).

leaflet A single division of a compound leaf.

legume A simple, single-compartmented fruit opening along two sides, as in the bean family.

lip One of a two-lipped corolla or calyx. In a bilaterally symmetrical flower with joined petals or sepals, the upper or lower part of the corolla or calyx. The odd petal in an orchid flower.

midrib The main vein of a leaf.

moniliform With lateral constrictions, giving a beadlike appearance.

monocotyledon (monocot) Plant having one seed leaf during embryonic development.

monoecious Having unisexual flowers, with male and female flowers on the same plant.

mouth brooder Fish species in which the males carry the fertilized eggs in their mouths until they hatch.

nematocyst One of the tiny stinging cells of jellyfish.

nerve A simple vein or slender rib.

node The place on a stem where one or more leaves are produced.

oblanceolate Lanceolate, but with the broadest part toward the apex; inverted lanceolate.

oblong Longer than wide, with sides nearly parallel or somewhat curving.

obovate Egg-shaped in outline, broadest above the middle or near the apex; inverted ovate.

obpyramidal Upside-down pyramid. Widest at top and tapering to a narrow base.

opposite Applied to 2 leaves at the same node, on opposite sides of the stem.

oval Broadly elliptic.

ovary The lower part of the pistil, where the seeds develop.

ovate Egg-shaped in outline, with the broader end toward the base.

ovipositor A specialized organ at the end of the abdomen of female insects for depositing eggs.

palmate Lobed or divided to radiate

from one point, as the fingers from the palm of the hand.

panicle A loose, elongate cluster of flowers, branched several times.

pappus The highly modified calyx of flowers of the Asteraceae or sunflower family. It is borne on the apex of the ovary.

parasite A plant or animal that gets its food from another living organism.

ectoral spine A single, solid projection at the front edge of the fins on the sides of some fishes.

edicel The stalk of a single flower in a cluster.

eduncle A stalk that bears either a solitary flower or a flower cluster.

eltate Attached to the inside margin of the stalk.

erennial A plant of 3 or more years duration.

erfect flower A bisexual flower; one containing both stamens and pistils.

erfoliate Applied to a sessile leaf or bract whose base completely surrounds the stem so that the stem seems to pass through the leaf or bract.

erianth The corolla and calyx of a flower.

etal One of the parts that compose the corolla.

etiole The stalk of a leaf.

innate Compound, with the leaflets on opposite sides of a common, stemlike axis.

istil The female reproductive organ of a flower, composed of the stigma, style, and ovary.

it viper Any of several species of venomous snakes with specialized heat-sensing organs (pits) on either side of the head.

umose Feathery. Specifically, bearing fine, spreading hairs.

od A dry fruit that splits open along 2 sides at maturity.

ollen The fertilizing grains borne in the anthers.

prostrate Lying flat upon the ground.

pubescent Hairy. Specifically, bearing short, downlike hairs.

raceme A simple, elongate cluster of stalked flowers, arranged singly along a central stalk.

ray The branch of an umbel. Also, a marginal strap-shaped flower in the sunflower family.

reflexed Bent back or downward.

reticulate With a netted appearance.

rhizatomous Having rhizomes.

rhizome A horizontal underground stem, often thick and short.

rootstock An underground stem.

rosette A basal cluster of leaves in a circular form, as in a dandelion.

samara An indehiscent fruit with winglike projections.

saprophyte A plant that gets its food from dead organic matter.

scandent Climbing, but without tendrils.

scape A flower stalk that arises from the ground without leaves.

scapose Having flowers on a scape.

seed The ripened ovule.

sepal A division of the calyx.

serrated Having sharp teeth that point forward.

sessile Without a stalk.

sheath A tubular envelope.

simple Of one piece; not compound.

spadix A thick, club-shaped stalk bearing small, crowded, stalkless flowers, as in the arum family.

spathe A hooded or leaflike sheath partly enclosing the spadix, as in the arum family. In certain other plants, a leaflike structure at the base of a flower or flower cluster.

spatulate Referring to leaves, spoon-shaped, narrowed downward from the rounded summit.

spike An elongate flower cluster in which sessile flowers are arranged along a central stem.

spine A stiff, pointed projection on a plant or animal that may or may not

FLOWER PARTS

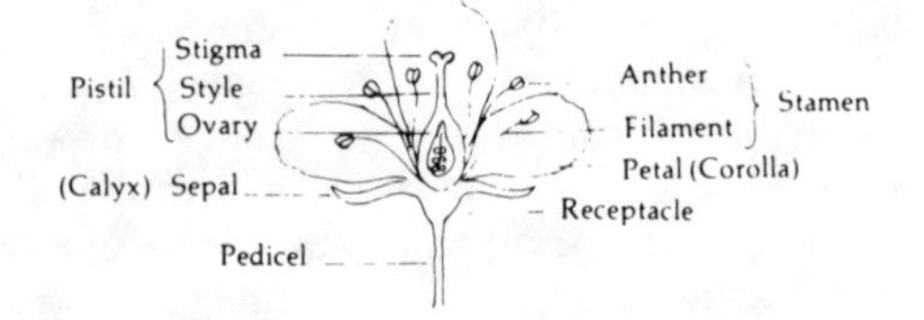

FLOWER TYPES

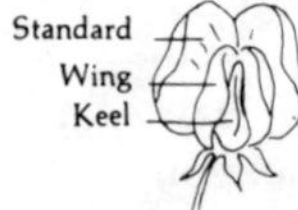

Labiate
(Mint)

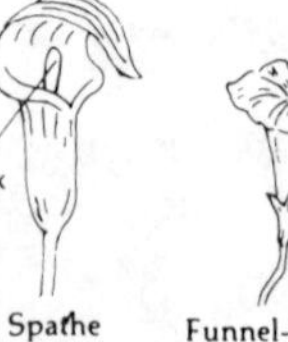

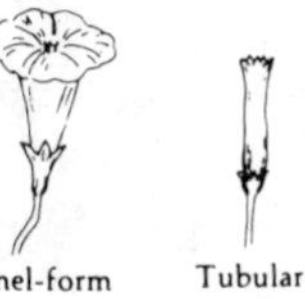

INFLORESCENCES

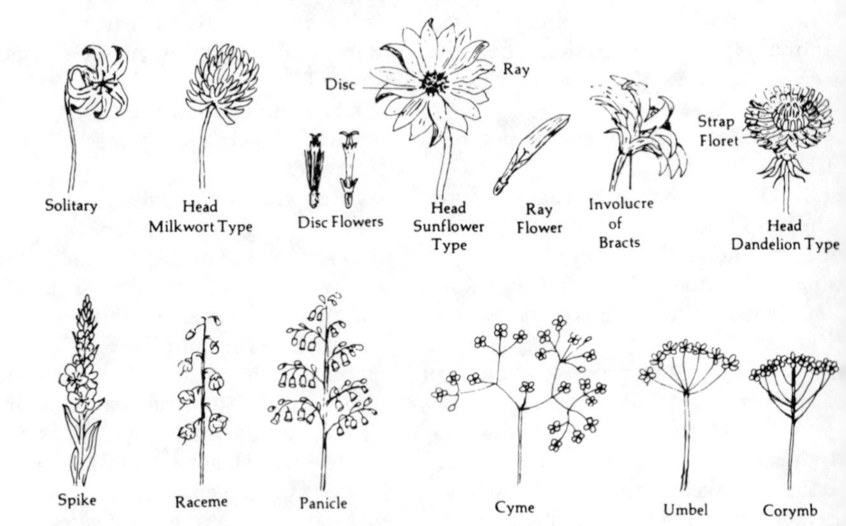

Figure I

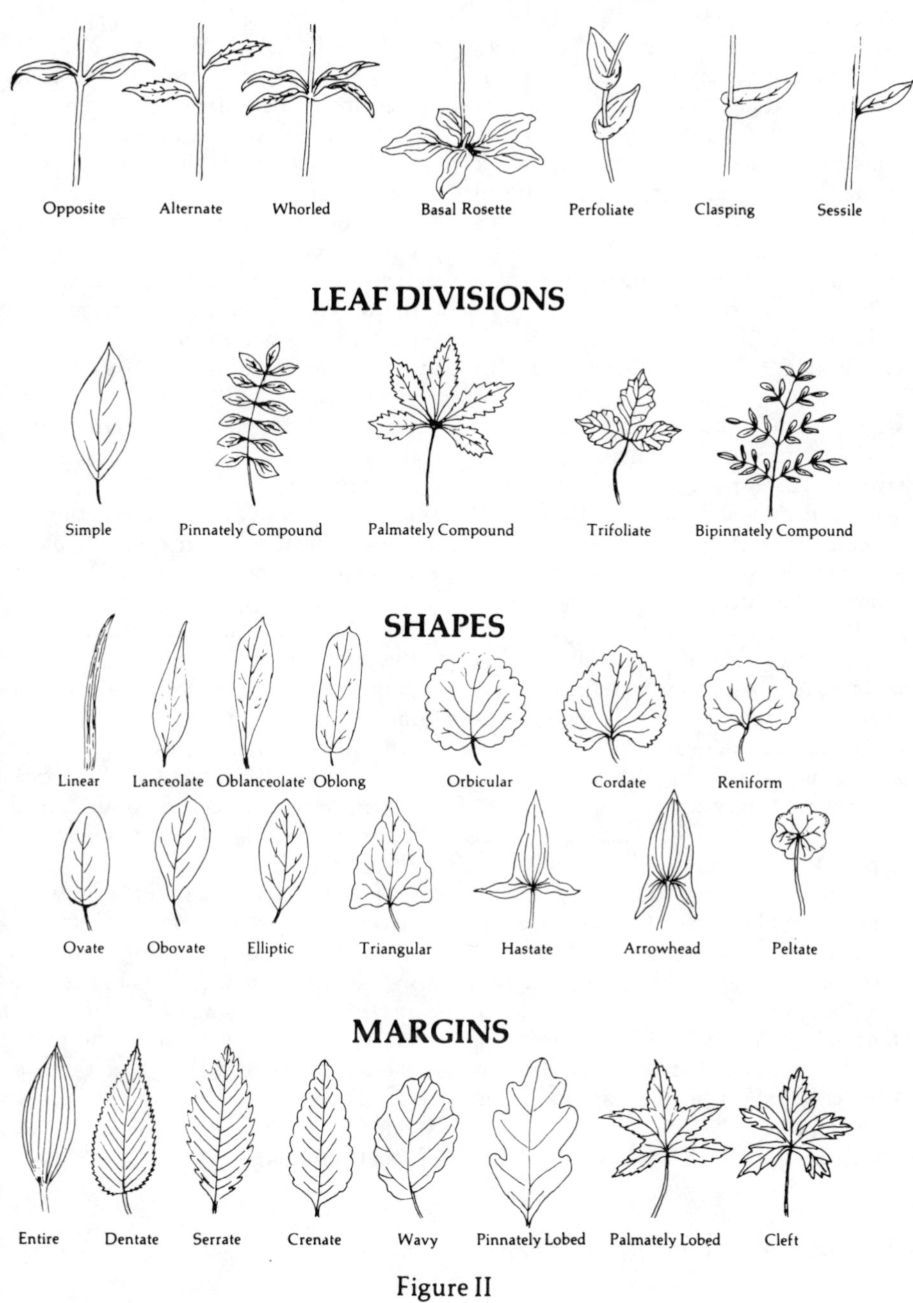
LEAF ARRANGEMENTS
Opposite
Alternate
Whorled
Basal Rosette
Perfoliate
Clasping
Sessile
LEAF DIVISIONS
Simple
Pinnately Compound
Palmately Compound
Trifoliate
Bipinnately Compound
SHAPES
Linear
Lanceolate
Oblanceolate
Oblong
Orbicular
Cordate
Reniform
Ovate
Obovate
Elliptic
Triangular
Hastate
Arrowhead
Peltate
MARGINS
Entire
Dentate
Serrate
Crenate
Wavy
Pinnately Lobed
Palmately Lobed
Cleft

Figure II

be capable of injecting venom, depending upon the species.

spore An asexual reproductive cell, produced by mushrooms and other organisms, that can withstand unfavorable environmental conditions for an indefinite period.

stamen The male reproductive organ of a flower, composed of the anther and the filament.

standard The broad upper petal of a flower of the bean family.

stellate Starlike.

sterile Unproductive. Without stamens or pistils.

stigma The terminal part of the pistil, which receives the pollen.

stinger A sharp organ, often on the posterior end of an animal and usually capable of injecting venom.

stipitate Borne on a short stalk.

stipule An outgrowth from the base of a leaf, usually occurring in pairs, one from either side of the leaf base.

stolon Any basal branch inclined to root.

style The part of the pistil between the stigma and the ovary.

subtending Standing beneath or in close proximity to.

taproot The main, descending root of a plant.

tendril A slender, coiling organ of a climbing plant, used for twining about a support.

terminal Borne at the tip of a stem or stalk.

toothed A nontechnical term referring to leaf margins that bear projections or indentations. See illustrations of leaf margins.

trifoliate Bearing 3 leaflets.

tuber A thick, short, underground stem, possessing buds or "eyes."

umbel A flower cluster, usually flat or convex, in which all the flower stalks arise from one point. In a compound umbel, small umbels grow at the end of each of the radiating stems.

unisexual flower A flower that contains only one kind of sex organ, either stamens or pistils.

universal veil A membranous sheath that covers some mushrooms as they push their way up through the soil.

urticating hairs Stinging hairs on some caterpillars.

veil The membranous covering over the gills of *Amanita* and some other mushrooms.

vein A thread of fibrous conducting tissue in a leaf or other organ; a blood vessel that returns blood to the heart in animals.

volva The bulbous, sometimes cup-like, remnant of the universal veil that persists at the base of the stalk of *Amanita* mushrooms.

weed A plant that aggressively invades areas of cultivation or other areas where the soil has been disturbed. Used particularly to refer to obnoxious or unwanted plants.

whorl An arrangement of 3 or more leaves or flowers in a circle about a stem, all attached at a single node.

wing A thin, rigid membrane extending from the surface of a stem, leaf stalk, fruit, or seed. Also, one of the lateral petals of a flower of the bean family.

woolly Bearing soft, curly, sometimes tangled hairs.

Medical and Chemical Terms

aesculin A coumarin glycoside found in the plant genus *Aesculus*.

aglycones An organic compound, especially of flavonoids without an attached sugar.

alkaloid Complex organic compounds

and bitter bases containing nitrogen, and often oxygen, and that are found in many seed plants. (See also discussion in chapter 2.)

anaphylactic shock A systemic reaction by some individuals to a specific antigen (such as from a wasp or bee) after previous sensitization. The results can be serious, even fatal, complications, such as respiratory failure, itching, and edema.

andromedotoxin An amorphous substance that may be white or crystalline that is considered to be resinoid.

antigen A substance, usually a protein or carbohydrate, that acts as a toxin in the body and stimulates the production of an antibody.

apocannoside Glycoside of *Apocynum*.

apocynin A resin found in *Apocynum*.

aporphine An alkaloid found in *Dicentra*.

araliin A glycoside found in *Aralia*.

aristinic acid A compound found in the plant genus *Aristolochia*.

aristolochin A resin found in *Aristolochia*.

atropine A mixture of compounds that inhibit the actions of acetylcholine in the parasympathetic nervous system, thus relaxing smooth muscles (e.g., dilation of the pupil of the eye).

azaridine An alkaloid found in *Melia*.

calcium oxalate crystals Sharp, puncturing crystals of the insoluble compound, oxalic acid.

calycanthine An alkaloid found in *Calycanthus*.

cascara sagrada Dried bark of buckthorn (*Rhamnus*) used as a laxative.

cephalanthin A glycoside found in *Cephalanthus*.

cephalin A glycoside found in *Cephalanthus*.

cicutoxin An unsaturated alcohol found in *Circuta*.

cocculidine An alkaloid found in *Cocculus*.

coclifoline An alkaloid found in *Cocculus*.

crinamine An alkaloid found in *Crinum*.

crinidine An alkaloid found in *Crinum*.

cyanogenic glycoside A glycoside capable of producing cyanide (as hydrogen cyanide).

cymarine A cardiac glycoside in the steroid group.

delphinine An alkaloid found in *Delphinium*.

diterpenoid alkaloids Alkaloids, including delphinine, found in larkspur (*Delphinium*).

dopamine A form of the amino acid dopa, which is found in the adrenal glands and is used in the treatment of Parkinson's disease.

dugaldin A glycoside found in *Helenium*.

epinephrine Adrenaline; a colorless, crystalline slightly basic sympathomimetic hormone that is the principal blood-pressure-raising hormone secreted by the adrenal medulla; epinephrine is used medically as a heart stimulant, a vasoconstrictor to control skin hemorrhages, and a muscle relaxant in bronchial asthma.

erythema Abnormal redness of the skin due to capillary congestion.

formic acid A colorless pungent, blister-forming, liquid acid found in the venom of ants and in many plants.

gastroenteric Referring to the stomach or intestines.

gastroenteritis Inflammation of the lining membrane of the stomach and the intestines.

gastrointestinal Referring to the digestive system; gastroenteric.

gastrointestinal tract Intestines and stomach.

glycoside Any of many sugar derivatives that contain a nonsugar group attached through an oxygen or nitrogen bond. Splitting of the bond by the process of hydrolysis results in a sugar, such as glucose.

haemotoxic (hemotoxic) venom A highly complex protein injected by some venomous snakes, especially certain pit vipers, that primarily causes destruction of animal cells and tissues, hemorrhaging, and results in various other blood- and tissue-related problems (in contrast to neurotoxic venom, which presumably affects nervous tissue and function). Because of the symptomatic complexity of snake venoms and the lack of detailed knowledge regarding their physiological effects, some medical doctors challenge the utility of the terms *haemotoxic* and *neurotoxic*.

hederagenin A saponic glycoside of English ivy (*Hederea helix*).

hederin A saponic glycoside of English ivy (*Hederea helix*).

heterocyclic ring A ring composed of atoms of more than one kind.

hieracifoline A pyrrolizidine alkaloid found in fireweed (*Erechtites*).

histamine A compound that is found in ergot and many animal tissues that is probably responsible for dilation and increased permeability of blood vessels which play a major role in allergic reactions.

hyaluronidase An enzyme that splits and lowers the viscosity of hyaluronic acid, facilitating the spread of fluids through tissues.

hydrogen cyanide (hydrocyanide) A poisonous, usually gaseous, compound that has the odor of bitter almonds.

hypotension Abnormally low blood pressure.

lobeline A crystalline alkaloid present in the plant genus *Lobelia* and used chiefly as a respiratory stimulant and as a smoking deterrent.

lycroine An alkaloid found in *Crinum*.

margosine An alkaloid found in *Melia*.

myocardial depression Weakening of the pumping action of the heart.

necrosis Localized death of living tissue; a common consequence of some venomous snakebites, such as the cottonmouth.

neurotoxic venom Toxic to the nerves or nervous tissue. (See haemotoxic venom.)

noradrenalin See norepinephrine.

norepinephrine A hormone present with adrenaline in the adrenal gland. It has an intense vasoconstrictor action, and is given by slow intravenous injection in shock and peripheral failure.

oleoresin A natural plant product containing chiefly essential oil and resin.

oxalates Salts of various elements, often in a crystalline form.

paraisine An alkaloid found in *Melia*.

phenol Any of various acid compounds composed of 6-carbon heterocyclic rings.

phospholipase Any of several enzymes that hydrolyze lecithins (compounds containing phosphorus widely distributed in plants and animals that form colloidal solutions in water and serve as a wetting agent).

phytolaccine A resin found in *Phytolacca*.

phytolaccotoxin An alkaloid found in *Phytolacca*.

posterior Situated behind; the hind parts of the body of livestock.

posterior paralysis Paralysis of the rear parts of livestock.

proteolytic enzymes Enzymes that decompose proteins, with the formation of simpler and soluble compounds.

protoberberine An alkaloid found in *Dicentra*.

pruritis Intense itching of the skin without eruptions.

pyrrolizidine alkaloid Alkaloids composed of two 5-carbon rings that cause severe liver damage.

resin An organic compound excreted from glands on the surface of some plants (e.g., pine trees) that will dissolve in ether or alcohol but not water. Resins normally range from light yellow to brown in color and are translucent.

resinoid Complex compounds that, upon extraction, are solid or semisolid at room temperature, brittle, easily melted or burned, soluble in a variety of organic solvents but insoluble in water.

saponin; saponic glycosides Any of various toxic glycosides that occur in plants and are characterized by production of a soapy lather.

scopolamine A poisonous alkaloid found in the roots of various plants, especially the nightshades.

serotonin A phenolic amine that is a powerful vasoconstrictor and is found especially in the blood serum and gastric mucosa of mammals.

serpentarine A resin found in *Aristolochia*.

strychnine A bitter poisonous alkaloid found in sweetshrubs (*Calycanthus*) and used as a poison, especially for rodents, and medicinally as a stimulant to the central nervous system. In excess, the chemical causes convulsions.

strychnine-like convulsions Convulsions similar to those caused by strychnine.

tannins Any of various soluble astringent complex phenolic substances of plant origin used in tanning, dyeing, the making of ink, and in medicine.

tetrahydrocannabinols Any of various resinous compounds that are psychoactive and found in marijuana (*Cannabis*).

toxalbumin Proteins of high toxicity that break down critical natural proteins and cause accumulation of ammonia; also called phytotoxins.

urushiol An oily toxic irritant principal present in poison ivy and some related plants, and in oriental lacquers derived from such plants.

zygacine An alkaloid in *Zigadenus*.

zygadenine An alkaloid in *Zigadenus*.

Selected References

Some of the following publications provided source material for this volume. They are also recommended for further reading.

Akre, R., et al. 1980. *Yellowjackets of America North of Mexico*. U.S. Department of Agriculture, Agriculture Handbook No. 552.

American Fisheries Society. 1970. *A List of Common and Scientific Names of Fishes from the United States and Canada* (3rd edition). Reeve M. Bailey, editor. American Fisheries Society Special Publication No. 6. Washington, D.C.

Bartram, W. 1928. *Travels, 1773–1778*. Mark Van Doren (ed.). New York: Dover.

Birkhead, W. 1972. Toxicity of Stings of Ariid and Ictalurid Catfishes. *Copeia*, 1972: (4): 790–807.

Boschung, H. 1977. *How to Know the Poisonous Snakes of Alabama*. Alabama Museum of Natural History. Nature Notebook (3). Tuscaloosa: University of Alabama.

———. 1978. *The Sharks of the Gulf of Mexico, Exclusive of the Small Benthic Species*. Alabama Museum of Natural History. Nature Notebook (4). Tuscaloosa: University of Alabama.

———. 1979. *The Batoids of the Gulf of Mexico, Exclusive of the Bathybenthic Species*. Alabama Museum of Natural History. Nature Notebook (5). Tuscaloosa: University of Alabama.

Boschung, H., et al. *The Audubon Society Field Guide to North American Fishes, Whales, and Dolphins*. New York: Chanticleer Press.

Bucherl, W. 1968. *Venomous Animals and Their Venoms*. W. Bucherl, E. Buckley, and V. Deulofeu (eds.). New York: Academic Press.

———. 1968. *Venomous Animals and Their Venoms*. W. Bucherl and E. Buckley (eds.). Volume 1, *Venomous Invertebrates*. New York: Academic Press.

———. 1971. *Venomous Animals and Their Venoms*. W. Bucherl and E. Buckley (eds.). Volume 3. *Venomous Invertebrates*. New York: Academic Press.

Campbell, G. 1983. *An Illustrated Guide to Some Poisonous Plants and Animals of Florida*. Englewood, FL: Pineapple Press.

Caras, R. 1975. *Dangerous to Man*. New York: Holt, Rinehart and Winston.

———. 1974. *Venomous Animals of the World*. Englewood Cliffs, NJ: Prentice-Hall.

Cary, C., E. Miller, and G. Johnstone. 1924. *Poisonous Plants of Alabama*. Auburn, AL: Extension Service, Circular 71.

Chermock, R. 1952. *A Key to the Amphibians and Reptiles of Alabama*. Geological Survey. Tuscaloosa: Alabama Museum. Paper 38.

———. 1977. *Edible and Poisonous Mushrooms: Nature Notebook*. Alabama Museum of Natural History. Tuscaloosa: University of Alabama.

Conant, R. 1975. *A Field Guide to Reptiles and Amphibians of Eastern and Central North America*. Boston: Houghton Mifflin.

Covell, C. 1984. *A Field Guide to the Moths of Eastern North America*. Boston: Houghton Mifflin.

Craig and Faust's Clinical Parasitology. 1970. Eighth Edition. Philadelphia: Lea & Febiger.

Dean, B., A. Mason, and J. Thomas. 1973. *Wildflowers of Alabama and Adjoining States*. Tuscaloosa: University of Alabama Press.

Duncan, W., and T. Jones. 1949. *Poisonous Plants of Georgia. Bulletin*. Athens: University of Georgia School of Veterinary Medicine, 49(13):1–46.

Ferguson, D., in R. Dominick, et al. 1971–72. *The Moths of America North of Mexico*, fasc. 20.2, Bombycoidea (Saturniidae). London: E. W. Classey Ltd. and R.B.D. Publications Inc.

Foelix, R. 1982. *Biology of Spiders*. (Translation of Biologie der Spinnent, 1979). Cambridge: Harvard University Press.

Freeman, J., and H. Moore. 1974. *Livestock-Poisoning Vascular Plants of Alabama*. Bulletin 460. Auburn, AL: Agricultural Experiment Station, Auburn University.

Gadd, L. 1980. *Deadly Beautiful: The World's Most Poisonous Animals and Plants*. New York: Macmillan.

Gennaro, J. 1963. *Observation on Treatment of Snakebite in America: Venomous and Poisonous Animals and Noxious Plants of the Pacific Region*. New York: Macmillan.

Gertsch, W. 1979. *American Spiders*. New York: Van Nostrand Reinhold.

Gibbons, W. 1983. *Their Blood Runs Cold: Adventures with Reptiles and Amphibians*. Tuscaloosa: University of Alabama Press.

Gleason, H., and A. Cronquist. 1963. *Manual of Vascular Plants of Northeastern United States and Adjacent Canada*. New York: Van Nostrand Reinhold.

Guba, E. 1977. *Wild Mushrooms: Food and Poison*. Kingston, MA: Pilgrim.

Habermehl, G. 1981. *Venomous Animals and Their Toxins*. New York: Springer-Verlag.

Hall, T. 1945. *Key to Woody Plants in the Alabama Section of the Tennessee Valley*. Muscle Shoals, AL: Tennessee Valley Authority.

Halstead, B. 1980. *Dangerous Marine Animals That Bite, Sting, Shock, Are Nonedible*. Centreville, MD: Cornell Maritime Press.

———. 1965. *Poisonous and Venomous Marine Animals of the World*. Volume One: *Invertebrates*. Washington, D.C.: U.S. Government Printing Office.

———. 1970. *Poisonous and Venomous Marine Animals of the World*. Volume Three: *Vertebrates*. Washington, D.C.: U.S. Government Printing Office.

Hardin, J., and H. Arena. 1974. *Human Poisoning from Native and Cultivated Plants*. Second edition. Durham, NC: Duke University Press.

Harmon, R., and C. Pollars. 1948. *Bibliography of Animal Venoms*. Gainesville: University of Florida Press.

Helm, T. 1976. *Dangerous Sea Creatures*. New York: Funk and Wagnalls.

Hoese, H., and R. Moore. 1977. *Fishes of the Gulf of Mexico*. College Station: Texas A & M University Press.

Jenkins, D. 1986. *Amanita of North America*. Eureka, CA: Mad River Press.

Kingsbury, J. 1964. *Poisonous Plants of the U.S. and Canada*. Englewood Cliffs, NJ: Prentice-Hall.

Krispyn, J., and H. Hermann. 1977. *The Social Wasps of Georgia: Hornets, Yellowjackets, and Polistine Paper Wasps*. Bulletin 207. Athens: University of Georgia Agricultural Experiment Station.

Lewis, J. 1981. *The Biology of Centipedes*. Cambridge, MA: Cambridge University Press.

Linzey, D. Ca. 1979. *Snakes of Alabama*. Huntsville, AL: Strode.

Miller, O. 1979. *Mushrooms of North America*. New York: E. P. Dutton.

Mohr, C. (1827–1901). *Plantlife of Alabama*. Tuscaloosa, AL: Alabama Geological Survey.

Mount, R. 1975. *The Reptiles and Amphibians of Alabama*. Auburn, AL: Auburn University Agricultural Experiment Station.

Muenscher, W. 1940. *Poisonous Plants of the United States*. New York: Macmillan.

Nicander: The Poems and Poetical Fragments. 1953. Edited with a translation and notes by A. Gos and A. Scholfield. Cambridge, MA: Cambridge University Press.

Perkins, K., and W. Payne. *Guide to the Poisonous and Irritant Plants of Florida*. Circular 441. Florida Cooperative Extension Service. Gainesville: University of Florida.

Piek, T. (ed.). 1986. *Venoms of the Hymenoptera: Biochemical, Pharmacological and Behavioral Aspects*. Orlando, FL: Academic Press.

Radford, A., H. Ahles, and C. Bell. 1974. *Manual of the Vascular Flora of the Carolinas*. Chapel Hill: University of North Carolina Press.

Shipp, B. 1986. *Guide to Fishes of the Gulf of Mexico*. Dauphin Island, AL: Dauphin Island Sea Laboratory.

Smith, K. (ed.). 1973. *Insects and Other Arthropods of Medical Importance*. London: British Museum (Natural History).

Stehr, F. 1987. *Immature Insects*. Dubuque, IA: Kendall/Hunt.

Stephens, H. 1980. *Poisonous Plants of the Central United States*. Lawrence: The Regents Press of Kansas.

Tu, A. 1984. *Vol. 2. Insect Poisons, Allergens, and Other Invertebrate Venoms*. New York: Marcel Dekker.

Wahlquist, H. 1966. *A Field Key to the Batoid Fishes (Sawfishes, Guitarfishes, Skates and Rays) of Florida and Adjacent Waters*. State of Florida. Board of Conservation, Division of Salt Water Fisheries, Technical Series No. 50. St. Petersburg: Florida Board of Conservation Marine Laboratory Maritime Base, Bayboror Harbor.

Wheeler, A. 1975. *Fishes of the World: An Illustrated Dictionary*. New York: Macmillan.

Index

References to plate numbers are to the section of color plates in the middle of the book.